AF389524

Plant Breeding Supporting the Sustainable Field Crop Production

Plant Breeding Supporting the Sustainable Field Crop Production

Editor

Balázs Varga

MDPI • Basel • Beijing • Wuhan • Barcelona • Belgrade • Manchester • Tokyo • Cluj • Tianjin

Editor
Balázs Varga
Cereal Breeding Department
Centre for Agricultural
Research
Martonvásár
Hungary

Editorial Office
MDPI
St. Alban-Anlage 66
4052 Basel, Switzerland

This is a reprint of articles from the Special Issue published online in the open access journal *Sustainability* (ISSN 2071-1050) (available at: www.mdpi.com/journal/sustainability/special_issues/field_crp_production).

For citation purposes, cite each article independently as indicated on the article page online and as indicated below:

LastName, A.A.; LastName, B.B.; LastName, C.C. Article Title. *Journal Name* **Year**, *Volume Number*, Page Range.

ISBN 978-3-0365-7017-4 (Hbk)
ISBN 978-3-0365-7016-7 (PDF)

Contents

About the Editor

Balázs Varga

Balázs Varga is a senior scientist at the Agricultural institute, Centre for Agricultural Research in Martonvásár, Hungary. His research aims to improve the abiotic stress tolerance of small-grain cereals. Additionally, he focuses on the determination of the water demand and water-use-efficiency of cereal species in lysimeters and the detection of the root growth of the plant during the vegetation period by applying CI-600 in situ root imager and WinRHIZO Pro systems.

sustainability

Editorial

Plant Breeding Supporting the Sustainable Field Crop Production

Balázs Varga

Agricultural Institute, Centre for Agricultural Research, Eötvös Loránd Research Network,
H-2462 Martonvásár, Hungary; varga.balazs@atk.hu

Citation: Varga, B. Plant Breeding
Supporting the Sustainable Field
Crop Production. *Sustainability* **2023**,
15, 4040. https://doi.org/10.3390/
su15054040

Received: 17 February 2023
Accepted: 21 February 2023
Published: 23 February 2023

The population of Earth exceed eight billion in 2022 and it is growing even faster. The most important challenge for humankind will be providing food and ensuring catering not only in developed countries, but in underdeveloped regions as well. Agriculture is one of the main pollution sources, and non-appropriate farm management could degrade the environment seriously [1]. Therefore, in the future, the productivity of the sector must be improved to be able to produce enough food, but reducing the ecological and environmental footprints must be in the focus. Sustainability means efficient usage of inputs in the long term and reducing the impacts of production on the environment and landscape. Reserving the capabilities of the cropping areas, efficient and reasonable utilization of the nutrients, water, and other inputs become especially important [2]. The problem could be approached from many aspects, but the two major topics are cultivation practices and plant materials. Various water and nutrient-sawing technologies are available, but these could be used efficiently if the harvested plants could contribute to implementing the goal of sustainability [3]. Abiotic and biotic stress tolerance would be key requirements because various environmental factors endanger food security such as drought and heat, and the spreading of invasive weeds, pests, and diseases harm the yield.

Plant breeding supports new varieties and hybrids for farmers. The potential yielding capacity of the genotypes is higher and higher, but the real productivity is limited by various factors. The maximization of the yield required intensive technologies, active pest and disease management, and a high amount of inorganic fertilizers. Improving the abiotic and biotic stress tolerance and the adaptability of the genotypes can contribute to achieving sustainable farming. Yield maximization is not feasible in sustainable farming but the role of the breeding would be to provide plant materials that could produce enough food even in organic and integrated farm management systems [4].

This Special Issue aims to highlight how plant breeding could contribute to strengthening sustainability in field crop production by integrating the application of modern technologies and tools.

Water shortage is one of the most important impacts of the changing climate in many parts of the world. Typically, the arid and semi-arid regions are affected, but the intensity of the extreme drought becomes more intense and frequent in other areas. Even the Carpathian Basin in Central Europe is exposed to drought in the Spring and Summer [5].

Roots have a key role in the water and nutrition cycle in the soil-plant-atmosphere system. The root system, its development and turnover during the vegetation period is a less studied topic because of its complexity. However, the root structure could be the driver of drought resistance. Drought resistance is a complex phenomenon, and it develops through the interaction of various plant properties that are determined by several genes, including dwarfing genes, such as *Rht1*, *Rht2*, *Rht8*, etc. [6]. Dwarfing genes reduce plant height, can increase the harvest index, and improve lodging resistance. The presence of these genes in the genome not only reduces the plant height but can also negatively impact the intensity of root development and root morphology. The impacts of the water withdrawal and the elevated CO_2 level on the root structure of three winter wheat cultivars carrying different dwarfing genes were studied by using in situ root scanning technology [7]. The water shortage induced intensive root development in the deeper soil layers and under elevated CO_2 concentration, the distribution of the root system was more homogeneous in the whole soil profile. The maximum root length was detected at the beginning of the heading, and this period took longer under elevated CO_2 concentration [7].

Seedlings are the most sensitive to the stresses and damage in this phenophase, which could lead to serious yield losses. Winter wheat seedling's vigour is an important parameter that was examined by Khaeim et al. [8] under drought stress in combination with high temperatures. In agronomic practice, the determination of the germination percentage is an important parameter for calculating the sowing rate. A new methodology had been developed to optimize the circumstances of the germination process in the lab [8]. The ideal temperature for seedling development was 20 C. Germination of various percentages can occur in a broad range of water quantities commencing at 0.65 mL, which represents 75% of thousand kernel weight, but the optimal range for germination is 4.45–7.00 mL, representing 525–825% of the thousand kernel weight [8].

Besides water, the other substrate of photosynthesis is carbon dioxide. It was confirmed that the elevation of the CO_2 could be positive to plant development and production and the CO_2 enrichment could counterbalance the negative impact of drought [9]. The responses of various cereal varieties to elevated CO_2 concentration were studied under optimum and limited water availability [10]. The CO_2 treatments were combined with the simulation of drought in two phenophases of cereals and the productivity of the plants as well as the water use efficiency was determined. Genotypic variations had been determined and it was concluded that the examined winter barley react positively to the CO_2 enrichment, but negative responses had been detected for oat [10].

Salinity is a serious phenomenon that limits field crop production [11]. The high salt content in the soil, high evaporation rate, or non-appropriate quality of irrigation water could induce the enrichment of salt in the upper soil layers. Improving the growing potential of wheat on solonetz would be important in Serbia [12]. Ten winter wheat and one triticale cultivar were tested in a field experiment and phosphogypsum was applied in two doses for soil reclamation. Stable genotypes were selected that could be grown successfully on solonetz and it was determined that the effect of the applied amelioration measure depended on the meteorological conditions of the growing season [12].

Rice is the second leading staple food in the world. Rice is more sensitive to environmental stresses compared to other cereals. The plants are very sensitive to cold stress at the seedling stage and their critical period is the reproductive stage [13]. Phenotypic observations combined with genotypic characterisation of near-isogenic breeding lines were evaluated by Akter et al. [14] aiming to support rice breeding with potential crossing partners to improve cold tolerance. A high variability in cold susceptibility was determined in the test assessment and nine lines could be identified that showed strong tolerance to cold. The results of the authors highlighted that cold tolerance can be in pair with a high- yielding capacity, therefore, an accurate pre-breeding could propose efficient parental lines for breeding programs [14].

Changing climatic conditions favour to spreading of invasive weed species [15]. Weeds usually have a stronger adaptive capacity to the harsh environment than cultivated plants, therefor more intense weed competition occurs under stress conditions. Another important problem would be that weeds could become resistant to herbicides and the management of these weeds would be a key issue in the infested areas [16]. The resistance of rigid ryegrass to pyroxsulam was studied by Kutasy et al. [17]. This species is one of the most serious herbicide-resistant annual grass species which cause problems in agricultural land worldwide. In [17] RNA-seq approach had been applied to identify sequences coding resistance to riazolopyrimidines. It was concluded that diagnosing the presence of target-site resistance and multiple target-site resistance mutations could improve the efficiency of weed management and the effectiveness of the traits becomes more predictable [17]. Nowadays, glyphosate is the most widely applied herbicide in the world even glyphosate resistance had been described for many weed species [18] and even GM crops [19] can degrade the active ingredients of these herbicides. Glyphosate and its metabolite (aminomethylphosphonic acid) accumulate in the soil and environment. Both chemicals are considered to be carcinogens and mutagens. An evaluation of sunflower established whether the plants can uptake from the contaminated soil herbicide residues and what kind of morphological changes in the above and belowground organs of the plants could induce [20]. It was confirmed that the plant absorbed glyphosate from the soil, and it became detectable in the roots and leaves. Higher concentrations of glyphosate induced a reduction in root length and thickness, but the dry biomass of the plants increased [20]. The presence of glyphosate in the biomass could harm human health directly and cause problems when the plants were used as fodder for animals.

The changing environmental conditions favour the breakdown of serious plant diseases [21]. Net blotch disease is becoming an important leaf pathogen of cereals, especially barley. Antioxidant enzymes are elements of the defensive system of plants. Abiotic and biotic stressors induce the accumulation of reactive oxygen forms in plant cells and antioxidants such as superoxide dismutase or peroxidase, and these enzymes have important roles in controlling and signalling metabolic and developmental processes. The connection between the infection severity of Pyrenophora teres f. teres and the level of the superoxide dismutase was analysed in barley genotypes [22]. A significant increase in superoxide dismutase activity can be determined in the inoculated population but the reactions were genotype and isolate-dependent. The results of Kunos et al. [22] proposed an early detection method for net blotch disease based on the measurements of the superoxide dismutase activity in the leaves.

Plant resistance breeding has century-old history, but it is still relevant. The changes in the environmental conditions could contribute spreading of plant pathogens because the vigour of plants could be lowered by non-favourable conditions and could serve favourable ecological conditions for epidemics [23]. Pepper is one of the most important horticultural crops in Hungary. The general defense response of pepper was analyzed by Szarka et al. [24]. The leaves were inoculated artificially with Xanthomonas vesicatoria and Pseudomonas phaseolicola and the symptoms were evaluated. Specific hypersensitive response genes do not protect but destroy the cells affected by pathogens [25]. The nonpathogen-specific GDR plays the role of the plant immune system due to its low stimulus threshold and high reaction rate [26]. Tissue retention capacity can be increased by breeding to a level where the GDR alone can provide adequate protection for the plant. In this case, the presence of specific hypersensitive reaction genes is unnecessary; sometimes it is even a burden to the plant [26]. Another important pathogen bacteria of pepper is bacterial spot disease caused by Xanthomonas hortorum p. gardneri, which causes star-shaped necrotic lesions on the foliage, stem, and fruit. The infection decreases the yield potential of plants as well as the quality of the fruits. The bacteria infect greenhouse- and field-grown plantations as well. Resistance breeding programs successfully developed commercial pepper lines with hypersensitive and quantitative resistance to various Xanthomonas species. The new sources of resistance identified in [26] provide a basis for further work on breeding disease-resistant varieties. It was documented that the different genotypes of Capsicum. baccatum differed in their response to wilt caused by Xanthomonas, which also demonstrated that the tested collection offered a valuable public source of resistance for pepper breeders to develop varieties resistant to bacterial spots.

Nematodes are important pests in horticulture. Application of root-specific resistance genes would be an efficient and sustainable way of protecting against nematodes. Tóth et al. [27] proposed an efficient methodology for transforming target genes by Agrobacterium rhizogenes using tomatoes as a model plant. The transformation efficiency was over 90% and the presence of *NeoR/KanR* and *DsRed* genes in the transformed plants were confirmed by PCR. The long-term effectiveness of the transformed genes had been confirmed after three months as well. The proposed method improves the ability to study root-specific genes and could be used for molecular studies of root–pathogen interactions [27].

Conflicts of Interest: The authors declare no conflict of interest.

References

1. Adegbeye, M.; Reddy, P.R.K.; Obaisi, A.; Elghandour, M.; Oyebamiji, K.; Salem, A.; Morakinyo-Fasipe, O.; Cipriano-Salazar, M.; Camacho-Díaz, L. Sustainable agriculture options for production, greenhouse gasses and pollution alleviation, and nutrient recycling in emerging and transitional nations—An overview. *J. Clean. Prod.* **2020**, *242*, 118319. [CrossRef]
2. Ullah, H.; Santiago Arenas, R.; Ferdous, Z.; Attia, A.; Datta, A. Improving water use efficiency, nitrogen use efficiency, and radiation use efficiency in field crops under drought stress: A review. *Adv. Agron.* **2019**, *156*, 109–157. [CrossRef]
3. Mojid, M.A.; Mainuddin, M. Water-Saving Agricultural Technologies: Regional Hydrology Outcomes and Knowledge Gaps in the Eastern Gangetic Plains—A Review. *Water* **2021**, *13*, 636. [CrossRef]
4. Le Campion, A.; Oury, F.-X.; Heumez, E.; Rolland, B. Conventional versus organic farming systems: Dissecting comparisons to improve cereal organic breeding strategies. *Org. Agric.* **2019**, *10*, 63–74. [CrossRef]
5. Mezősi, G.; Bata, T.; Meyer, B.C.; Blanka, V.; Ladányi, Z. Climate Change Impacts on Environmental Hazards on the Great Hungarian Plain, Carpathian Basin. *Int. J. Disaster Risk Sci.* **2014**, *5*, 136–146. [CrossRef]
6. Dowla, M.N.U.; Edwards, I.; O'Hara, G.; Islam, S.; Ma, W. Developing Wheat for Improved Yield and Adaptation under a Changing Climate: Optimization of a Few Key Genes. *Engineering* **2018**, *4*, 514–522. [CrossRef]
7. Varga, B.; Farkas, Z.; Varga-László, E.; Vida, G.; Veisz, O. Elevated Atmospheric CO_2 Concentration Influences the Rooting Habits of Winter-Wheat (*Triticum aestivum* L.) Varieties. *Sustainability* **2022**, *14*, 3304. [CrossRef]
8. Khaeim, H.; Kende, Z.; Balla, I.; Gyuricza, C.; Eser, A.; Tarnawa, Á. The Effect of Temperature and Water Stresses on Seed Germination and Seedling Growth of Wheat (*Triticum aestivum* L.). *Sustainability* **2022**, *14*, 3887. [CrossRef]
9. Mitterbauer, E.; Enders, M.; Bender, J.; Erbs, M.; Habekuß, A.; Kilian, B.; Ordon, F.; Weigel, H.-J. Growth response of 98 barley (*Hordeum vulgare* L.) genotypes to elevated CO_2 and identification of related quantitative trait loci using genome-wide association studies. *Plant Breed.* **2017**, *136*, 483–497. [CrossRef]
10. Farkas, Z.; Anda, A.; Vida, G.; Veisz, O.; Varga, B. CO_2 Responses of Winter Wheat, Barley and Oat Cultivars under Optimum and Limited Irrigation. *Sustainability* **2021**, *13*, 9931. [CrossRef]
11. Machado, R.M.A.; Serralheiro, R.P. Soil Salinity: Effect on Vegetable Crop Growth. Management Practices to Prevent and Mitigate Soil Salinization. *Horticulturae* **2017**, *3*, 30. [CrossRef]
12. Banjac, B.; Mladenov, V.; Petrović, S.; Matković-Stojšin, M.; Krstić, Đ.; Vujić, S.; Mačkić, K.; Kuzmanović, B.; Banjac, D.; Jakšić, S.; et al. Phenotypic Variability of Wheat and Environmental Share in Soil Salinity Stress [3S] Conditions. *Sustainability* **2022**, *14*, 8598. [CrossRef]
13. Zhang, Q.; Chen, Q.; Wang, S.; Hong, Y.; Wang, Z. Rice and Cold Stress: Methods for Its Evaluation and Summary of Cold Tolerance-Related Quantitative Trait Loci. 2014. [Online]. Available online: http://www.thericejournal.com/content/7/1/24 (accessed on 9 September 2014).
14. Akter, N.; Biswas, P.S.; Syed, A.; Ivy, N.A.; Alsuhaibani, A.M.; Gaber, A.; Hossain, A. Phenotypic and Molecular Characterization of Rice Genotypes' Tolerance to Cold Stress at the Seedling Stage. *Sustainability* **2022**, *14*, 4871. [CrossRef]
15. Bajwa, A.; Kaur, S.; Franks, S.; Clements, D.R.; Jones, V.L. Article 664034 Citation: Clements DR and Jones VL (2021) Rapid Evolution of Invasive Weeds under Climate Change: Present Evidence and Future Research Needs. *Front. Agron.* **2021**, *3*, 664034. [CrossRef]
16. Heap, I. Global perspective of herbicide-resistant weeds. *Pest Manag. Sci.* **2014**, *70*, 1306–1315. [CrossRef]
17. Kutasy, B.; Takács, Z.; Kovács, J.; Bogaj, V.; Razak, S.; Hegedűs, G.; Decsi, K.; Székvári, K.; Virág, E. Pro197Thr Substitution in *Ahas* Gene Causing Resistance to Pyroxsulam Herbicide in Rigid Ryegrass (*Lolium Rigidum* Gaud.). *Sustainability* **2021**, *13*, 6648. [CrossRef]
18. Heap, I.; Duke, S.O. Overview of glyphosate-resistant weeds worldwide. *Pest Manag. Sci.* **2017**, *74*, 1040–1049. [CrossRef]
19. Cerdeira, A.L.; Duke, S.O. The Current Status and Environmental Impacts of Glyphosate-Resistant Crops. *J. Environ. Qual.* **2006**, *35*, 1633–1658. [CrossRef]
20. Farkas, D.; Horotán, K.; Orlóci, L.; Neményi, A.; Kisvarga, S. New Methods for Testing/Determining the Environmental Exposure to Glyphosate in Sunflower (*Helianthus annuus* L.) Plants. *Sustainability* **2022**, *14*, 588. [CrossRef]
21. Jeger, M.J. The impact of climate change on disease in wild plant populations and communities. *Plant Pathol.* **2021**, *71*, 111–130. [CrossRef]
22. Kunos, K.M.V.; Cséplő, M.; Seress, D.; Eser, A.; Kende, Z.; Uhrin, A.; Bányai, J.; Bakonyi, J.; Pál, M. The Stimulation of Superoxide Dismutase Enzyme Activity and Its Relation with the *Pyrenophora teres* f. *teres* Infection in Different Barley Genotypes | Enhanced Reader. *Sustainability* **2022**, *14*, 2597. [CrossRef]
23. Mokhena, T.; Mochane, M.; Tshwafo, M.; Linganiso, L.; Thekisoe, O.; Songca, S. Impact of Climate Change on Plant Diseases and IPM Strategies. IntechOpen. Available online: https://www.intechopen.com/books/advanced-biometric-technologies/liveness-detection-in-biometrics (accessed on 29 August 2019).
24. Szarka, J.; Timár, Z.; Hári, R.; Palotás, G.; Péterfi, B. General Defense Response under Biotic Stress and Its Genetics at Pepper (*Capsicum annuum* L.). *Sustainability* **2022**, *14*, 6458. [CrossRef]

25.   Balint-Kurti, P.; Balint-Kurti, P. The plant hypersensitive response: Concepts, control and consequences. *Mol. Plant Pathol.* **2019**, *20*, 1163–1178. [CrossRef] [PubMed]
26.   Tóth, Z.G.; Tóth, M.; Fekete, S.; Szabó, Z.; Tóth, Z. Screening Wild Pepper Germplasm for Resistance to *Xanthomonas hortorum* pv. *gardneri*. *Sustainability* **2023**, *15*, 908. [CrossRef]
27.   Tóth, M.; Tóth, Z.G.; Fekete, S.; Szabó, Z.; Tóth, Z. Improved and Highly Efficient *Agrobacterium rhizogenes*-Mediated Genetic Transformation Protocol: Efficient Tools for Functional Analysis of Root-Specific Resistance Genes for *Solanum lycopersicum* cv. Micro-Tom. *Sustainability* **2022**, *14*, 6525. [CrossRef]

sustainability

Article

Elevated Atmospheric CO$_2$ Concentration Influences the Rooting Habits of Winter-Wheat (*Triticum aestivum* L.) Varieties

Balázs Varga *, Zsuzsanna Farkas, Emese Varga-László *, Gyula Vida and Ottó Veisz

Agricultural Institute, Centre for Agricultural Research, Eötvös Loránd Research Network, 2462 Martonvásár, Hungary; farkas.zsuzsanna@atk.hu (Z.F.); vida.gyula@atk.hu (G.V.); veisz.otto@atk.hu (O.V.)
* Correspondence: varga.balazs@atk.hu (B.V.); emese.laszlo82@gmail.com (E.V.-L.)

Abstract: The intensity and the frequency of extreme drought are increasing worldwide. An elevated atmospheric CO$_2$ concentration could counterbalance the negative impacts of water shortage; however, wheat genotypes show high variability in terms of CO$_2$ reactions. The development of the root system is a key parameter of abiotic stress resistance. In our study, biomass and grain production, as well as the root growth of three winter-wheat varieties were examined under optimum watering and simulated drought stress in a combination with ambient and elevated atmospheric CO$_2$ concentrations. The root growth was monitored by a CI-600 in situ root imager and the photos were analyzed by RootSnap software. As a result of the water shortage, the yield-related parameters decreased, but the most substantial yield reduction was first detected in Mv Karizma. The water shortage influenced the depth of the intensive root development, while under water-limited conditions, the root formation occurred in the deeper soil layers. The most intensive root development was observed until the heading, and the maximum root length was recorded at the beginning of the heading. The period of root development took longer under elevated CO$_2$ concentration. The elevated CO$_2$ concentration induced an accelerated root development in almost every soil layer, but generally, the CO$_2$ fertilization induced in the root length of all genotypes and under each treatment.

Keywords: cereals; climate change; water shortage; carbon dioxide; root development

Citation: Varga, B.; Farkas, Z.; Varga-László, E.; Vida, G.; Veisz, O. Elevated Atmospheric CO$_2$ Concentration Influences the Rooting Habits of Winter-Wheat (*Triticum aestivum* L.) Varieties. *Sustainability* **2022**, *14*, 3304. https://doi.org/10.3390/su14063304

Academic Editor: Roberto Mancinelli

Received: 24 February 2022
Accepted: 10 March 2022
Published: 11 March 2022

Publisher's Note: MDPI stays neutral with regard to jurisdictional claims in published maps and institutional affiliations.

1. Introduction

One of the most important drivers of climatic changes is the increasing atmospheric CO$_2$ concentration [1]. The rising temperature is a well-known worldwide phenomenon; however, the changes in precipitation vary greatly by region [2–4]. Still, it can be stated that the frequency and the intensity of drought have been increasing in the most important agricultural areas of the world, and this has resulted in reduced harvested yield and has caused uncertainties in this strategic sector [5]. Many studies have focused on the negative effects of a water-limited environment regarding winter-wheat productivity [2,5,6]. Apart from yield reduction, due to intensive drought, water sources can be utilized less efficiently [7,8]. Several recently published studies confirmed that the elevated CO$_2$ could more or less counterbalance the negative effects of water shortage through the intensification of photosynthesis in C4 plants [9,10] and the regulation of the stomata closure in C3 plants [11,12]. The majority of these studies focused on the aboveground biomass, especially on the harvested yield; however, it must be highlighted that the root system plays an important role in the water and nutrient uptake of the plants [13–15]. Rooting depth, as well as the structure of the root system and the depth of intensive root development, are the crucial factors that can influence plants' water uptake. Still, a well-developed root structure is not the sole determinant of drought tolerance. Roots are often more varied than shoots and are affected by changes in the climate, soil conditions, plant varieties and soil nutrient and water availability throughout the growing season [16]. The accurate examination of the belowground parts of a plant is more complicated than the analysis of the aboveground biomass. Conventionally, destructive and non-repeatable approaches have been applied in order to examine the rooting habits of plants, such as soil coring [17,18], the use of mini rhizotrons [19] or pots of different sizes, but all of these methods have their limitations. First of all, the root development and turnover cannot be measured throughout the vegetation period within the same plant stand. Recently, various methods have been developed to detect the properties of the belowground biomass. They mainly consist of imaging and an image-processing phase. Another option could be to use impedance spectroscopy; however, only a few parameters of the root system can be determined by this approach. This tool is not suitable for measuring the parameters of the individual roots or for inspecting the depth of the intensive root development during the plant-growth period [20,21]. MRI (Magnetic Resonance Imaging) or CT (X-ray Computed Tomography) technologies are highly sophisticated methodologies, but they are exceedingly expensive and could be applied only in pot experiments [22–24]. However, under real field conditions, as well as in a model experiment, root-scanning technology can be efficiently used throughout vegetation. The transparent polycarbonate tubes can be dug or drilled into the soil at different positions and lengths to detect the rooting habits of the plants from germination until harvest. The advantages of this approach are that the measurements can be carried out in the same position and the potential influences of the various environmental factors on the rooting can be excluded [25]. Wheat landraces are better adapted to changing climatic conditions and to stress environments than modern cultivars due to their population's genetic structure, buffering capacity, and morpho-physiological traits, such as rooting habits conferring adaptability to stress environments [26]. However, even among the recently bred cultivars, there is high variability in terms of abiotic stress tolerance [27,28]. Drought resistance is a complex phenomenon, and it develops through the interaction of various plant properties that are determined by several genes, including dwarfing genes, such as *Rht1, Rht2, Rht8,* etc. [29]. Using dwarfing genes to reduce plant height increases the harvest index, improves lodging resistance and increases grain yield. Their application has been one of the major strategies in developing modern bread-wheat cultivars [30]. The presence of these genes in modern wheat varieties is necessary because due to the cultivation technology, shorter plants are more resistant to lodging. The presence of the effective dwarfing genes in the genome not only reduces the plant height but also can negatively affect the intensity of the root development [29].

The objectives of the study were (1) to determine the dynamics of root development during the vegetation period of winter-wheat varieties, (2) to examine how plants react through the modification of the root development to the drought stress in the different soil layers, and (3) to quantify the genotypic responses to elevated atmospheric CO_2 concentrations through the root development under optimum and limited water availability. The results of the experiments could contribute to the efficiency of plant-breeding activities for improved drought tolerance. Moreover, from a practical point of view, the experiments emphasize that the rooting habits of the different varieties should also be considered by the farmers.

2. Materials and Methods

2.1. Experimental Layout

Three registered Hungarian winter-wheat (*Triticum aestivum* L.) varieties, Mv Pálma, Mv Karéj and Mv Karizma carrying *Rht8, Rht1* and *Rht2* dwarfing genes (30), respectively, were selected for the model experiment, which was carried out in two similar climate-controlled greenhouse chambers of the Agricultural Institute, Centre for Agricultural Research, Hungary. Plants were vernalized at 4 °C for 6 weeks and the germinated seeds were planted into plastic containers (120 cm × 90 cm × 100 cm) filled with ca. 1000 liters of a 3:1:1 (*v*/*v*) mixture of soil, sand and humus. Water-soluble fertilizer (14% N, 7% P2O5, 21% K2O, 1% Mg, 1% B, Cu, Mn, Fe, Zn; Volldünger Classic; Kwizda Agro Ltd., Vienna, Austria) was added bi-weekly to both water treatments based on the manufacturer's recommendations. Plant density was 450 plants/m², which is similar to the commonly applied sowing rate in local agricultural practice (Figures 1b and 2). Each container consisted of 8 rows and each row of 28 plants.

Figure 1. (**a**) The position of the polycarbonate tubes in the container; (**b**) The plant stand after the planting.

Figure 2. The experimental design in the ambient chamber. The tubes with the black end caps are the polycarbonate tubes for scanning the root system.

The air temperature and the additional light intensity of the greenhouse chambers were automatically regulated. The air temperature was increased from the initial 10–12 °C to 24–26 °C over 16 weeks, while air humidity was maintained between 60% and 80% and was regulated by ventilating the greenhouse chambers' air [31]. Whenever it was necessary, the natural-light intensity was enhanced by artificial illumination to 500 μmol m^{-2} s^{-1} at the beginning of the vegetation period, which was gradually increased to 700 μmol m^{-2} s^{-1}. Thiovit Jet fungicide (Syngenta AG, Basel, Switzerland) (active ingredient: sulfur) and Karate 2.5 WG insecticide (Syngenta AG, Basel, Switzerland) (active ingredient: lambda-cyhalothrin) were applied two times (BBCH 23 and 37) [32] following the distributor's recommendations by the dosage against powdery mildew and aphids, respectively. The atmospheric CO_2 concentration in the control (ambient) chamber was maintained at ~400 ppm and the gas concentration was enhanced to 750 ppm in the other chamber by using a network of perforated pipes placed at a height of 0.5 m above the plants. The uniform distribution was achieved through ventilation.

The containers were separated into two parts of equal size (optimum watered and drought-stressed) by using water-insulating PVC foil (thickness 1 mm).

The water-holding capacity of the soil was determined by using the gravimetric method before starting the experiment and the control treatments were watered until optimum (60%) soil-water content [SWC]. The water content of the soil was monitored by 5 TA sensors (Decagon Devices Ltd., Pullman, WA, USA) at 3 depths (30, 60 and 90 cm). Water-stressed plants did not receive additional watering after the planting until the volumetric soil-water content dropped below 8–10% (average of the three depths). The plants in the drought-stress treatment were irrigated first at the BBCH 51 stage, 68 days after the planting. Afterwards, halved water doses compared to the control were applied to the stress-treated stands.

2.2. Measurements

Without overlapping, three transparent polycarbonate tubes were set up in the containers in a horizontal position, at 30, 60 and 90 cm soil depths (Figure 1a). The root development/turnover was monitored every two weeks in the same positions of the tubes by a CI-600 in situ root imager (CID-Bioscience Ltd., Camas, WA, USA). The plant phenophases were ranked according to the BBCH scale (Table 1) [32].

Table 1. List of phenophases when the root length measurements were carried.

BBCH Codes	Explanation
BBCH 17	Leaf development: 7 or more leaves unfolded
BBCH 29	End of tillering: the maximum number of tillers detectable
BBCH 37	Flag leaf just visible, still rolled
BBCH 51	Beginning of heading: the tip of inflorescence emerged from the sheath, the first spikelet just visible
BBCH 69	End of flowering: all spikelets have completed flowering but some dehydrated anthers may remain
BBCH 77	Late milk stage
BBCH 83	Ripening, early dough

RootSnap software (CID-Bioscience Ltd., Camas, WA, USA) was applied for image processing and to determine the root length (Figure 3). Simultaneously, soil temperature and soil-water content were monitored continuously in the three layers by 5 TE sensors and EM50 data loggers (Decagon Devices Ltd., Pullman, WA, USA).

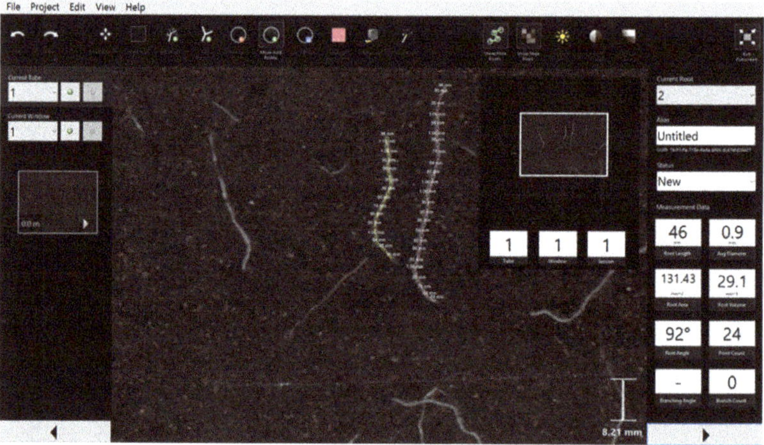

Figure 3. An image taken by the CI-600 root scanner under analysis by the RootSnap software.

At the end of the vegetation period, all plants were manually harvested row by row (8 replications). The weight of the total aboveground biomass, grain weight and thousand-kernel weight was measured by a digital balance (ME 1002E, Mettler-Toledo Ltd., Worthington, OH, USA), and the harvest index (Equation (1)) was calculated.

$$\text{Harvest index} = \frac{\text{grain yield (g)}}{\text{total aboveground biomass (g)}} * 100 \tag{1}$$

Relative changes in the root length in response to elevated carbon dioxide levels were calculated as:

$$CO_2 \text{ responses} = \frac{Ex}{A} \tag{2}$$

where A is the root length at the 400 ppm CO_2 level, Ex is the root length at the 700 ppm CO_2 level.

2.3. Analysis

The experimental design ($2 \times 2 \times 3$) consisted of two CO_2 levels, two watering levels and three genotypes. The rooting habits of a closed plant stand were analyzed. The scanning was always carried out in three replications in the same position. The effects of the tested factors on yield parameters were determined by multi-way ANOVA and means were compared by Tukey's HSD test ($p \leq 0.05$). The significant differences in the length parameters between the water treatments and CO_2 levels were evaluated by Student's t-test and the significant differences between the root lengths measured in the plant growth stages were analyzed by one-way ANOVA followed by Tukey's HSD test.

3. Results

3.1. Effects of Water Shortage and Elevated Atmospheric CO_2 Concentration on Yield Parameters

The biomass (BM), grain yield (GY), thousand-kernel weight (TKW) and harvest index (HI) of the individual treatments are summarized in Table 2.

Table 2. Responses of winter-wheat varieties to elevated CO_2 and water shortage.

Factors			Variables			
Genotypes	CO_2 Levels	Watering	BM (g)	GY (g)	TKW (g)	HI (%)
	NC	C	67.4	26.06	36.87	38.57
Mv Pálma	NC	D	62.0	23.37	41.57	37.54
	EC	C	73.6	30.47	35.63	41.95
	EC	D	65.3	25.65	40.53	39.62
	NC	C	73.8	27.47	43.00	37.18
Mv Karéj	NC	D	66.4	24.45	42.60	37.05
	EC	C	75.6	28.44	41.47	37.90
	EC	D	67.7	25.83	40.52	38.15
	NC	C	77.4	32.12	41.77	41.76
Mv Karizma	NC	D	70.6	28.96	37.87	41.17
	EC	C	79.4	33.70	41.46	42.71
	EC	D	75.9	32.20	35.11	42.64
$HSD_{5\%}$ values	Watering		1.817	1.223	n.s.	n.s.
	CO_2		n.s.	1.079	6.42	7.69
	Genotype		1.198	0.998	6.59	8.08

BM: aboveground biomass; GY: grain yield; TKW thousand-kernel weight; HI: harvest index; NC: ambient CO_2; EC: 750 ppm CO_2; C: controlled watering; D: drought stress, n.s. the effects of the factor are not significant (the presented data are means of eight replications).

The ANOVA shows that the effects of the watering levels, genotypes and the CO_2 concentrations significantly influenced the grain yield, but their interactions were not statistically significant. The effects of the watering and the genotypes were significant on the aboveground biomass, but the ANOVA shows that the interactions of the factors remained insignificant (Tables S1 and S2). Different CO_2 levels and irrigation regimes were assumed to have contradictory effects on the yield-related parameters. Water shortage resulted in decreased BM and GY both under ambient and elevated CO_2 concentrations. The ratio of the yield reduction did not differ between the two CO_2 concentrations. Compared to the well-watered plant stands, the biomass of Mv Pálma and Mv Karéj decreased by 8.02% and 10.03% under ambient CO_2 concentration and 11.2% and 10.45% under elevated CO_2, respectively, in the drought-stress treatment (Table 2).

Generally, the simulated drought stress did not significantly affect the TKW and the HI, but the effects of the genotypes and watering levels were statistically significant in terms of both parameters (Tables S3 and S4). In the case of Mv Pálma, water shortage led to a 12.75% and 13.75% increase in TKW compared with the non-stressed controls under 400 ppm and 750 ppm CO_2 concentration, respectively. The highest yielding capacity was observed for Mv Karizma, and this genotype showed the lowest yield reduction under the simulated drought stress. In Mv Karizma, the elevated CO_2 concentration played a role in counterbalancing the negative effects of the water shortage. Taking the average of the three examined varieties, the CO_2 enrichment increased the biomass (by 4.57% and 4.97% under normal watering and drought stress, respectively) as well as the grain yield (by 8.1% and 8.96% under normal watering and drought stress, respectively) (Table 2). Improved harvest indices (4.3% and 4.0% under normal watering and drought-stress conditions, respectively) could be determined as a result of the observed tendencies in biomass and grain weight, which would be highly favorable for nutrient- and water-utilization efficiency. In Mv Karéj, no differences were observed between the two water treatments in terms of CO_2 reactions. The interaction of water shortage and CO_2 fertilization showed opposite tendencies in the other two varieties. The increase in the yield parameters was more intense under the control water supply in Mv Pálma, while the positive effect of the elevated CO_2 compared with the control was more significant under drought-stress conditions in Mv Karizma (Table 2).

3.2. Dynamics of the Root Development of Winter-Wheat Varieties under Optimum Watering and Drought-Stressed Conditions Grown at Ambient and Elevated CO_2 Concentrations

Under ambient CO_2 concentration, the root length of Mv Pálma continuously developed in the well-watered treatment after planting until it reached its maximum level at the BBCH 51 stage in the two upper soil layers (30 and 60 cm), while a faster root development was detected in the drought-stressed treatment when the root length did not significantly increase after the BBCH 29 and BBCH 37 stages at 30 and 60 cm, respectively. (Table 3). Significantly higher root length was observed at 30 cm between the BBCH 29 and 51 stages under drought-stressed conditions, but the root length did not significantly differ between the water treatment at 60 cm except at the BBCH 21 and 83 stages. At 90 cm, the root length of Mv Pálma reached its maximum (does not significantly increase further) at BBCH 69 under optimum irrigation and at BBCH 51 under drought-stressed conditions, and significantly higher root length was measured under drought stress between BBCH 37 and BBCH 83 than in the control treatment (Table 2). Under elevated CO_2, the root length of Mv Pálma was consequently higher in the well-irrigated treatment at 30 cm between BBCH 21 and BBCH 77 than under drought stress (Table 3). The root length did not significantly increase at 30 or 90 cm after the BBCH 37 stage, but at 60 cm, the root length did not significantly increase after the BBCH 29 stage under drought-stressed conditions.

Under ambient CO_2 concentration, the root length of Mv Karéj reached its maximum in the upper soil layer (30 cm) at the BBCH 29 stage in both water treatments, and the root length was significantly higher under optimum watering at BBCH 51 and BBCH 77 than in the drought-stressed treatment (Table 4). In the middle layer (60 cm), the root length did not significantly increase after BBCH 37 and BBCH 29 in the control and drought-stressed treatments, respectively, and a significant difference between the water treatments can be observed only at the BBCH 29 phenophase. In Mv Karéj, the maximum root length was observed at the BBCH 51 stage at 90 cm in both water treatments, but significantly higher values were measured under drought-stressed conditions from BBCH 37 compared to the control treatment.

Table 3. Root length of Mv Pálma in different stages of vegetation under well-watered and drought-stressed conditions in different soil layers by ambient and elevated atmospheric CO_2 concentrations.

	400 ppm						750 ppm					
	C30	D30	C60	D60	C90	D90	C30	D30	C60	D60	C90	D90
BBCH 17	362[a4]	160[b5]	n.r.	n.r.	n.r.	n.r.	503[a4]	518[a3]	n.r.	n.r.	n.r.	n.r.
BBCH 21	884[a3]	897[a4]	262[a4]	276[a4]	n.r.	n.r.	1133[a3]	851[b2]	634[a4]	656[a2]	n.r.	n.r.
BBCH 29	1215[b2]	1516[a1]	858[b3]	1225[a23]	457[a4]	666[a3]	1380[a2]	596[b3]	933[a23]	1021[a1]	1029[b3]	1180[a4]
BBCH 37	1249[b2]	1532[a1]	1385[a1]	1452[a1]	951[b3]	2228[a2]	1778[a1]	1063[b1]	1185[a1]	1079[a1]	1947[a1]	1952[a1]
BBCH 51	1276[b1]	1452[a1]	1537[a1]	1492[a1]	1194[b2]	3190[a1]	1649[a1]	1097[b1]	1231[a1]	1146[a1]	1871[a1,2]	1843[a1,2]
BBCH 69	1253[a2]	1242[a2,3]	1234[a2]	1337[a1]	1540[b1]	2877[a1]	1477[a2]	1087[b1]	1079[a1,2]	1027[a1]	1852[a1,2]	1548[b3]
BBCH 77	1506[a1]	1137[b3]	927[a3]	1034[a3]	1008[b2,3]	2730[a1]	1334[a2]	876[b2]	821[a3]	674[b2]	1646[a2]	1637[a2,3]
BBCH 83	1293[a2]	1439[a1,2]	1209[b2]	1387[a1,2]	1340[b1,2]	3200[a1]	1026[a3]	991[a1,2]	862[a3]	995[a1]	1887[a1,2]	1882[a1,2]

C30, D30, C60, D60, C90 and D90 indicate the soil layer at 30 cm under controlled watering, the soil layer at 30 cm under drought stress, the soil layer at 60 cm under controlled watering, the soil layer at 60 cm under drought stress, the soil layer at 90 cm under controlled watering and the soil layer at 90 cm under drought stress, respectively. Lowercase letters indicate significant differences between drought and control treatments within the same soil layer and CO_2 treatment (Student's *t*-test) ($p < 0.05$) and the numbers in superscripts indicate significant differences between phenophases (HSD test ($p \leq 0.05$) (n = 3), n.r., no roots were observed.

Table 4. Root length of Mv Karéj at different stages of vegetation under well-watered and drought-stressed conditions in different soil layers by ambient and elevated atmospheric CO_2 concentrations.

	400 ppm						750 ppm					
	C30	D30	C60	D60	C90	D90	C30	D30	C60	D60	C90	D90
BBCH 17	357[a4]	412[a4]	n.r.	n.r.	n.r.	n.r.	60[b6]	208[a4]	n.r.	0	0	0
BBCH 21	1450[a3]	1425[a2]	288[a3]	237[a3]	n.r.	n.r.	588[b5]	1060[a3]	544[a3]	208[b4]	0	0
BBCH 29	1982[a1,2]	1818[a1]	982[b2]	1280[a1,2]	470[a3]	785[a3]	1240[a4]	1257[a2,3]	1121[a2]	472[b3]	750[a3]	559[b3]
BBCH 37	2111[a1]	1890[a1]	1476[a1]	1375[a1]	2326[b2]	2832[a2]	1539[a1]	1550[a1]	1398[a1]	746[b1,2]	2341[a1,2]	1562[b1,2]
BBCH 51	2167[a1]	1582[b2]	1659[a1]	1474[a1]	3205[b1]	3804[a1]	1657[a1]	1548[a1]	1396[a1]	862[b1]	2703[a1]	1551[b1,2]
BBCH 69	1870[a2]	1575[a2]	1484[a1]	1334[a1,2]	2867[b1]	3747[a1]	1551[a1,2]	1338[b1]	1242[a1,2]	809[b1,2]	2387[a1,2]	1664[b1]
BBCH 77	1578[a3]	1153[b3]	1122[a2]	1122[a2]	2454[b2]	3480[a1]	1411[a2,3]	1298[a2]	1041[a2]	673[b2]	2038[a2]	1361[b2]
BBCH 83	2173[a1]	1933[a1]	1623[a1]	1466[a1]	3232[b1]	3802[a1]	1382[a3]	969[a3]	1117[a2]	662[b2]	2404[a1,2]	1716[b1]

C30, D30, C60, D60, C90 and D90 indicate the soil layer at 30 cm under controlled watering, the soil layer at 30 cm under drought stress, the soil layer at 60 cm under controlled watering, the soil layer at 60 cm under drought stress, the soil layer at 90 cm under controlled watering and the soil layer at 90 cm under drought stress, respectively. Lowercase letters indicate significant differences between drought and control treatments within the same soil layer and CO_2 treatment (Student's *t*-test) ($p < 0.05$) and the numbers in superscripts indicate significant differences between phenophases (HSD test ($p \leq 0.05$) (n = 3), n.r., no roots were observed.

Under 750 ppm CO_2, the root length of Mv Karéj reached its maximum at the BBCH 37 stage at each soil layer under optimum and drought-stressed conditions. No significant differences were observed between the water treatments at the 30 cm soil layer after the BBCH 21 stage, but the CO_2 enrichment resulted in a significant decrease in root length in each phenophase at 60 and 90 cm (Table 4).

The root development of Mv Karizma showed the highest variability among the studied varieties under ambient CO_2 concentration in different soil layers and irrigation regimes. Intensive root development was observed from planting until the BBCH 29 stage at 30 cm, but the root length did not increase further during vegetation. Oppositely, a significant reduction in root length was determined from the BBCH 69 stage (Table 5). The root length at 30 cm was significantly higher under optimum watering than under drought stress in each phenophase. At 60 cm, the root-length development was faster under drought-stress conditions than under optimum irrigation, and the measured data did not significantly increase after BBCH 37 and BBCH 51 under drought stress and optimum watering, respectively. The water shortage induced a significantly more developed root system in Mv Karizma for each phenophase at 60 cm (Table 5). The root system developed intensively until the BBCH 51 stage at 90 cm under both watering regimes, then a root turnover can be observed in the BBCH 69 and 77 stages, but another intensive root formation was detected at the BBCH 83 stage. Significantly higher root-length values were measured in Mv Karizma under drought-stressed conditions at 90 cm than in the control treatment from BBCH 29 until BBCH 83, except BBCH 37 at 400 ppm CO_2 concentration. Under elevated CO_2 levels (750 ppm), a delay can be observed at 30 cm in the time that the root length takes to reach its maximum compared to the ambient treatment. The maximum values were observed at the BBCH 37 stage in both water treatments. The water supply had no significant effects on the root length of Mv Karizma during vegetation at 30 cm under 750 ppm CO_2 between the BBCH29 and BBCH77 stages. At 60 cm, the highest root length was observed at the end of vegetation under optimum watering, while the highest data were measured at the BBCH 37 and BBCH 51 stages under limited water supply. The water shortage induced a significant increase in root length in each phenophase of Mv Karizma under CO_2 enrichment. The trends in root development at 90 cm were similar to that at 60 cm, and the water shortage resulted in an increase in root length between the BBCH 37 and BBCH 77 stages (Table 5).

Table 5. Root length of Mv Karizma at different stages of vegetation under well-watered and drought-stressed conditions in different soil layers by ambient and elevated atmospheric CO_2 concentrations.

	400 ppm						750 ppm					
	C30	D30	C60	D60	C90	D90	C30	D30	C60	D60	C90	D90
BBCH 17	375[a4]	156[b5]	n.r.	n.r.	n.r.	n.r.	200[a4]	338[b4]	n.r.	n.r.	n.r.	n.r.
BBCH 21	2013[a3]	1057[b4]	114[b4]	153[a4]	n.r.	n.r.	641[a3]	956[b3]	294[b5]	525[a4]	n.r.	n.r.
BBCH 29	2729[a1,2]	1711[b1,2]	561[b3]	1843[a3]	528[a4]	195[b4]	1170[a2]	1067[a2,3]	742[b4]	1355[a3]	684[b3]	568[a3]
BBCH 37	2820[a1]	1829[b1,2]	810[b2]	2356[a1,2]	2847[a2]	3251[a3]	1381[a1]	1322[a1]	1366[b2,3]	2041[a1]	1997[b2]	3345[a1]
BBCH 51	2931[a1]	1853[b1]	1142[b1]	2560[a1]	3469[b1]	4411[a1]	1338[a1,2]	1181[a1,2]	1592[b2]	2188[a1]	2483[b2]	3242[a1]
BBCH 69	2470[a2]	1609[b2]	1156[b1]	2086[a2,3]	2433[b3]	3744[a2]	1244[a1,2]	1155[a1,2]	1432[b2,3]	1660[a2]	2407[b2]	3178[a1]
BBCH 77	2128[a3]	1476[b2,3]	991[b1,2]	2012[a2,3]	2629[b2,3]	3481[a3]	1180[a1,2]	1013[a2]	1271[b3]	1665[a2]	2196[b2]	2535[a2]
BBCH 83	1963[a3]	1766[b1,2]	1182[b1]	2474[a1]	3365[b1]	4468[a1]	1342[a1,2]	1031[b2]	1845[a1]	1792[a2]	2864[a1]	3050[a1]

C30, D30, C60, D60, C90 and D90 indicate the soil layer at 30 cm under controlled watering, the soil layer at 30 cm under drought stress, the soil layer at 60 cm under controlled watering, the soil layer at 60 cm under drought stress, the soil layer at 90 cm under controlled watering and the soil layer at 90 cm under drought stress, respectively. Lowercase letters indicate significant differences between drought and control treatments within the same soil layer and CO_2 treatment (Student's *t*-test) ($p < 0.05$) and the numbers in superscripts indicate significant differences between phenophases (HSD test ($p \leq 0.05$) (n = 3), n.r., no roots were observed.

3.3. CO_2 Reactions of Winter-Wheat Genotypes during Vegetation at Different Soil Layers under Well-Watered and Drought-Stressed Conditions

The responses of Mv Pálma to the elevated CO_2 showed a variability during the vegetation period and significant alterations were determined between the well-watered and drought-stressed treatments (Figure 4). The CO_2 response was positive and significant from the sowing until the BBCH 21 stage at 30 and 60 cm, and this tendency can be observed for 90 cm at the BBCH 29 stage. Generally, the CO_2 enrichment induced faster root development in Mv Pálma.

Positive and statistically significant CO_2 reactions were observed between the BBCH 37 and BBCH 83 growth stages under optimum watering at 90 cm, and between BBCH 51 and BBCH 69 at 30 cm (Figure 4). The CO_2 fertilization resulted in a significant decrease in the root length of Mv Pálma under drought-stressed conditions at the BBCH 37 stage, and at the later stages of vegetation in each soil layer except in the BBCH 51 stage at 60 cm (Figure 4).

Overall, the CO_2 responses of Mv Karéj were negative in each phenophase, and the reactions were not influenced by the watering (Figure 5). The negative CO_2 responses were more intensive at the end of vegetation, which indicates that the CO_2 fertilization influenced the water balance of plants and induced faster root turnover. The unfavorable impacts of the CO_2 enrichment were more intensive under drought-stressed conditions on the root length, but this process could be a component of the survival strategy of this genotype (Figure 5).

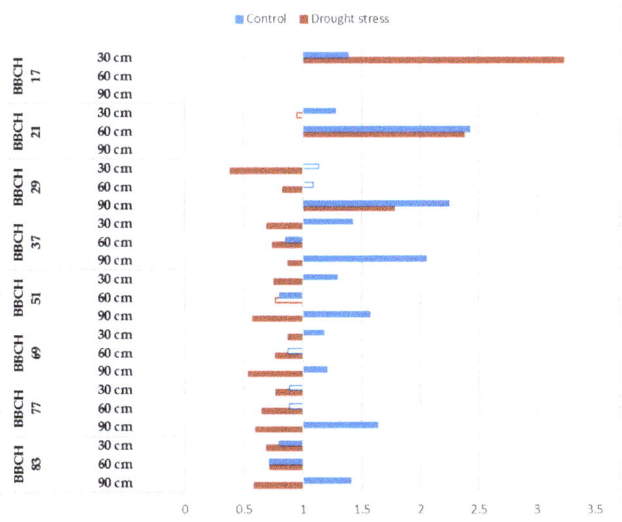

Figure 4. Relative changes in root length of Mv Pálma to elevated CO_2 concentration during the vegetation period at 30, 60 and 90 cm soil layer under optimum watering (control) and drought-stressed conditions. Full bars represent significant differences compared to the control (400 ppm) $p \leq 0.05$ level (n = 3).

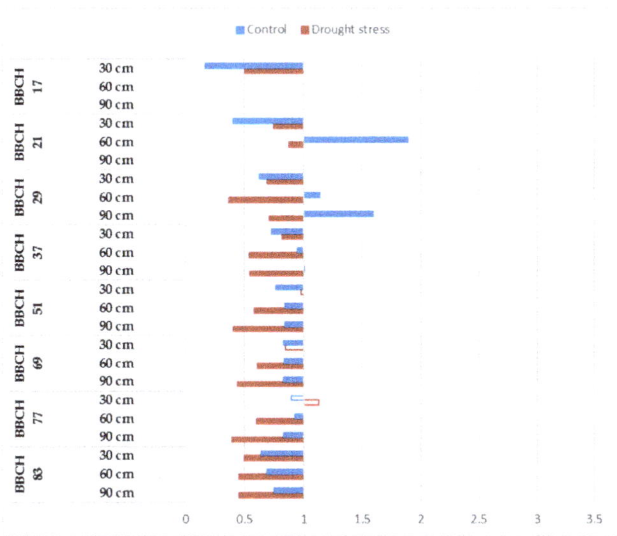

Figure 5. Relative changes in root length of Mv Karéj to elevated CO_2 concentration during the vegetation period at 30, 60 and 90 cm soil layer under optimum watering (control) and drought-stressed conditions. Full bars represent significant differences compared to the control (400 ppm) $p \leq 0.05$ level (n = 3).

The CO_2 reactions of Mv Karizma showed variability during vegetation, and this parameter was influenced significantly by the watering (Figure 6). Roots can be observed only at 30 cm at the BBCH 17 stage, and significant positive CO_2 responses were observed in terms of the root development under drought-stressed conditions, while the opposite tendency was observed under optimum irrigation. The negative impacts of the CO_2 enrichment were detected at 30 cm under optimum irrigation in the BBCH 21 stage, but significant positive responses to both watering levels were observed at 60 cm. The reactions of Mv Karizma to CO_2 enrichment was consequently negative at 30 cm, and this trend was not influenced by the intensity of watering, while the CO_2 reactions were tendentially negative between the BBCH29 and BBCH 83 stages under drought-stressed conditions and positive at 60 cm under controlled irrigation. Negative CO_2 reactions were determined at 90 cm at both watering levels, but this tendency was significant at first when plants reached the BBCH 51 stage (Figure 6).

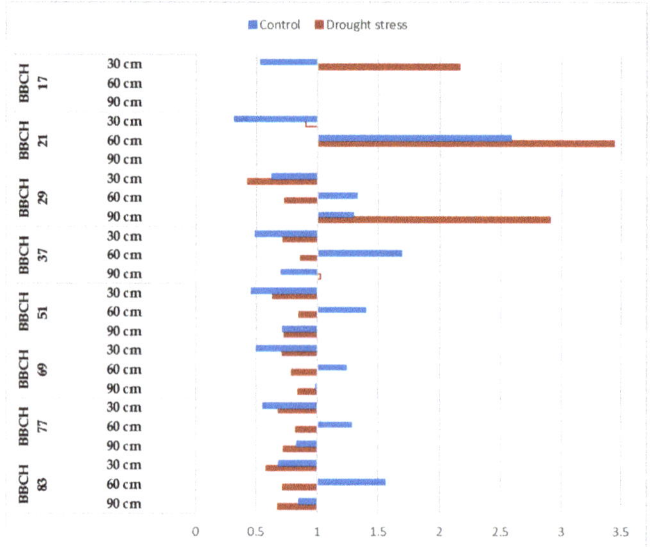

Figure 6. Relative changes in root length of Mv Karizma to elevated CO_2 concentration during the vegetation period at 30, 60 and 90 cm soil layer under optimum watering (control) and drought-stressed conditions. Full bars represent significant differences compared to the control (400 ppm) $p \leq 0.05$ level (n = 3).

4. Discussion

CO_2 molecules in the atmosphere are essential substrates of photosynthesis; therefore, in general, increasing the concentration of CO_2 leads to improved assimilation and crop productivity. Previously, the effects of elevated CO_2 have been widely investigated in various field crops, such as wheat [33,34], maize [35,36], rice [37], sorghum [35], etc. Primarily, these studies determined that CO_2 fertilization improved biomass and grain production, which is also confirmed by our experiment. It has already been correspondingly established that there are differences in the CO_2 reactions of cereal varieties, especially under stress conditions [37–39]. In our experiment, Mv Karizma showed the most intense responses to the increased CO_2 level. Previously conducted studies also highlighted that the increasing CO_2 concentration could partly counterbalance the negative impact of abiotic stresses, such as drought or heat [7,40]. In our experiment, the CO_2 response was variety dependent and the most favorable reactions were observed for Mv Karizma. The yield reduction under drought stress dropped by approximately one half under elevated CO_2 compared to ambient conditions. The physiological background of this result might be that the high CO_2

concentration modifies the intensity of the photosynthesis [39,41], decreases the stomatal conductance and increases the evaporation [42]. Based on this, it can be concluded that although the influence of elevated CO_2 on the aboveground parts of the plants is well-known, the development of the root system should also offset the increased water loss and nutrient uptake. There are differences in the rooting habits of wheat varieties, even under ambient conditions [43,44]; furthermore, the cultivars also give diverse responses to CO_2 [45,46]. Considering the above, it would be highly important to reveal how these effects interact under various irrigation regimes.

The novel aspect of our study is that the effects of the elevated CO_2 concentration were combined with the drought stress, and the differences in the CO_2 responses of the tested varieties were also determined.

Asseng et al. [47] described in a minirhizotron study that the fastest root-growth development was observed between the visible terminal spikelet stage (BBCH 51) and anthesis (BBCH 65), and the maximum root length was measured beyond 90 days after sowing. Based on our results, the most intensive root growth occurred in the vegetative phase, while the date of the highest root length was influenced by the atmospheric CO_2 as well. Under ambient conditions, the maximum root length could be observed at the BBCH 37 stage, but the elevated CO_2 resulted in a constant increase in the root length until the BBCH 51 phase. Uddin et al. [48] described that elevated CO_2 resulted in the intensive root-length formation of wheat plants, especially in the upper soil layer. In our study, however, significant differences were observed between the examined varieties in terms of CO_2 reactions. Our study partly confirmed this trend: although accelerated root development was detected in each soil layer induced by the CO_2 fertilization, overall, the root length was reduced under elevated CO_2 concentration. Mitchell et al. [49] proposed that the potential reason for this phenomenon could be that the surplus assimilates availability for extra root growth under elevated CO_2. Manschadi et al. [50] found in a wheat experiment, in root-observation chambers, that genotypes that were developed for dry conditions had longer root systems in deep soil layers. In our study, the intensive root formation in the deeper soil layers (60 and 90 cm) under drought-stressed conditions indicate the good adaptive capacity of the examined genotypes, especially that of Mv Pálma and Mv Karizma, to a water-limited environment.

In the root characteristics, no differences were detected in near-isogenic lines in the case of the reduced height genes (*Rht1* and *Rht2*) in Western Australia [51], whereas in Argentina, an increase in the total root length and root weight was associated with reduced height [52]. In our experiment, Mv Karizma had the greatest root length, especially under dry conditions, in the deepest soil layer. This might be due to the presence of the less-efficient dwarfing gene (*Rht2*) in its genome. Mv Pálma had the less-developed root system, even under elevated CO_2 concentration, which could be in accordance with the presence of the *Rht8* gene.

5. Conclusions

As a consequence of the water shortage, the depth of the intensive root development remained in the deeper soil layers and under elevated CO_2 concentration, the distribution of the root system was more homogeneous in the whole soil profile than under ambient conditions. The elevated CO_2 concentration induced an accelerated root formation, but considering the whole vegetation period, the CO_2 fertilization had a reducing effect on the root length.

The most intensive root development was detected in the vegetative stage of the plants. The maximum root length was observed at the beginning of the heading, and this period took longer under elevated CO_2 concentration. After the heading, the development of the new roots became slower and intensive root turnover was observed. Increased root formation was determined at the maturity stage in almost every treatment, which could be in association with the re-growth of the stubble under field conditions.

The water shortage generally stimulated root growth, but the depth of the intensive root development proved to be variety dependent and this phenomenon was more intensive under ambient CO_2 concentration.

Supplementary Materials: The following supporting information can be downloaded at: https://www.mdpi.com/article/10.3390/su14063304/s1, Table S1: The effects of the tested factors and their interactions on the aboveground biomass, Table S2: The effects of the tested factors and their interactions on the grain yield, Table S3: The effects of the tested factors and their interactions on the thousand-kernel weight, Table S4: The effects of the tested factors and their interactions on the harvest index.

Author Contributions: Conceptualization, B.V.; methodology, B.V.; validation, G.V., O.V. data curation, B.V., E.V.-L.; writing—original draft preparation, B.V., Z.F.; writing—review and editing G.V., O.V.; visualization, E.V.-L.; funding acquisition, O.V. All authors have read and agreed to the published version of the manuscript.

Funding: This research was funded by the Hungarian Government and the European Union, with the co-funding of the European Regional Development Fund within the framework of the Széchenyi 2020 Program's grant number GINOP-2.3.2-15-2016-00029.

Institutional Review Board Statement: Not applicable.

Informed Consent Statement: Not applicable.

Data Availability Statement: The data presented in this study are available on request from the corresponding author.

Conflicts of Interest: The authors declare no conflict of interest. The funders had no role in the design of the study; in the collection, analyses, or interpretation of data; in the writing of the manuscript, or in the decision to publish the results.

References

1. Intergovernmental Panel on Climate Change. *Climate Change 2013: The Physical Science Basis–Working Group I Contribution to the IPCC Fifth Assessment Report*; Cambridge University Press: Cambridge, UK, 2013.
2. Matiu, M.; Ankrest, D.P.; Menzel, A. Interactions between temperature and drought in global and regional crop yield variability during 1961–2014. *PLoS ONE* **2017**, *12*, e0178339. [CrossRef] [PubMed]
3. Lobell, D.B.; Hammer, G.L.; McLean, G.; Messina, C.; Roberts, M.J.; Schlenker, W. The critical role of extreme heat for maize production in the United States. *Nat. Clim. Chang.* **2013**, *3*, 497–501. [CrossRef]
4. Good, P.; Both, B.B.B.; Chadwick, R.; Hawkins, E.; Jonko, A.; Lowe, J.A. Large differences in regional precipitation change between a first and second 2 K of global warming. *Nat. Commun.* **2016**, *7*, 13667. [CrossRef] [PubMed]
5. Leng, G.; Hall, J. Crop yield sensitivity of global major agricultural countries to droughts and the projected changes in the future. *Sci. Total Environ.* **2019**, *654*, 811–821. [CrossRef]
6. Lesk, C.; Rowhani, P.; Ramankutty, N. Influence of extreme weather disasters on global crop production. *Nature* **2016**, *529*, 84–87. [CrossRef]
7. Varga, B.; Vida, G.; Varga-László, E.; Hoffmann, B.; Veisz, O. Combined effects of drought stress and elevated atmospheric CO_2 concentration on the yield parameters and water use properties of winter wheat (*Triticum aestivum* L.). *J. Agron. Crop. Sci.* **2017**, *203*, 192–205. [CrossRef]
8. Farkas, Z.; Varga-László, E.; Anda, A.; Veisz, O.; Varga, B. Effects of waterlogging, drought and their combination on yield and water-use efficiency of five Hungarian winter wheat varieties. *Water* **2020**, *12*, 1318. [CrossRef]
9. Chun, J.A.; Wang, Q.; Timlin, D.; Fleischer, D.H.; Reddy, V.R. Effect of elevated carbon dioxide and water stress on gas exchange and water use efficiency in corn. *Agric. For. Meteorol.* **2011**, *151*, 378–384. [CrossRef]
10. Meng, F.C.; Zhang, J.H.; Yao, F.M.; Hao, C. Interactive effects of elevated CO_2 concentration and irrigation on photosynthetic parameters and yield of maize in Northeast China. *PLoS ONE* **2014**, *5*, e98318. [CrossRef]
11. Wang, M.; Xie, B.; Fu, Y.; Dong, C.; Hui, L.; Guanghui, L.; Liu, H. Effects of different elevated CO_2 concentrations on chlorophyll contents gas exchange, water use efficiency, and PSII activity on C3 and C4 cereal crops in a closed artificial ecosystem. *Photosynth. Res.* **2015**, *126*, 351–362. [CrossRef]
12. Franzaring, J.; Holz, I.; Fangmeier, A. Responses of old and modern cereals to CO_2-fertilisation. *Crop Pasture Sci.* **2013**, *64*, 943–956. [CrossRef]
13. Rogers, E.D.; Benfrey, P.N. Regulation of plant root system architecture: Implication for crop advancement. *Curr. Opin. Biotechnol.* **2015**, *32*, 93–98. [CrossRef] [PubMed]

14. Ristova, D.; Busch, W. Natural variation of root traits: From development to nutrient uptake. *Plant Physiol.* **2014**, *166*, 518–527. [CrossRef] [PubMed]
15. Benlloch-Gonzalez, M.; Bochicchio, R.; Berger, J.; Bramley, H.; Palta, J.A. High temperature reduces the positive effects of elevated CO_2 on wheat root system growth. *Field Crop. Res.* **2014**, *165*, 71–79. [CrossRef]
16. Zuo, Q.; Jie, F.; Zhang, R.; Mend, L. A generalized function of wheat's root length density distribution. *Vadose Zone J.* **2004**, *3*, 271–277. [CrossRef]
17. Wasson, A.P.; Rebetzke, G.J.; Kirkegaard, J.A.; Christopher, J.; Richards, R.A.; Watt, M. Soil coring at multiple field environments can directly quantify variation in deep root traits to select wheat genotypes for breeding. *J. Exp. Bot.* **2014**, *65*, 6231–6249. [CrossRef]
18. Zhao, G.; Liu, J.; Ciu, J.; Wang, H.; Wen, G. Revealing the mechanism of the force dragging the soft bag in the dynamic process pf deep soil coring. *Powder Technol.* **2019**, *344*, 251–259. [CrossRef]
19. Cai, G.; Vanderbought, J.; Klotzche, A.; van der Kruk, J.; Neumann, J.; Hermes, N.; Vereecken, H. Construction of Minirhizotron Facilities for Investigating Root Zone Processes. *Vadose Zone J.* **2016**, *15*, vzj2016.05.0043. [CrossRef]
20. Weigand, M.; Kemna, A. Imaging and functional characterization of crop root systems using spectroscopic electrical impedance measurements. *Plant Soil* **2019**, *435*, 201–224. [CrossRef]
21. Weigand, M.; Kemna, A. Multi-frequency electrical impedance tomography as a non-invasive tool to characterize and monitor the crop root system. *Biogeosciences* **2017**, *14*, 921–939. [CrossRef]
22. Xu, Z.; Valdes, C.; Clarke, J. Existing and potential statistical and computational approaches for the analysis of 3D CT images of plant roots. *Agronomy* **2018**, *8*, 71. [CrossRef]
23. Atkinson, J.A.; Pound, M.P.; Bennett, M.J.; Wells, D.M. Uncovering the hidden half of the plants using new advances in root phenotyping. *Curr. Opin. Biotechnol.* **2019**, *55*, 1–8. [CrossRef] [PubMed]
24. Tracy, S.R.; Nagel, K.A.; Postman, J.A.; Fassbender, H.; Wasson, A.; Watt, M. Crop improvement from phenotyping roots: Highlights reveal expanding opportunities. *Trends Plant Sci.* **2020**, *25*, 105–118. [CrossRef] [PubMed]
25. Barnett, S.; Zhao, S.; Ballard, R.; Franco, C. Selection of microbes for control of Rhizoctonia root rot on wheat using a high throughput pathosystem. *Biol. Control.* **2017**, *113*, 45–57. [CrossRef]
26. Jaradat, A.A. Wheat landraces: A mini-review. *Emir. J. Food Agric.* **2013**, *25*, 20–29. [CrossRef]
27. Abido, W.E.A.; Zsombik, L. Effect of water stress on germination of some Hungarian wheat landraces varieties. *Acta Ecol. Sin.* **2018**, *38*, 422–428. [CrossRef]
28. Nagy, É.; Lehoczki-Krsjak, S.; Lantos, C.; Pauk, J. Phenotyping for testing drought tolerance on wheat varieties of different origins. *S. Afr. J. Bot.* **2018**, *116*, 216–221. [CrossRef]
29. Dowla, M.A.N.; Edwards, I.; O'Hara, G.; Islam, S.; Ma, W. Developing Wheat for Improved Yield and Adaptation Under a Changing Climate: Optimization of a Few Key Genes. *Engineering* **2018**, *4*, 514–522. [CrossRef]
30. Gulyás, G.; Bognár, Z.; Láng, L.; Bedő, Z. Distribution of dwarfing genes (Rht-B1b and Rht-D1b) in Martonvásár wheat breeding materials. *Acta Agron. Hung.* **2011**, *59*, 249–254. [CrossRef]
31. Tischner, T.; Kőszegi, B.; Veisz, O. Climatic programmes used on Martonvásár phytotron most effectively in recent years. *Acta Agron. Hung.* **1997**, *45*, 85–104.
32. Lancashire, P.D.; Bleiholder, H.; van den Boom, T.; Langlüdekke, P.; Stauss, R.; Weber, E.; Witzenberger, A. A uniform decimal code for growth stages of crops and weeds. *Ann. Appl. Biol.* **1991**, *119*, 561–601. [CrossRef]
33. Kimball, B.A.; Pinter, P.J.; Garcia, R.L.; LaMorte, R.L.; Wall, G.W.; Hunsaker, D.J. Productivity and water use of wheat under free-air CO_2 enrichment. *Glob. Chang. Biol.* **1995**, *1*, 429–442. [CrossRef]
34. Högy, P.; Wiese, H.; Koehler, P.; Schwadorf, K.; Breuer, J.; Franzaring, J. Effects of elevated CO_2 on grain yield and quality of wheat: Results from a 3-year free-air CO_2 enrichment experiment. *Plant Biol.* **2009**, *11*, 60–69. [CrossRef]
35. Manderscheid, R.; Erbs, M.; Weigel, H.-J. Key physiological parameters related to differences in biomass production of maize and four sorghum cultivars under drought and free-air CO_2 enrichment. *Procedia Environ. Sci.* **2015**, *29*, 89–90. [CrossRef]
36. Qiao, Y.; Miao, S.; Li, Q.; Jin, J.; Lou, X.; Tang, C. Elevated CO_2 and temperature increase grain oil concentration but their impacts on grain yield differ between soybean and maize grown in a temperate region. *Sci. Total Environ.* **2019**, *666*, 405–413. [CrossRef] [PubMed]
37. Hu, S.; Wang, Y.; Yang, L. Response of rice yield traits to elevated atmospheric CO_2 concentration and its interaction with cultivar, nitrogen application rate and temperature: A meta-analysis of 20 years FACE studies. *Sci. Total Environ.* **2020**, *764*, 142797. [CrossRef]
38. Weigel, H.-J.; Manderscheid, R. Crop growth responses to free-air CO_2 enrichment and nitrogen fertilization: Rotating barley, ryegrass, sugar beet and wheat. *Eur. J. Agron.* **2012**, *43*, 97–107. [CrossRef]
39. Zheng, Y.; He, C.; Guo, L.; Hao, L.; Cheng, D.; Li, F.; Peng, Z.; Xu, M. Soil water status triggers CO_2 fertilization effect on the growth of winter wheat (*Triticum aestivum*). *Agric. For. Meteorol.* **2020**, *291*, 108097. [CrossRef]
40. Zhang, X.; Shi, Z.; Jiang, D.; Högy, P.; Fangmeier, A. Independent and combined effects of elevated CO_2 and post-anthesis heat stress on protein quantity and quality in spring wheat grains. *Food Chem.* **2019**, *277*, 524–530. [CrossRef]
41. Sinha, P.G.; Saradhi, P.P.; Upverty, D.C.; Bhatnagar, A.K. Effect of elevated CO_2 concentration on photosynthesis and flowering in three wheat species belonging to different ploidies. *Agric Ecosyst. Environ.* **2011**, *142*, 432–436. [CrossRef]

42. Manderscheid, R.; Dier, M.; Erbst, M.; Sickora, J.; Weigel, H.-J. Nitrogen supply–A determinant in water use efficiency of winter wheat grown under free-air CO_2 enrichment. *Agric Water Manag.* **2018**, *210*, 70–77. [CrossRef]

43. Severini, A.D.; Wasson, A.P.; Evans, J.R.; Richards, R.A.; Watt, M. Root phenotypes at maturity in diverse wheat and triticale genotypes grown in three field experiments: Relationships to shoot selection, biomass, grain yield, flowering time, and environment. *Field Crop. Res.* **2020**, *255*, 107870. [CrossRef]

44. Dreccer, M.F.; Condon, A.G.; Macdonald, B.; Rebetzke, G.J.; Awasi, M.-A.; Borgognone, M.G.; Peake, A.; Pinare-Chavez, F.J.; Hungt, A.; Jackway, P.; et al. Genotypic variation for lodging tolerance in spring wheat: Wider and deeper root plates, a feature of low lodging, high yielding germplasm. *Field Crop Res.* **2020**, *258*, 1078942. [CrossRef]

45. Hansen, E.M.O.; Hauggaard-Nielsen, H.; Launay, M.; Rose, P.; Mikkelsen, T.N. The impact of ozone exposure, temperature and CO_2 on the growth and yield of three spring wheat varieties. *Environ. Exp. Bot.* **2019**, *168*, 103868. [CrossRef]

46. Erice, G.; Sanz-Sáez, Á.; González-Torralba, J.; Méndez-Espinoza, A.M.; Urretavizcaya, I.; Nieto, M.T.; Serret, M.D.; Araus, J.L.; Irigoyen, J.J.; Aranjuelo, I. Impact of elevated CO_2 and drought on yiled and quality traits of a historical (Blanqueta) and a modern (Sula), durum wheat. *J. Cereal. Sci.* **2019**, *87*, 194–201. [CrossRef]

47. Asseng, S.; Ritchie, J.T.; Smucker, A.J.M.; Robertson, M.J. Root growth and water uptake during water deficit and recovering in wheat. *Plant Soil* **1998**, *201*, 265–273. [CrossRef]

48. Uddin, S.; Löw, M.; Parvin, S.; Fitzgerald, G.; Bahrami, H.; Tausz-Posch, S.; Armstrong, R.; O'Leary, G.; Tausz, M. Water use and growth responses of dryland wheat grown under elevated [CO_2] are associated with root length in deeper, but not upper soil layer. *Field Crops Res.* **2018**, *224*, 170–181. [CrossRef]

49. Mitchell, J.H.; Chapman, S.C.; Rebetzke, G.J.; Bonnett, D.G.; Fukai, S. Evaluation of a reduced-tilling (tin) gene in wheat lines grown across different production environments. *Crop Pasture Sci* **2012**, *63*, 128–141. [CrossRef]

50. Manschadi, A.M.; Christopher, J.; Devoil, P.; Hammer, G.L. The role of root architectural traits in adaptation of wheat to water-limited environments. *Funct. Plant Biol.* **2006**, *33*, 823–837. [CrossRef]

51. Siddique, K.H.M.; Belford, R.K.; Tennant, D. Root: Shoot ratios of old and modern, tall and semi-dwarf wheat in a Mediterranean environment. *Plant Soil* **1990**, *121*, 89–98. [CrossRef]

52. Miralles, D.J.; Slafer, G.A.; Lynch, V. Rooting patterns in near-isogenic lines of spring wheat for dwarfism. *Plant Soil* **1997**, *197*, 79–86. [CrossRef]

 *sustainability*

Article

Screening Wild Pepper Germplasm for Resistance to *Xanthomonas hortorum* pv. *gardneri*

Zoltán Gábor Tóth [1], Máté Tóth [1], Sándor Fekete [2], Zoltán Szabó [1] and Zoltán Tóth [1,*]

[1] Institute of Genetics and Biotechnology, Applied Plant Genomics Group, Hungarian University of Agriculture and Life Sciences, Szent-Györgyi Albert Str. 4, 2100 Gödöllő, Hungary

[2] GTIG, Török Ignác Secondary School of Gödöllő, Petőfi Sándor Str. 12-14, 2100 Gödöllő, Hungary

* Correspondence: toth.zoltan.gen@uni-mate.hu

Abstract: Bacterial spot disease on peppers is caused by four species of the genus *Xanthomonas*. This disease causes black spot lesions not only on the leaves but also on the fruit, leading to yield and quality loss. *Xanthomonas* species cause major disease outbreaks in tropical, subtropical and humid continental regions worldwide. Bacterial blight caused by xanthomonads occurs on both greenhouse- and field-grown peppers and is particularly important in areas characterized by hot and humid environmental conditions. As pesticides are currently not sufficiently effective in the control of bacterial spot, the development of pepper varieties resistant to *Xanthomonas* species, including *X. hortorum* pv. *gardneri*, is of primary importance for sustainable production. In our research, 119 lines of *Capsicum baccatum* from the USDA ARS gene bank (Griffin, GA) and MATE (Hungarian University of Agriculture and Life Sciences) were tested against strains of *X. hortorum* pv. *gardneri* under greenhouse conditions. Four accessions of the wild pepper species *C. baccatum* appeared to be resistant to seven strains of *X. hortorum* pv. *gardneri* in greenhouse trials. The resistant genotypes of *X. hortorum* pv. *gardneri* identified in this study can be used for the resistance gene pyramidation against different bacterial spotted *Xanthomonas* species in pepper.

Keywords: sustainability; pepper breeding; bacterial spot resistance; *Xanthomonas hortorum* pv. *gardneri*

Citation: Tóth, Z.G.; Tóth, M.; Fekete, S.; Szabó, Z.; Tóth, Z. Screening Wild Pepper Germplasm for Resistance to *Xanthomonas hortorum* pv. *gardneri*. *Sustainability* **2023**, *15*, 908. https://doi.org/10.3390/su15020908

Academic Editor: Balázs Varga

Received: 29 November 2022
Revised: 26 December 2022
Accepted: 30 December 2022
Published: 4 January 2023

1. Introduction

Pepper (*Capsicum* spp.) is one of the most important vegetables and spices worldwide because of its versatility, its high nutritional value, as well as its role in reducing human micronutrient deficiencies [1]. The chili pepper genus is currently known to consist of 43 species [2] that are native to temperate and tropical Central and South America, Mexico and the West Indies, five of which are domesticated: *C. annuum*, *C. chinense*, *C. frutescens*, *C. baccatum*, and *C. pubescens* [3]. Despite the crucial agronomic importance of this crop, diseases are still one of the most important damaging factors in pepper production, and only few can be controlled by chemical treatments. However, environmental and consumer concerns, as well as the risk of crop pathogen insensitivity, underline the importance of developing resistant pepper varieties as a more effective way of reducing the impact of disease [4,5]. *Capsicum* germplasm collections represent an immense genetic wealth resource this purpose, but screening for disease resistance is often a lengthy, labor-intensive, expensive and mostly complex process.

Bacterial spot disease of peppers occurs in both processing and fresh market pepper fields, particularly where growing conditions are characterized by high humidity. Known for its aggressiveness, the pathogen *X. hortorum* pv. *gardneri* causes large, star-shaped necrotic lesions on the foliage, stem and fruit. Diseased leaves drop prematurely, resulting in extensive defoliation, thus reducing plant productivity and exposing fruits to potential sunscalding. Stem lesions occur as narrow, light brown, longitudinally raised cankers. Fruit with such large lesions are non-marketable in both the fresh and processed markets.

Bacterial spot is caused by four bacterial species of the genus *Xanthomonas*: *X. euvesicatoria*, *X. vesicatoria*, *X. perforans* and *X. hortorum* pv. *gardneri* [6]. *X. hortorum* pv. *gardneri* was first described in 1957 [7]. The pathogen *X. hortorum* pv. *gardneri* was recognized as a causative agent of bacterial spot epidemics in the year 2000. Currently, there are few chemicals available that are effective against these pathogens, and even these are ineffective under high disease pressure and have a negative impact on the environment. Since the overuse of copper has led to the development of resistance in the bacterial population, there is a need to find new sustainable and environmentally friendly alternatives [8,9]. The growing number of bacterial spot diseases caused by emerging varieties of xanthomonads makes it increasingly urgent for growers to find practical and sustainable solutions. The use of varieties resistant to bacterial leaf spot offers a potential tool for disease control in field conditions. The spread of new *Xanthomonas* species over the last few years has made the discovery of new sources of resistance a major priority. Thus, host resistance is considered to be a critical element of disease management strategies.

Resistance breeding programs successfully developed commercial pepper lines with hypersensitive and quantitative resistance to various *Xanthomonas* species [10]. Genes regulating resistance have been found in *X. euvesicatoria* [11], *X. perforans* species T3 [12], *X. perforans* species T4 [13], and *X. vesicatoria* [14], and six dominant resistance genes have been identified in pepper to date, namely *Bs1*, *Bs2*, *Bs3*, *Bs4C*, *Bs7* and *BsT* [15]. However, the persistence of resistance can change rapidly due to the changing geographical distribution of the pathogen and the rapid emergence of new pathogenic variants [16]. In the case of *X. hortorum* pv. *gardneri*, only one resistance gene source has been described so far, although in this case the infection was caused by a single *Xantomonas* strain (USVLXG1) and resistance was not tested with other *Xanthomonas* isolates [17]. The introduction of resistance genes with different mechanisms of action could provide a durable and sustainable solution to this ongoing host–pathogen arms race.

Commercial pepper varieties carrying recessive resistance genes, namely *bs5* and *bs6*, have not proven effective against *X. hortorum* pv. *gardneri*. Based on current publications, the recessive resistance gene *bs8* identified in *C. annuum* accession PI 163192 is the only one that was shown to confer resistance to *X. hortorum* pv. *gardneri* [18]. With the increasing frequency of *Xanthomonas* outbreaks around the world, there is a growing rationale to identify new sources of resistance in pepper to emerging bacterial species, including *X. hortorum* pv. *gardneri* and to incorporate these new resistance genes into breeding programs.

Preliminary results are presented below, which are aimed at screening wild pepper cultivars for heritable resistance to *X. hortorum* pv. *gardneri*. The aim of our current work was to identify new sources of resistance to *X. hortorum* pv. *gardneri* in pepper and to identify potential candidate genes involved in resistance to bacterial spot disease. A diverse USDA panel of pepper accessions were screened for resistance to *X. hortorum* pv. *gardneri*, and highly resistant accessions were identified in our work.

2. Materials and Methods

2.1. Plant Materials

The work was carried out on a collection of *C. baccatum* accessions originating from the USDA ARS genebank. We tested 119 accessions including 97 *C. baccatum* var. *pendulum*, 2 *C. baccatum* var. *pratermissum*, 7 *C. baccatum* var. *baccatum*, and 2 *C. baccatum* var. *umbilicatum*, and in 11 cases the variant was not specified (Table S1). In order to improve germination, the seeds were pre-soaked for four hours before they were sterilized with 10% calcium hypochlorite ($Ca(ClO)_2$) solution for twenty minutes with a few drops of Tween20. The seeds were rinsed five times with sterile Milli-Q Water. After that, all of the seeds were placed on MS medium [19] with 20 g sucrose (MS20) in sterilized glass storage jars until the cotyledon stage. Afterwards, the plants were placed into pots filled with Klasmann Traysubstrate soil mixed with 50 g NPK and microelement fertilizer (14% N + 7% P + 21% K + 1% Mg + 1% microelements (B, Cu, Mn, Fe, Zn) to 70 L of soil.

The plants were then grown under controlled greenhouse conditions at $24 \pm 2\,^\circ$C, with 14/10 h photoperiods, and 50–70% relative humidity until they reached the six-mature-leaf stage. A cultivated variety *C. annuum* Fö "Fehérözön" was used as a susceptible control.

2.2. Bacterial Isolates and Plant Infiltration

To find a resistant phenotype, two *Xg* strains were used for plant infiltration—LMG962 from the Belgian Coordinated Collections of Microorganisms (BCCM) and SRB, which is a field isolate from Serbia [7]. Bacteria were cultivated in YDC [20] medium for two days in an incubator at a temperature of 30 °C. After two days of incubation, bacteria were collected and diluted with sterile Milli-Q Water to a final concentration of 10^6 cfu/mL. The infiltration was performed in planta with a needle-free syringe into the intercellular spaces on the abaxial side of the first true leaves. After infiltration, the plants were placed in plant growth chambers at 25 °C with a 16/8 h photoperiod, at 50–70% relative humidity.

The response to infection was monitored continuously, but the final evaluation was made after 7 days. The plants that showed complete resistance to both isolates were also tested with five further *Xg* strains: Xg51, Xg152, Xg153, Xg156, Xg177 (Table 1), which were donated by Brian Staskawicz (Innovative Genomics Institute University of California, Berkeley).

Table 1. *Xg* strains used in this study and their origin.

Species	Strain Name	Origin	Host Isolated	Year Isolated	Isolation ID	Collector
	LMG962	Republic of Yugoslavia	tomato	1957	-	Sutic D.
	SRB	Republic of Yugoslavia	tomato	1957	-	Sutic D.
X. hortorum pv. *gardneri*	Xg51	-[1]	-	-	-	T. Minsavage
	Xg152	-	-	-	-	-
	Xg153	Gibsonburg, OH	tomato	2010	SM194-10	S. Miller
	Xg156	Blissfield, MI	tomato	2010	SM177-10	S. Miller
	Xg177	Sandusky Co., OH	tomato	2012	SM795-12	S. Miller

[1] means that the exact origin, host, year, isolation ID or collector of the isolate has not been established.

2.3. Bacterial Growth Test

In order to determine how the presence of resistance affected bacterial growth in resistant plants, leaf disc samples were collected from the infiltrated area at different times. For the test of the LMG962 isolate, four resistant and three susceptible plants were used in two replicates at 25 °C and 30 °C. The photoperiod and humidity were the same as for the accession testing. Sampling was performed at 0, 2, 4, 6, 8, 10, 12 days post inoculation (dpi) from the inoculated area. In the case of resistant plants, re-isolation was also performed 30 days after infection to monitor possible changes in bacterial counts. The isolated leaf discs were macerated in 100 µL sterilized water and prepared in dilution lines and were scattered on YDC medium. After two or three days, the colonies were counted. The cfu/cm^2 of leaves was calculated by counting the number of adult colonies.

3. Results

3.1. Evaluation of C. baccatum Accessions for the Susceptibility and Resistance to X. hortorum pv. gardneri

One hundred and nineteen *C. baccatum* accessions were inoculated with two highly virulent strains (LMG962 and SRB) of *X. hortorum* pv. *gardneri* and evaluated for symptoms. Disease evaluation started one week after inoculation and the earliest typical disease symptoms were observed on the leaves 3–6 days after inoculation on the infected pepper plants. The average incubation period varied between 4 and 7 days until the initial abnormal discoloration appeared, while the length of time required to develop complete leaf necrosis in sensitive individuals varied between 6 and 10 days.

Susceptible symptoms typically started with chlorotic lesions, which developed into angular, water-soaked lesions and eventually into brown or black necrotic lesions. Phenotyping of the control line "Fehérözön" (highly susceptible) conformed to the expected disease reactions. In our studies, we found that *C. baccatum* accessions differed in terms of the earliness and intensity of symptoms. The vast majority of the accessions were susceptible to both strains, showing a similar phenotype to "Fehérözön" used as a control. A small part of the accessions showed partial resistance to one of the *X. hortorum* pv. *gardneri* strain. Based on the evaluation of their reactions, four of the 119 *C. baccatum* genotypes showed strong resistance to both *Xg* isolates used in the first selection round (Figure 1).

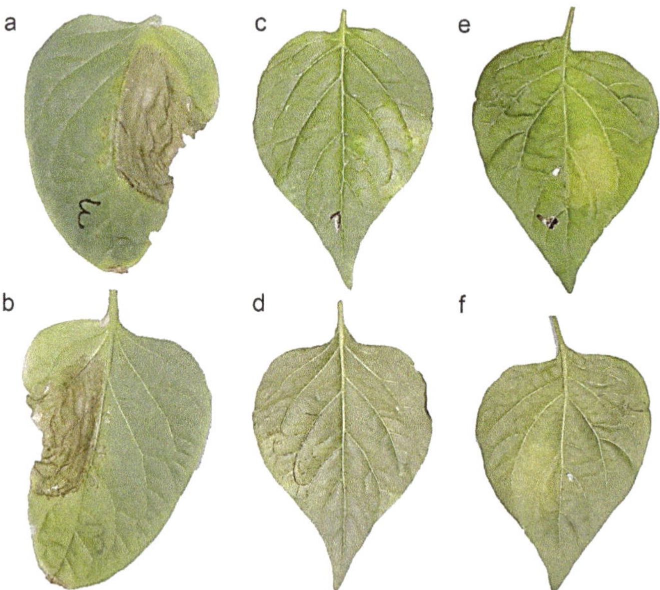

Figure 1. Symptoms of bacterial spot 7 days after inoculation with strain LMG962 in susceptible Fö (**a,b**) and the resistant phenotype on Cbp2 plants infected with strain LMG962 (**c,d**) and SRB (**e,f**). Adaxial and abaxial view.

In the second phase, all four previously resistant phenotypes of *C. baccatum* accessions were challenged with five additional *Xg* isolates. Based on our results, all four accessions also showed strong resistance (Table 2) to all five additional *Xg* isolates (Xg51, Xg152, Xg153, Xg156, Xg 177).

Table 2. *C. baccatum* var. *pendulum* accessions showing partial or complete resistance to the *X. hortorum* pv. *gardneri* strains used.

Species	Variety	Accession	SRB	LMG962	XG51	XG152	XG153	XG156	XG177
		Cbp2	+	+	+	+	+	+	+
		Cbp3	+	+	+	+	+	+	+
		Cbp4	+	+	+	+	+	+	+
		PI 441541	+	−					
		PI 441542	+	−					
C.baccatum	*pendulum*	PI 370010	−	+					
		PI 441543	+	-					
		PI 441578	+	−					
		PI 441552	−	+					
		PI 441533	−	+					
		PI 441520	+	−					
		Cbp1	+	+	+	+	+	+	+

+: resistant response to infection, −: sensitive response to infection.

In the resistant plants, the infected area turned light green, but at 30 dpi, it was no longer possible to distinguish between infiltrated and untreated areas. Small necrotic spots appeared on the leaves at 10–12 dpi typically along the leaf veins, but these lesions increased very little if at all after 30 days (Figure 2). No difference was observed between the emerging phenotype in resistant plants according to the bacterial strain used for inoculation.

Figure 2. Symptoms of bacterial spot 30 days after inoculation with strain LMG962, SRB and Xg51 in a resistant accession plant.

3.2. Evaluation of the Bacterial Growth Test

We determined the effect of the identified resistance gene on bacterial growth in the plant and the correlation of visual symptoms with bacterial population levels. The growth in the populations of different bacterial races was monitored for 12 days after inoculation and incubation at 25 °C and 30 °C. "Fehérözön" was used as a negative control. Compared to the almost equal initial values, the bacterial growth curve in planta at 25 °C indicated that resistant plants reached a maximum of 4.75 log at 4 dpi and then showed a minimal decrease with no significant change in the following 6–12 days. In contrast, in the case of the sensitive plants, a value of 6.7 log at 4 dpi was obtained and a steady increase was detected until leaf drop at 8 dpi (Figure 3). In the majority of cases, the infected leaves of the susceptible plants had fallen off by the eighth day and sampling could not be continued. At 30 °C, the resistant plants reached a maximum value of 5.55 log at 6 dpi and then a slight decrease was observed. Susceptible plants showed a value of 7.1 log at 6 dpi until leaf drop (Figure 4). In susceptible plants, the temperature had no significant effect on the growth curve of *Xg* bacteria, but significantly affected the tolerance of plants to infection, because inoculated leaves fell off earlier at higher temperatures. Plants kept at 25 °C were also sampled at 30 dpi and the results show that no significant change in the cfu value in the leaf occurred. In the case of plants kept at 30 °C, it was not possible to collect a sample after one month, due to early aging of the plant and leaf fall.

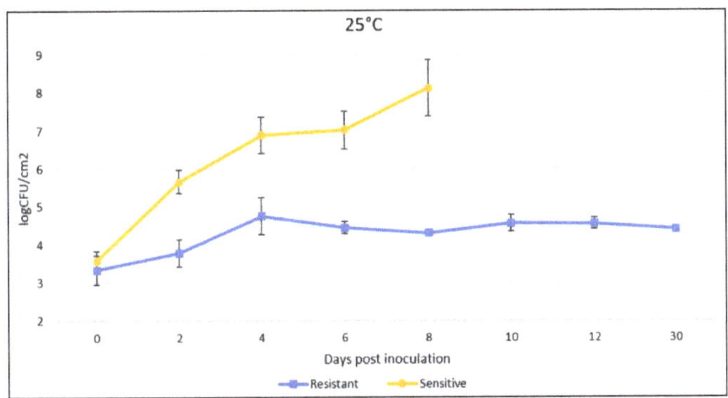

Figure 3. Time course of bacterial population growth at 25 °C after infiltration of leaves of resistant and susceptible plants by LMG962 strains.

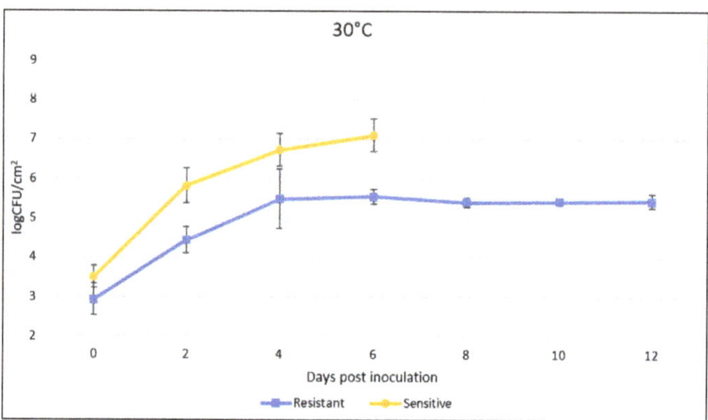

Figure 4. Time course of bacterial population growth at 30 °C after infiltration of leaves of resistant and susceptible plants by LMG962 strains.

4. Discussion

C. baccatum is the most popular and widely cultivated chili pepper in South America. There is a wild form called *C. baccatum* var. *baccatum* and a domesticated form (*C. baccatum* var. *pendulum*) [21]. Brazil is considered to be the center of the genetic diversity of *Capsicum* spp., and at least 116 accessions already described are found in Brazil [22,23]. It is expected that new species [2] and new resistance genes for pepper breeding will be found in this important and untapped genetic resource. Pathogens are increasingly threatening the cultivation and yield of most *Capsicum* species worldwide. In this context, the search for new sources of resistance remains a major challenge today and *Capsicum* germplasm collections represent a promising target for this purpose.

Previous searches for resistance to *X. hortorum* pv. *gardneri* concentrated on the *Xg* isolates which are most prevalent in the US. In contrast to previous research, in our present work we also used the isolates most commonly found in Europe and the US. The results of this screening study in which pepper accessions were inoculated with seven *Xg* strains indicated that there were significant differences in resistance between PI lines from the USDA ARS pepper germplasm collection. Screening was carried out in two stages. As a first step, 119 accessions from the *C. baccatum* collection were challenged with two different strains of *X. hortorum* pv. *gardneri* to determine the extent and types of resistance in *C. baccatum*. Sensitive symptoms of bacterial spot were observed in 107 accessions and *Xg* was recovered from all plants, indicating that none of the accessions carried a source of resistance to this pathogen. Eight accessions showed partial resistance to one of the *Xg* isolates and strong resistance was identified in four PI lines (Cbp1, Cbp2, Cbp3 and Cbp4). In the second stage, screening was carried out on Cbp1, Cbp2, Cbp3 and Cbp4 accessions using the five potentially important isolates in Europe and the US. Complete screening was carried out under controlled conditions in growth chambers using a method (leaf screening of young plants with a concentrated bacterial suspension) that clearly indicated the presence of the pathogen and the rapid onset of symptoms.

The screening results showed two main types of resistance (Table 2) in the tested plant population. Species-specific resistance was the most common and was present in 6% of all accessions tested. Resistance to all species used (potential species-nonspecific resistance) was very rare and was present in 3% of accessions tested.

5. Conclusions

Species shifts in bacterial spot pathogen populations are very common [10,24] and pose a serious problem and challenge for pepper breeding programs due to the decline in resistance persistence [25]. Therefore, there is an increasing need to identify new sources of resistance to these pathogens. The resistance germplasm of *X. hortorum* pv. *gardneri* identified in this study can be of significant importance for both breeders and researchers and is highly relevant for introgression into pepper lines and in terms of gene pyramidation efforts against various *Xanthomonas* species. The resistance genes that we identified in *C. baccatum* might also have significant potential for resistance inhibition in other pepper species such as *C. annuum*. Interspecific crosses provide the opportunity to pass genes from one species to another, but in most cases, this can present several difficulties. A successful wide hybridization among species is highly dependent on the genetic distance between species. According to our current knowledge, two options are available for introgressing *C. baccatum* genes into *C. annuum*. The first option is using *C. chinense* and *C. frutescens* as bridge species [26], while the second option is to carry out direct crosses between *C. annuum* and *C. baccatum* combined with in vitro embryo rescue [27]. Our future goal is to use both methods to introduce our newly identified valuable resistance gene into *C. annuum*. Using the four resistant parental lines, crosses are underway with the aim of creating F1 and F2 progeny to determine the inheritance of disease resistance.

The new sources of resistance identified in this study provide a basis for further work on breeding disease-resistant varieties that are resilient to the changing population structure of pathogens. The current study showed that the different genotypes of *C. baccatum* differed in their response to wilt caused by the pathogen *Xanthomonas*, which also demonstrated that the pepper PI collection offered a valuable public source of resistance for pepper breeders to develop varieties resistant to bacterial spot. In conclusion, the information in this study can be summarized in that the use of a resistant genotype, together with cultural practices and sanitary control measures, is considered the most feasible and sustainable way to control *X. hortorum* pv. *gardneri* wilt.

Supplementary Materials: The following supporting information can be downloaded at: https://www.mdpi.com/article/10.3390/su15020908/s1, Table S1: Screening of *C. baccatum* germplasm to *X. hortorum* pv. *gardneri*.

Author Contributions: Conceptualization, Z.G.T. and Z.T.; methodology, Z.G.T. and Z.T.; software, Z.T.; validation, Z.G.T., Z.S. and Z.T.; formal analysis, Z.G.T., M.T., Z.S. and Z.T.; investigation, Z.G.T. and Z.T.; resources, Z.G.T. and Z.T.; data curation, Z.G.T. and Z.T.; writing—original draft preparation, Z.G.T. and Z.T.; writing—review and editing, Z.G.T., S.F., Z.S. and Z.T.; visualization, Z.T.; supervision, S.F. and Z.T.; project administration, Z.G.T. and Z.T. All authors have read and agreed to the published version of the manuscript.

Funding: This research received no external funding.

Institutional Review Board Statement: Not applicable.

Informed Consent Statement: Not applicable.

Data Availability Statement: Not applicable.

Conflicts of Interest: The authors declare no conflict of interest.

References

1. FAOSTAT, Food and Agriculture Organization (FAO). 2021. Available online: http://www.fao.org/faostat/en/#home (accessed on 15 April 2021).
2. Barboza, G.E.; García, C.C.; Scaldaferro, M.; Bohs, L. An amazing new *Capsicum* (*Solanaceae*) species from the Andean-Amazonian Piedmont. *PhytoKeys* **2020**, *167*, 13–29. [CrossRef] [PubMed]
3. USDA, ARS, National Genetic Resources Program. Germplasm Resources Information Network (GRIN). National Germplasm Resources Laboratory, Beltsville, Maryland. Available online: http://www.arsgrin.gov/cgi-bin/npgs/html/exsplist.pl (accessed on 23 May 2015).
4. Mundt, C.C. Durable resistance: A key to sustainable management of pathogens and pests. *Infect. Genet. Evol.* **2014**, *27*, 446–455. [CrossRef] [PubMed]
5. Collinge, D.B.; Sarrocco, S. Transgenic approaches for plant disease control: Status and prospects. *Plant Pathol.* **2022**, *71*, 207–225. [CrossRef]
6. Jones, J.B.; Lacy, G.H.; Bouzar, H.; Stall, R.E.; Schaad, N.W. Reclassification of the xanthomonads associated with bacterial spot disease of tomato and pepper. *Syst. Appl. Microbiol.* **2004**, *27*, 755–762. [CrossRef]

7. Šutic, D. Bakterioze crvenog patlidzana (Tomato bacteriosis). *Rev. Appl. Mycol.* **1957**, *36*, 734–735.
8. Khanal, S.; Hind, S.R.; Babadoost, M. Occurrence of Copper-Resistant *Xanthomonas perforans* and *X. gardneri* in Illinois Tomato Fields. *Plant Health Prog.* **2020**, *21*, 338–344. [CrossRef]
9. El-Fiki, I.A.I.; Youssef, M.; Eman, H.O. Controlling The Bacterial Leaf Spot Disease in Pepper Caused by *Xanthomonas vesicatoria* Using Natural Bacteritoxicants. *Egypt. Acad. J. Biol. Sci. F. Toxicol. Pest Control.* **2022**, *14*, 229–245. [CrossRef]
10. Stall, R.E.; Jones, J.B.; Minsavage, G.V. Durability of resistance in tomato and pepper to *xanthomonads* causing bacterial spot. *Annu. Rev. Phytopathol.* **2009**, *47*, 265–284. [CrossRef]
11. Yang, W.C.; Sacks, E.J.; Ivey, M.L.L.; Miller, S.A.; Francis, D.M. Resistance in Lycopersicon esculentum intraspecific crosses to race T1 strains of *Xanthomonas campestris* pv. *vesicatoria* causing bacterial spot of tomato. *Phytopathology* **2005**, *95*, 519–527. [CrossRef]
12. Wang, H.; Hutton, S.F.; Robbins, M.D.; Sim, S.C.; Scott, J.W.; Yang, W.; Jones, J.B.; Francis, D.M. Molecular mapping of hypersensitive resistance from tomato 'Hawaii 7981' to *Xanthomonas perforans* race T3. *Phytopathology* **2011**, *101*, 1217–1223. [CrossRef]
13. Sharlach, M.; Dahlbeck, D.; Liu, L.; Chiu, J.; Jiménez-Gómez, J.M.; Kimura, S.; Koenig, D.; Maloof, J.N.; Sinha, N.; Minsavage, G.V.; et al. Fine genetic mapping of *RXopJ4*, a bacterial spot disease resistance locus from *Solanum pennellii* LA716. *Theor. Appl. Genet.* **2013**, *126*, 601–609. [CrossRef] [PubMed]
14. Scott, J.W.; Francis, D.M.; Miller, S.A.; Somodi, G.C.; Jones, J.B. Tomato bacterial spot resistance derived from PI 114490; inheritance to race T2 and relationship across three pathogen races. *J. Am. Soc. Hortic. Sci.* **2003**, *128*, 698–703. [CrossRef]
15. Potnis, N.; Krasileva, K.; Chow, V.; Almeida, N.F.; Patil, P.B.; Ryan, R.P.; Sharlach, M.; Behlau, F.; Dow, J.M.; Momol, M.T.; et al. Comparative genomics reveals diversity among xanthomonads infecting tomato and pepper. *BMC Genom.* **2011**, *12*, 146. [CrossRef]
16. Potnis, N.; Timilsina, S.; Strayer, A.; Shantharaj, D.; Barak, J.D.; Paret, M.L.; Vallad, G.E.; Jones, J.B. Bacterial spot of tomato and pepper: Diverse *Xanthomonas* species with a wide variety of virulence factors posing a worldwide challenge. *Mol. Plant Pathol.* **2015**, *16*, 907–920. [CrossRef] [PubMed]
17. Potnis, N.; Branham, S.E.; Jones, J.B.; Wechter, W.P. Genome-Wide Association Study of Resistance to *Xanthomonas gardneri* in the USDA Pepper (Capsicum) Collection. *Phytopathology* **2019**, *109*, 1217–1225. [CrossRef] [PubMed]
18. Sharma, A.; Minsavage, G.V.; Gill, U.S.; Hutton, S.F.; Jones, J.B. Identification and Mapping of *bs8*, a Novel Locus Conferring Resistance to Bacterial Spot Caused by *Xanthomonas gardneri*. *Phytopathology* **2022**, *112*, 1640–1650. [CrossRef]
19. Murashige, T.; Skoog, F. A Revised Medium for Rapid Growth and Bio Assays with Tobacco Tissue Cultures. *Physiol. Plant.* **1962**, *15*, 473–497. [CrossRef]
20. Schaad, N.W. *Laboratory Guide for the Identification of Plant Pathogenic Bacteria*; The American Phytopathological Society: St. Paul, MN, USA, 1988; pp. 81–82.
21. Dewitt, D.; Bosland, P.W. *The Complete Chile Pepper Book—A Gardener's Guide to Choosing, Growing, Preserving and Cooking*; Timber Press: Portland, OR, USA, 2009; p. 336.
22. Moscone, E.A.; Scaldaferro, M.A.; Grabiele, M.; Cecchini, N.M.; García, Y.S.; Jarret, R.; Daviña, J.R.; Ducasse, D.A.; Barboza, G.E.; Ehrendorfer, F. The evolution of chili peppers (*Capsicum-Solanaceae*): A cytogenetic perspective. *Acta Hortic.* **2007**, *745*, 137–169. [CrossRef]
23. Cardoso, R.; Ruas, C.F.; Giacomin, R.M.; Ruas, P.M.; Ruas, E.A.; Barbieri, R.L.; Rodrigues, R.; Gonçalves, L.S.A. Genetic variability in Brazilian *Capsicum baccatum* germplasm collection assessed by morphological fruit traits and AFLP markers. *PLoS ONE* **2018**, *13*, e0196468. [CrossRef]
24. Horvath, D.M.; Stall, R.E.; Jones, J.B.; Pauly, M.H.; Vallad, G.E.; Dahlbeck, D.; Staskawicz, B.J.; Scott, J.W. Transgenic resistance confers effective field level control of bacterial spot disease in tomato. *PLoS ONE* **2012**, *7*, e42036. [CrossRef]
25. Gassmann, W.; Dahlbeck, D.; Cjesnokova, O.; Minsavage, G.V.; Jones, J.B.; Staskawicz, B.J. Molecular evolution of virulence in natural field strains of *Xanthomonas campestris* pv. *vesicatoria*. *J. Bacteriol.* **2000**, *182*, 7053–7059. [CrossRef] [PubMed]
26. Manzur, J.P.; Fita, A.; Prohens, J.; Rodríguez-Burruezo, A. Successful Wide Hybridization and Introgression Breeding in a Diverse Set of Common Peppers (*Capsicum annuum*) Using Different Cultivated Ají (*C. baccatum*) Accessions as Donor Parents. *PLoS ONE* **2015**, *10*, e0144142. [CrossRef] [PubMed]
27. Potnis, N.; Minsavage, G.; Smith, J.K.; Hurlbert, J.C.; Norman, D.; Rodrigues, R.; Stall, R.E.; Jones, J.B. Avirulence proteins AvrBs7 from *Xanthomonas gardneri* and AvrBs1.1 from *Xanthomonas euvesicatoria* contribute to a novel gene-for-gene interaction in pepper. *Mol. Plant-Microbe Interact.* **2012**, *25*, 307–320. [CrossRef] [PubMed]

 *sustainability*

Article

Phenotypic Variability of Wheat and Environmental Share in Soil Salinity Stress [3S] Conditions

Borislav Banjac [1], Velimir Mladenov [1,*], Sofija Petrović [1], Mirela Matković-Stojšin [2], Đorđe Krstić [1], Svetlana Vujić [1], Ksenija Mačkić [1], Boris Kuzmanović [1], Dušana Banjac [3], Snežana Jakšić [3], Danilo Begić [1] and Rada Šućur [1]

[1] Faculty of Agriculture, University of Novi Sad, Sq. Dositeja Obradovića 8, 21000 Novi Sad, Serbia; borislav.banjac@polj.edu.rs (B.B.); sonjap@polj.uns.ac.rs (S.P.); djordjek@polj.uns.ac.rs (Đ.K.); antanasovic.svetlana@polj.uns.ac.rs (S.V.); ksenija.mackic@polj.uns.ac.rs (K.M.); kuzmanovic.boris@gmail.com (B.K.); danilo.begic@polj.uns.ac.rs (D.B.); rada.sucur@polj.edu.rs (R.Š.)
[2] Institute Tamiš, Novoseljanski put 33, 26000 Pančevo, Serbia; matkovic.stojsin@institut-tamis.rs
[3] Institute of Field and Vegetable Crops, Maksima Gorkog 30, 21000 Novi Sad, Serbia; dusana.banjac@nsseme.com (D.B.); snezana.jaksic@nsseme.com (S.J.)
* Correspondence: velimir.mladenov@polj.edu.rs

Abstract: Through choosing bread wheat genotypes that can be cultivated in less productive areas, one can increase the economic worth of those lands, and increase the area under cultivation for this strategic crop. As a result, more food sources will be available for the growing global population. The phenotypic variation of ear mass and grain mass *per* ear, as well as the genotype × environment interaction, were studied in 11 wheat (*Triticum aestivum* L.) cultivars and 1 triticale (*Triticosecale* W.) cultivar grown under soil salinity stress (3S) during three vegetation seasons. The results of the experiment set on the control variant (solonetz) were compared to the results obtained from soil reclaimed by phosphogypsum in the amount of 25 t × ha^{-1} and 50 t × ha^{-1}. Using the AMMI analysis of variance, there was found to be a statistically significant influence of additive and non-additive sources of variation on the phenotypic variation of the analyzed traits. Although the local landrace Banatka and the old variety Bankut 1205 did not have high enough genetic capacity to exhibit high values of ear mass, they were well-adapted to 3S. The highest average values of grain mass per ear and the lowest average values of the coefficient of variation were obtained in all test variants under microclimatic condition B. On soil reclaimed by 25 t × ha^{-1} and 50 t × ha^{-1} of phosphogypsum, in microclimate C, the genotypes showed the highest stability. The most stable genotypes were Rapsodija and Renesansa. Under 3S, genotype Simonida produced one of the most stable reactions for grain mass per ear.

Keywords: soil salinity stress; adaptation; environmental share; interaction; plant breeding; wheat

Citation: Banjac, B.; Mladenov, V.; Petrović, S.; Matković-Stojšin, M.; Krstić, Đ.; Vujić, S.; Mačkić, K.; Kuzmanović, B.; Banjac, D.; Jakšić, S.; et al. Phenotypic Variability of Wheat and Environmental Share in Soil Salinity Stress [3S] Conditions. *Sustainability* **2022**, *14*, 8598. https://doi.org/10.3390/su14148598

Academic Editor: Balázs Varga

Received: 13 June 2022
Accepted: 11 July 2022
Published: 14 July 2022

Publisher's Note: MDPI stays neutral with regard to jurisdictional claims in published maps and institutional affiliations.

1. Introduction

Wheat (*Triticum* sp.) is one of the most significant plant species worldwide. It has played an important role in the development of mankind, participating not only in human nutrition, but in the development of many human activities as well. There is a crucial need to improve the production of wheat for a growing population [1]. Wheat grain yield is a so-called super-trait, a highly quantitative trait that depends on many components that determine it, as well as on environmental factors [2–5].

Soil, as the medium for crop production, can be the limiting factor in crop establishment and in achieving an adequate yield [6]. This is mostly in reference to soils with high concentrations of different salt types. Representatives of such soils belong to halomorphic soils, among which is solonetz [7]. Due to their poor chemical and physical properties, these soils limit plant growth, and lead to reduced yields in arid and semi-arid areas [8–10]. For these reasons, there is a need to increase crop productivity in this type of soil, with the application of different management practices (primarily fertilization) by analyzing plants' responses to applied measures. These actions would lead to increased wheat production and the development of cultivars that could be grown successfully in such conditions. Sairam, et al. [11], emphasized that there are a few identified bread wheat genotypes tolerant to 3S. In Serbia, however, there is a lack of commercial wheat cultivars for production on salt-affected soils. Some of the responses to abiotic stress include chromatin changes; possible phenotypic alterations are temporary, and sometimes return to baseline levels when non-stress conditions have been restored, if possible [12].

Considering that grain is the result of the plant's tendency to reproduce, it can be said that the yield is a consequence of the total phenotype variation of a certain genotype for the purpose of reproduction. Since selection per yield is not possible in early generations of selection per se, phenotypic markers become more important in indirect selection per yield under abiotic stress conditions. This applies in particular to those traits that are highly quantitative, where the application of selection based on molecular markers (AS—Marker-Assisted Selection) is difficult. Therefore, the successful breeder's choice depends on the information on the genetic variability of each yield component [13]. Testing ear traits has a significant place in wheat breeding. Ear mass is the total mass of wheat generative plant part and, as a highly quantitative trait, it is inherited by the minor genes. This genes system allows significant phenotypic variation under the effect of environmental factors, which is most often reflected in significant genotype $\times$ environment interaction [4,14,15]. Grain mass per ear is a highly quantitative component of wheat phenotypic variability, and is a consequence of the minor genes' activity [16,17]. Therefore, the degree of their activity is conditioned by mutual interaction and the effect of the environment [18]. Grain mass per ear is an influential component of wheat yield when grown on solonetz. Generally, salt stress affects plant metabolic processes by impairing cell water potential, membrane function, and uptake of nutrients, and in total reduces predicted crop yields [19].

Wheat grain yield and its components are under not only the influence, but also the interaction, of genotype and environmental factors (G $\times$ E interaction). A widely used multivariate method for studying G $\times$ E interaction is AMMI (Additive Main Effects and Multiplicative Interaction) [4,20–22]. This multivariate data analysis firstly calculates genotype and environmental effects (additive source of variation), using analysis of variance (ANOVA), and then analyzes residual effects (G $\times$ E interaction), using principal component analysis (PCA) [23]. Therefore, the AMMI method is an effective tool, because it considers a large part of the G $\times$ E sum of the squares, and provides an adequate interpretation of the genotypes' stability [24].

Conducting the AMMI analysis in different agro-ecological environments, Petrović, et al. [25], Banjac, et al. [26], and Neisse, et al. [27], concluded that the multiplicative variation of ear characteristics and total grain yield was more pronounced than the additive variation, whereby genotypes with high specific stability in the given agro-ecological environments could be singled out as favorable. Therefore, experiments set up in different agro-ecological environments are of great importance in evaluating the stability of genotypes under varying environmental conditions [20,28].

The breeding process aims to improve the traits of existing cultivars, and to develop new genetic variability which will achieve the best possible economic effect. Therefore, the present study investigates the evaluation of the yield of wheat genotypes as a response to 3S. The objectives of this study were: (i) to evaluate ear mass and grain mass per ear of 11 wheat genotypes grown on salinity soil; (ii) to evaluate ear traits of those genotypes under different levels of reclamation; (iii) to compare the responses of important grain yield components to different growing conditions, and to identify genotypes with adequate traits for cultivation under stressed environments.

The findings of this research help to increase the economic worth of lower grade land by taking into account the ongoing degradation of arable lands caused by numerous variables of the modern age. This study helps to consider the possibility of growing wheat on solonetz after its repair by appropriate reclamation measures, and improves our understanding of how crops react to these types of stress, because solonetz does not provide favorable conditions for growing wheat, and is primarily used as a pasture. The study's findings painted a clear picture of wheat behavior under the influence of climate change. These findings could help to produce new genetic diversity, but they also suggest that growing wheat in a low bonitet class of soil has a bioremediation function that increases the land's economic worth, and also increases the area used for cultivating wheat.

2. Materials and Methods

2.1. Plant Material and Field Exams

Twelve genotypes were studied, including 10 cultivars, one local population of hexaploid (2n = 6× = 42) wheat (*Triticum aestivum* ssp. *aestivum* L.), and one cultivar of triticale (*Triticosecale* W.). The tested wheat cultivars included eight winter cultivars (Renesansa, Pobeda, Evropa 90, Novosadska rana 5, Dragana, Rapsodija, Simonida and Cipovka), while cultivar Nevesinjka was optional. Newer cultivars were also considered, in order to assess their utility in abiotic stress situations, which are mostly induced by soil type. Two older wheat genotypes, that were present in the Serbian area, were used in the experiment (Banatka and Bankut 1205, which originated from Hungary), as shown in Table 1. The sample was selected based on previous studies of the existing genetic variability [15,29,30]. Triticale cultivar Odisej was sown. Triticale was used as a test plant to determine how wheat tolerated the abiotic stress conditions, because it is a synthetic hybrid with a high degree of resistance to abiotic stress factors.

Table 1. Pedigree of examined genotypes (11 wheat genotypes and 1 triticale genotype).

Genotype Description	Genotype Name	Genotype Pedigree
Winter wheat cultivar	Renesansa	Jugoslavija/NS 55–25
Winter wheat cultivar	Pobeda	Sremica/Balkan
Winter wheat cultivar	Evropa 90	Talent/NSR2
Winter wheat cultivar	Novosadska rana 5	NSR1/Tisa//Partizanka/3/Mačvanka 1
Winter wheat cultivar	Dragana	Sremka 2/Francuska
Winter wheat cultivar	Rapsodija	Agri/Nacozari F76//Nizija
Winter wheat cultivar	Simonida	NS 63–25/Rodna//NS-3288
Winter wheat cultivar	Cipovka	NS 3288/Rodna
Local population; old winter wheat	Banatka	LV-Banat
Winter wheat cultivar; old winter wheat	Bankut 1205	Bankut 5/Marquis
Optional wheat cultivar	Nevesinjka	Dugoklasa/Jarka
Triticosecale cultivar	Odisej	LT 338.75/BL. 517

The experiment was set up in Banat (Autonomous Province of Vojvodina, Republic of Serbia) at Kumane site (45,539° N, 20,228° E, 72 m altitude), on stressed halomorphic soil of solonetz type, on an experimental area of 2 ha. The experiment was performed during three vegetation periods, marked as microclimate conditions A, B and C. A field trial was conducted according to randomized complete block design (RCBD), with three replications. The cultivars were sown in 155 m-long rows, using machines, with an interrow distance of 12.5 cm. Every cultivar was sown in 8 rows. During sowing, 134 kg $\times$ ha^{-1} of mineral fertilizer NPK 15:15:15 was applied. Depending on weather conditions, the crops were fertilized during vegetation seasons in late March or early April, using mineral fertilizer KAN in the amount of 200 kg $\times$ ha^{-1}. A total of 30 plants per treatment were analyzed. They were represented by the primary stem (10 primary stems $\times$ 3 repetitions) in order to evaluate the phenotypic variation of the yield components: ear mass and grain mass per ear. In all vegetation seasons, the harvest was performed and samples were taken when caryopsis was hard (could no longer be dented by thumb-nail) at physiological maturity, Zadoks growth stage 92 [31]. Solonetz is a halomorphic soil with more than 15% sodium ion-Na^+ adsorbed in the exchange complex. As a result, it is alkalized (pH > 9) and unfavorable for crops. The heavy mechanical composition of the compacted and impermeable Bt, na horizon severely limits solonetz's production capability, as shown in Table 2, Figure 1.

Table 2. Adsorbed cations content and salinity properties of solonetz at Kumane.

Horizon (Depth cm)	Adsorbed Cations						Salinity Properties		
	Ca^{++} (mg/100 g Soil)	Mg^{++} (mg/100 g Soil)	K^+ (mg/100 g Soil)	Na^+ (mg/100 g Soil)	Ca^{++} (%) *	Na^+ (%) *	ECe 25 °C ** (mS/cm)	Total Salts (%)	pH Soil Extract
Aoh/E, na (0–15)	128.26	37.92	74.68	20.69	53.14	7.91	0.62	0.03	5.41
Bt, na (15–111)	392.98	136.74	26.98	269.67	58.92	32.90	2.16	0.15	7.72
Bt, na C, na (111–156)	707.81	143.79	19.16	214.04	61.16	18.06	1.10	0.17	8.89
C, na (156–200)	658.51	136.50	15.64	152.65	63.80	14.49	1.03	0.12	8.79

* percentage in relation to the total ions content in the exchange complex of soil, ** ECe 25 °C = electrical conductivity of soil extract (ECe at 25 °C).

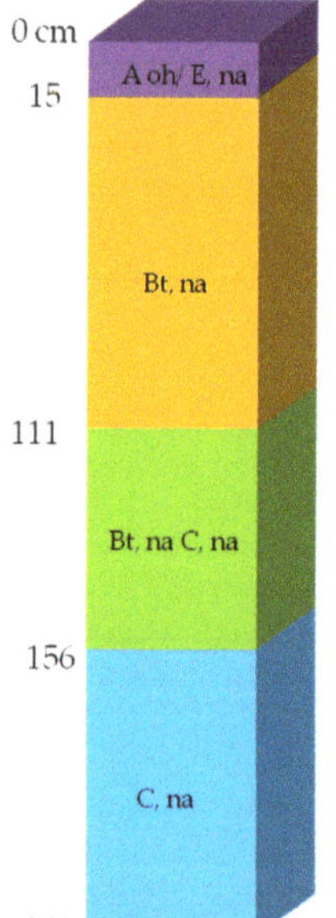

A oh/ E horizon (0–15 cm) – complex humic and eluvial soil horizon, in dry condition grayish yellow brown color 10YR 5/2, and in wet condition brownish black color 10YR 3/2, polyhedral structure, clay loam, non-calcareous, imbued with the root of primary vegetation.

Bt, na horizon (15–111 cm) – very powerful illuvial horizon, in the dry condition is dark grayish yellow color 2,5Y 5/2, and in wet condition 2,5Y 3/2 brownish black color, columnar structure, heavy clay, In top part of horizon is non-calcareous. By increasing depth, the content of calcium carbonates also increases, so the lower part of the horizon is less compact and brighter in color. Bt horizon is characterized by a strongly alkaline reaction, with pH values above 9.

Bt, na C, na horizon (111–156 cm) – transition horizon, in the dry condition is pale yellow color 2,5Y 8/3, and in wet condition is yellowish brown 2,5Y 5/3, polyhedral structure, heavy clay, strongly calcareous. Bt,na C,na transition horizon is is characterized by a strongly alkaline reaction, with pH values 9.38).

C, na horizon (156–200 cm) – loess deposits.

Figure 1. Soil profile description of solonetz soil type in Kumane.

The results from soil with two reclamation levels, from 25 t × ha^{-1} and 50 t × ha^{-1} of phosphogypsum, were processed in addition to the control results (soil without reclamation—natural pasture). The soil in the investigated plot was drained, to allow salts to leak into the neighboring drainage canals.

Each treatment was studied as a separate agro-ecological habitat for plant growth and development for one growing period. As a result, 9 alternative agro-ecological growing conditions were obtained, all of which were similar in terms of agro-technical circumstances, but differed in terms of phosphogypsum treatments (Table 3).

Table 3. Description of examined environments.

	Environments	
	code E1	solonetz; natural pasture
Microclimate condition A	code E2	Soil reclaimed by 25 t × ha^{-1} phosphogypsum
	code E3	Soil reclaimed by 50 t × ha^{-1} phosphogypsum
	code E4	solonetz; natural pasture
Microclimate condition B	code E5	Soil reclaimed by 25 t × ha^{-1} phosphogypsum
	code E6	Soil reclaimed by 50 t × ha^{-1} phosphogypsum
	code E7	solonetz; natural pasture
Microclimate condition C	code E8	Soil reclaimed by 25 t × ha^{-1} phosphogypsum
	code E9	Soil reclaimed by 50 t × ha^{-1} phosphogypsum

Apart from the unfavorable characteristics of solonetz soil, there were other abiotic stress conditions at the Kumane site. The joint action of the steppes, clearly expressed temperature changes and extremes, strong winds and water retention on the plot surface had an effect on the selection of this site. Weather conditions at the Kumane experimental field throughout the vegetation season, when wheat was produced, were, for the most part, typical of that environment. Sowing wheat was accompanied by a deficit of precipitation. Winters were marked by very cold weather, strong ground frosts and lack of snow cover. The extreme minimum temperature was −17.8 °C. May set a record for precipitation (microclimate condition B, 162.1 mm). Weak precipitation, relatively high air temperatures and frequent winds caused the drying of the surface layer of the soil, as shown in Figure 2.

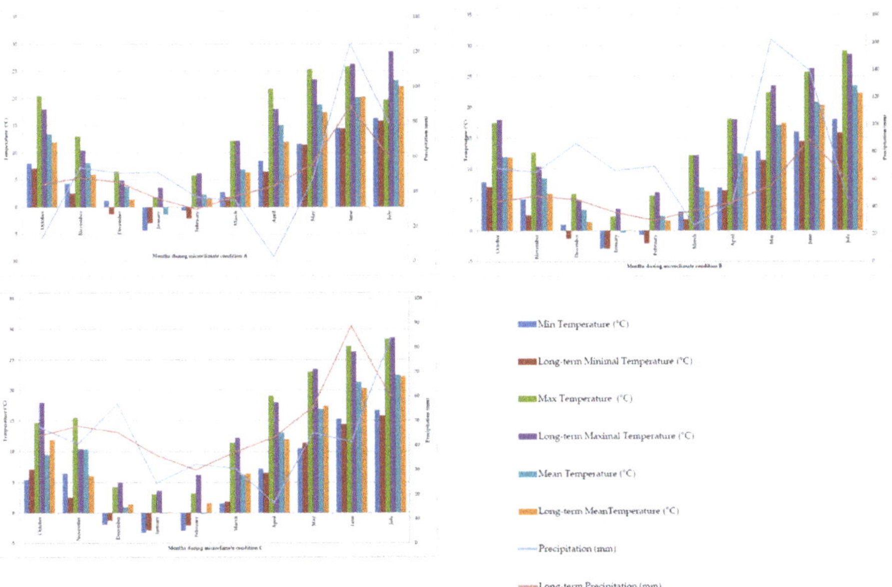

Figure 2. Minimum, maximum and mean values monthly temperature (°C) and monthly precipitation (mm) during the field trial at microclimate conditions A, B and C in the Kumane site.

2.2. Statistical Tools

For each investigated trait, the parameters of the descriptive statistics were calculated: mean value and coefficient of variation. Analysis of variation in the experiment, its quantification, and identification of sources of variation were performed using the AMMI model (Additive Main Effects and Multiplicative Interaction). Thus, AMMI ANOVA presented the main additive components, and then the multivariate source of variation (non-additive component of variance) was reflected. The genotype x environment interaction was further decomposed by a multivariate model PCA analysis [24].

Genotype $\times$ environment interaction was tested using the AMMI analysis by [23]. Data processing was performed in GenStat 9th Edition (trial ver.) VSN International Ltd., (www.vsn-intl.com/ accessed on 12 June 2022) [32]. All biplots were generated in Microsoft Excel, 2013.

The mean squares (MS) from analysis of variance were used to estimate components of the variance (genotypic variance σ^2_g, phenotypic variance σ^2_p, interaction variance $\sigma^2_{g\times y\times t}$, and ecological variance σ^2_e), as follows [33]:

Genotypic variance:

$$\sigma^2_g = \frac{MS1 - MS2}{r \times y \times t}$$

Ecological variance:

$$\sigma^2_e = MSe$$

Variance of interaction:

$$\sigma^2_{g\times y\times t} = \frac{MS2 - MS3}{r}$$

Phenotypic variance:

$$\sigma^2_p = \sigma^2_g + \sigma^2_{g\times y\times t} + \sigma^2_e$$

where: $MS1$ = mean square for genotype; $MS2$ = mean square for genotype $\times$ year $\times$ treatment; $MS3$ = mean square for error; r = replications; y = years; t = treatments.

Mean values ($\bar{x}$) were used for genetic analyses, to determine the genotypic coefficient of variation (CV_g) and the phenotypic coefficient of variation (CV_p), according to Singh and Chaudhary [34]:

$$CV_g(\%) = \frac{\sqrt{\sigma^2_g}}{\bar{x}} \times 100; \ CV_p(\%) = \frac{\sqrt{\sigma^2_p}}{\bar{x}} \times 100$$

Heritability in broad sense (h^2) for all traits was computed using the formula given as [35]:

$$h^2 = \frac{\sigma^2_g}{\sigma^2_p} \times 100$$

Heatmap analysis of Pearson moment correlation coefficients and correlation matrix analysis by the principal components method (PCA) were performed, in order to express the relationships between examined traits and grain yield, using the R Project for Statistical Computing, Version 4.2.0, 22 April 2022 ucrt [36].

3. Results

3.1. Ear Mass

The average value of ear mass per agro-ecological environments was 1.41 g, i.e., on soil without reclamation and two treatments during the experiment. The highest deviations from that average were recorded in cultivars Banatka ($\bar{x} = 0.90$ g) and Odisej ($\bar{x} = 2.25$ g), Table 4, Figure 2.

Table 4. Average values ($\bar{x}$) and coefficient of variation (V) of ear mass for examined genotypes (11 wheat genotypes and 1 triticale genotype) in nine agro-ecological growing conditions.

Genotype	Environments							
	Solonetz;		Soil Reclaimed by Phosphogypsum				Average Value	
	Natural Pasture		25 t $\times$ ha^{-1}		50 t $\times$ ha^{-1}			
	Codes E1; E4 and E7		Codes E2; E5 and E8		Codes E3; E6 and E9			
	$\bar{x}$ (g)	V (%)	$\bar{x}$ (g)	V (%)	$\bar{x}$ (g)	V (%)	$\bar{\bar{x}}$ (g)	V (%)
Renesansa	1.4	13.6	1.6	2.8	1.4	10.2	1.5	8.9
Pobeda	1.6	6.9	1.6	25.4	1.2	11.0	1.5	14.4
Evropa 90	1.7	5.3	1.3	3.0	1.6	15.8	1.5	8.0
NSR5	1.5	6.8	1.2	12.0	1.4	6.3	1.4	8.4
Dragana	1.8	9.5	1.0	9.8	1.1	17.8	1.3	12.4
Rapsodija	1.5	11.0	1.4	13.3	1.2	10.9	1.4	11.7
Simonida	1.4	11.0	1.3	10.4	1.3	7.5	1.3	9.6
Cipovka	1.5	9.6	1.3	13.2	1.3	5.3	1.4	9.4
Banatka	0.8	8.6	0.8	7.0	1.0	3.8	0.9	6.5
Bankut 1205	1.1	7.4	1.1	6.4	1.2	7.7	1.1	7.2
Nevesinjka	1.3	5.1	1.6	13.3	1.7	11.8	1.5	10.1
Odisej	2.1	5.5	2.2	12.9	2.4	4.4	2.2	7.6
Average value	1.5	8.4	1.4	10.8	1.4	9.4		
							LSD$_{0.05}$ = 0.180	
							LSD$_{0.01}$ = 0.236	

The analysis of variance showed that environmental share had a great influence on phenotype formation. This was a result of the significant share of the environment sum of squares—as an additive effect—and the genotype $\times$ environment interaction— which had a multivariate nature—in the total variation of the experiment. The genotype response to variations in the actions of environmental factors was reflected in a statistically highly significant value of mean squares of genotype $\times$ environment interaction. This interaction was 26.64% of the share of total variation in the experiment sum of squares (Table 5).

In addition to this significance, the variance analysis of ear mass for the total sample showed high significances of mean squares value for both genotypes and environments. Thereby, in the experiment total variation, the main effects of the variance analysis, genotype and environments, had a 59.97% share of the experiment sum of squares. A large share of the sum of squares within the main effects of the variance analysis belonged to agro-ecological factors (36.78%), while a smaller share belonged to the genotype sum of squares (23.19%), Table 5.

Although most of the total variability was explained by the first major component (IPCA$_1$ 36.71%), the statistical significance of the remainder indicated that, after the isolation of its influence, part of the variance remained unexplained; the other main components were therefore also analyzed. A total of six statistically significant main axes were distinguished, the second of which covered the largest part.

Almost all genotypes showed a stable reaction for the ear mass. Genotype Dragana showed less stability than the others. Triticale Odisej, with the largest ear mass, had the highest genotype $\times$ environment interaction, i.e., the lowest stability of all the genotypes. Although the local population, Banatka, proved to be one of the most stable, and the old cultivar, Bankut 1205, a fairly stable genotype, their average values of ear mass were the lowest, compared to the others. The position of the points for these two genotypes, which were grouped around the middle E1, indicated that these old cultivars were well-adapted to the unfavorable conditions of solonetz soil, but without any potential for high ear mass in tested conditions (Figure 3).

Table 5. AMMI analysis of variance for the ear mass of 11 wheat and 1 triticale cultivars examined across nine environments.

Source of Variation [1]	df [2]	MS [3]	F Value	F Table		The Share of Total Variation
				0.05	0.01	
Total	323	0.4	-	-	-	100
Treatments	107	1.1	** 13.03	1.00	1.00	86.60
Genotypes	11	2.7	** 33.93	1.83	2.32	23.19
Environments	8	5.9	** 75.54	1.94	2.51	36.78
Blocks	18	0.1	0.98	1.57	1.87	1.10
Interactions	88	0.4	** 4.87	1.00	1.00	26.64
$IPCA_1$ [4]	18	0.7	** 8.74	1.57	1.87	36.71
$IPCA_2$	16	0.5	** 6.39	1.57	1.87	23.85
$IPCA_3$	14	0.4	** 4.98	1.75	2.18	16.26
$IPCA_4$	12	0.3	** 3.22	1.75	2.18	9.00
$IPCA_5$	10	0.3	** 3.28	1.83	2.32	7.64
$IPCA_6$	8	0.2	* 2.45	1.94	2.51	4.56
$IPCA_7$	6	0.1	1.32	2.09	2.80	1.85
Residuals	4	0.01	0.13	2.37	3.32	-
Error	198	0.08	-	-	-	-

[1] All sources were tested in relation to the error; [2] degree of freedom; [3] mean of square; [4] extracted interaction axes; *. F value is statistical significant at 0.05 possibility; **. F value is statistical significant at 0.01 possibility.

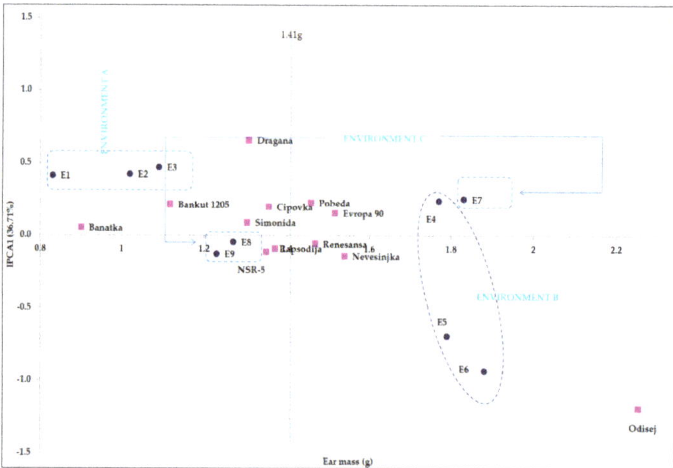

Figure 3. AMMI 1 biplot of 11 wheat and 1 triticale cultivars across nine environments for the estimation of G × E interaction for ear mass. Legend: codes E1, E4 and E7 = solonetz; natural pasture in microclimate conditions A, B and C; codes E2, E5 and E8 = soil reclaimed by 25 t × ha^{-1} phosphogypsum in microclimate conditions A, B and C; codes E3, E6 and E9 = soil reclaimed by 50 t × ha^{-1} phosphogypsum in microclimate conditions A, B and C.

The biplot clearly shows the group of points that represent different microclimate conditions. Points E1 (solonetz without reclamation), E2 (solonetz with applied 25 t $\times$ ha^{-1} of phosphogypsum) and E3 (solonetz with applied 50 t $\times$ ha^{-1} of phosphogypsum) form one group (Environment A). Microclimate condition A was characterized by low genotype $\times$ environment interaction and the great effect of the additive component. In this group, the lowest trait value was on solonetz without reclamation, so that it increased slightly with increasing doses of phosphogypsum. This indicated the visible effect of repaired solonetz in that season. During the second year of the experiment (points E4, E5 and E6), difference in the multivariate part of the variation was manifested, while it was absent in the additive part. This was the reason why the average values of ear mass in all treatments were approximately the same. In the third vegetation season, similar to the previous season, the predominant source of variation was in the additive component. During this season, the genotypes had the best results of the ear mass on solonetz without repair (E7). Considering the positions of points E8 and E9, it can be concluded that the cultivars were most stable on the soil with applied 25 t $\times$ ha^{-1} of phosphogypsum and 50 t $\times$ ha^{-1} phosphogypsum in microclimate condition C. The first vegetation season did not contribute to the decrease of genotype $\times$ environment interaction for the ear mass. During this season, regardless of treatment, the genotypes had the lowest ear mass averages, as shown in Figure 3.

3.2. Grain Mass per Ear

The average value of grain mass per ear during the three-year experiment ranged from $\bar{x} = 1.0$ g in the treatment of 25 t $\times$ ha^{-1} phosphogypsum to $\bar{x} = 1.1$ g in the soil without repair and in the treatment of 50 t $\times$ ha^{-1} phosphogypsum. The uniformity of variation of grain mass per ear was indicated by similar values of average coefficient of variation ($V = 10.2–10.7\%$) during the three experimental seasons, as shown in Table 6.

Table 6. Average value ($\bar{x}$) and coefficient of variation (V) of grain mass per ear for examined genotypes (11 wheat genotypes and 1 triticale genotype) in nine agro-ecological growing conditions.

Genotype	Environments							
	Solonetz;		Soil Reclaimed by Phosphogypsum				Average Value	
	Natural Pasture		25 t $\times$ ha^{-1}		50 t $\times$ ha^{-1}			
	Codes E1; E4 and E7		Codes E2; E5 and E8		Codes E3; E6 and E9			
	$\bar{x}$ (g)	V (%)	$\bar{x}$ (g)	V (%)	$\bar{x}$ (g)	V (%)	$\bar{\bar{x}}$ (g)	V (%)
Renesansa	1.1	12.4	1.2	12.3	1.2	15.3	1.2	13.3
Pobeda	1.3	5.3	1.2	21.2	0.9	9.4	1.1	12.0
Evropa 90	1.3	5.1	0.9	3.7	1.3	13.9	1.2	7.6
NSR5	1.1	12.5	1.0	10.7	1.2	9.1	1.1	10.8
Dragana	1.4	9.6	0.6	5.3	0.8	10.2	0.9	8.4
Rapsodija	1.1	13.3	1.1	9.8	1.0	14.1	1.1	12.4
Simonida	1.1	14.6	1.0	6.0	0.9	8.0	1.0	9.5
Cipovka	1.0	22.6	0.8	13.1	1.0	4.0	0.9	13.2
Banatka	0.6	10.0	0.6	8.9	0.7	19.7	0.6	12.9
Bankut 1205	0.8	2.1	0.8	7.7	0.8	8.2	0.8	6.0
Nevesinjka	1.0	6.4	1.3	14.3	1.3	11.6	1.2	10.8
Odisej	1.5	8.2	1.7	15.8	1.9	2.0	1.7	8.7
Average value	1.1	10.2	1.0	10.7	1.1	10.5		

$$\text{LSD}_{0.05} = 0.142$$
$$\text{LSD}_{0.01} = 0.187$$

In addition to the highly significant mean of squares of the environments, high statistical significance of the mean of squares of the genotype × environment interaction was also recorded. By analyzing the variance of grain mass per ear for the total sample, it was calculated that the main effects, genotype and environments, had a 55.74% share of the experiment sum of squares in the total experiment variation. Within the main effects of the variance analysis, most of the sum of squares belonged to the environmental share (31.89%), while a smaller share belonged to the genotype sum of squares (23.85%). Genotype × environment interaction had a 30.30% share in the experiment sum of squares, and showed high statistical significance. A total of six main components were distinguished, out of which the first five were statistically significant, as shown in Table 7. The $IPCA_1$ accounted for most of the interaction (37.58%), which is why the AI 1 biplot is also shown, in Figure 4.

Table 7. AMMI analysis of variance for the grain mass per ear of 11 wheat and 1 triticale cultivars, examined across nine environments.

Source of Variation [1]	df [2]	MS [3]	F Value	F Table 0.05	F Table 0.01	The Share of Total Variation
Total	323	0.3	-	-	-	100
Treatments	107	0.7	** 13.10	1.00	1.00	86.04
Genotypes	11	1.8	** 35.31	1.83	2.32	23.85
Environments	8	3.4	** 39.55	1.94	2.51	31.89
Blocks	18	0.1	* 1.64	1.57	1.87	1.82
Interactions	88	0.3	** 5.61	1.00	1.00	30.30
$IPCA_1$ [4]	18	0.5	** 10.30	1.57	1.87	37.58
$IPCA_2$	16	0.4	** 8.41	1.57	1.87	27.26
$IPCA_3$	14	0.3	** 5.28	1.75	2.18	14.95
$IPCA_4$	12	0.2	** 3.77	1.75	2.18	9.15
$IPCA_5$	10	0.2	** 2.84	1.83	2.32	5.76
$IPCA_6$	8	0.1	1.90	1.94	2.51	3.08
Residuals	10	0.06	1.08	1.83	2.32	-
Error	198	0.05	-	-	-	-

[1] All sources were tested in relation to the error; [2] degree of freedom; [3] mean of square; [4] extracted interaction axes; * F value is statistical significant at 0.05 possibility; **. F value is statistical significant at 0.01 possibility.

According to the achieved interaction values, i.e., the distance from zero axis, the genotypes were grouped towards stability. Genotypes Rapsodija, Renesansa, Bankut 1205 and Banatka showed the most stable response, relative to the first main component, and after them: Pobeda, Simonida, Cipovka and Evropa 90. Medium stable genotypes were Novosadska rana 5 and Nevesinjka, while genotypes Dragana and triticale Odisej were evaluated as the least stable, as shown in Figure 4.

The distribution of points of the agro-ecological environments indicates a great similarity in conditions for achieving grain mass stability per ear. However, 2 was singled out as the most stable. Even so, this does not make this environment the most favorable in relation to the others, since the genotypes that were part of it had a mean value of grain mass per ear lower than the total mean value of the experiment for this trait. Agro-ecological environment E6 had the highest interaction score, i.e., it was the environment where genotypes could not show their stable response.

Cultivars Rapsodija and Renesansa stood out as the most stable, compared to the first interaction axis, and with average values higher than the general average (Figure 4).

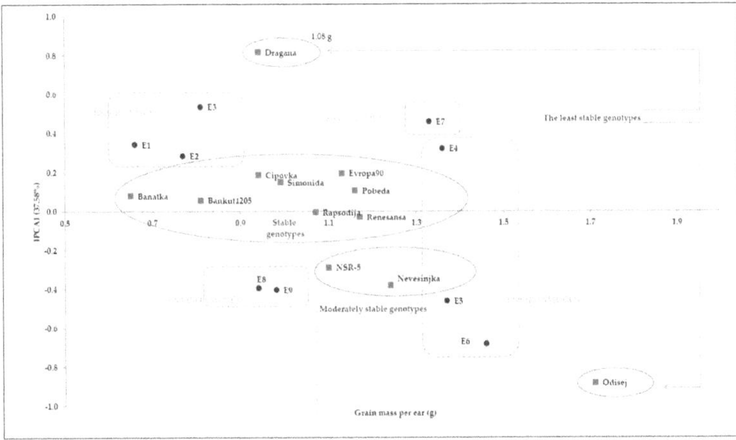

Figure 4. AMMI 1 biplot of 11 wheat and 1 triticale cultivars across nine environments for the estimation of G × E interaction for grain mass per ear. Legend: codes E1, E4 and E7= solonetz; natural pasture in microclimate conditions A, B and C; codes E2, E5 and E8 = soil reclaimed by 25 t × ha^{-1} phosphogypsum in microclimate conditions A, B and C; codes E3, E6 and E9 = soil reclaimed by 50 t × ha^{-1} phosphogypsum in microclimate conditions A, B and C.

3.3. Genetic Parameters

The phenotypic coefficient of variation (CVp) was higher than the genotypic coefficient of variation (CVg) for both analyzed traits. This indicates that the present variation was not only due to genotypes, but also due to the influence of the environment. For all tested microclimates combined, low broad sense heritability values were observed for both traits (31.98% for ear mass and 29.69% for grain mass per ear), as shown in Table 8.

Table 8. The mean value, estimates of variance components, genotypic and phenotypic variance, and heritability of ear mass and grain mass per ear.

Traits [a]	Mean Values	Estimates of Variance Components [b]				Genotypes Mean of Square	CVg (%)	CVp (%)	h^2 (%)
		σ^2_g	σ^2_p	σ^2_i	σ^2_e				
All tested microclimates combined									
EM	1.41	0.087	0.272	0.104	0.081	2.744 **	20.92	36.98	31.98
GMpE	1.08	0.057	0.192	0.079	0.056	1.837 **	22.11	40.57	29.69
Microclimate A									
EM	0.98	0.021	0.074	0.008	0.045	0.256 **	14.84	27.84	28.38
GMpE	0.75	0.008	0.054	0.014	0.032	0.150 **	12.00	31.19	14.81
Microclimate B									
EM	1.82	0.233	0.390	0.116	0.041	2.484 **	26.59	34.41	59.74
GMpE	1.40	0.155	0.262	0.079	0.028	1.665 **	28.14	36.59	59.16
Microclimate C									
EM	1.44	0.072	0.353	0.125	0.156	1.176 **	18.59	41.17	20.40
GMpE	1.08	0.046	0.256	0.105	0.105	0.837 **	19.82	46.76	17.97

[a] EM = ear mass (g); GMpE = grain mass per ear (g); [b] σ^2_g-genotypic variance, σ^2_p-phenotypic variance, σ^2_i - interaction variance, σ^2_e-ecological variance; CV_g = genotypic coefficient of variation; CV_p = phenotypic coefficient of variation; h^2 = heritability in broad sense. **. Tested value is statistical significant at 0.01 possibility.

Lower genotypic than phenotypic variance values are often present when genetic factors are examined in connection to microclimates. This outcome also contributed to the lower-than-expected heritability values in a more general sense. This was due to the significant treatment differences (solonetz, natural pasture and treatments by phosphogypsum), which resulted in high interaction values that were considered when evaluating heritability. However, microclimate B stood out, due to having the most favorable conditions for plant development and the expression of the examined traits. As a result, it can be inferred that more favorable microclimate conditions minimized the differences between the investigated treatments. Higher heritability values for the ear mass (59.74%) and the grain mass per ear (59.16%) were calculated in microclimate B than in microclimates A and C.

3.4. Correlations between Studied Parameters and Grain Yield of Wheat under 3S Conditions

In wheat breeding, analyzing the correlation dependency across yield components is crucial, as selection within one feature influences the change of another variable. Single Pearson coefficients are most frequently calculated for correlation analysis, and are shown in the heatmap (Figure 5). The association assessment is also shown by the biplot obtained by PCA analysis, as shown in Figure 6.

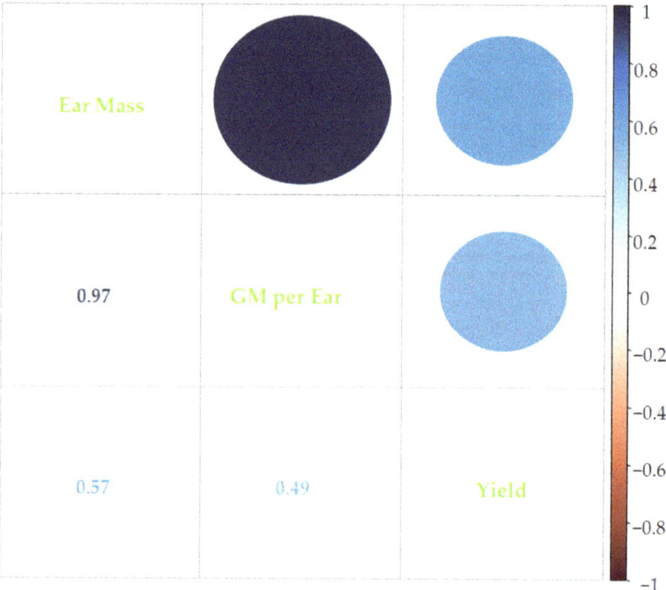

Figure 5. Heatmap of Pearson correlation coefficient for ear mass (EarMass), grain mass per ear (GM per Ear) and grain yield (Yield) for examined genotypes grown under 3S, during microclimate conditions A, B and C.

Given that there was a positive correlation between the studied parameters and grain yield, this indicates that there was a tendency for an increase in one component to result in an increase in another component—in this case, grain yield, which is important for breeding efforts. This is particularly important because the experiment was conducted in 3S conditions brought on by an increase in the amount of sodium ions in the soil.

The association between analyzed parameters and grain yield was estimated in more detail through the principal components method, presented by a biplot (Figure 6).

By comparing the values of the first (Dim1) and second (Dim2) principal components of PCA for ear parameters and grain yield, as well as genotypes, a biplot analysis was conducted to explore multivariate associations between the examined variables. The importance of the acquired data is shown by the fact that the first two axes together explained 99.2% of the variation. Acute angles that overlapped the vectors of the investigated traits indicated that there were positive correlations between them, which is also consistent with the reported correlation study (Figures 5 and 6).

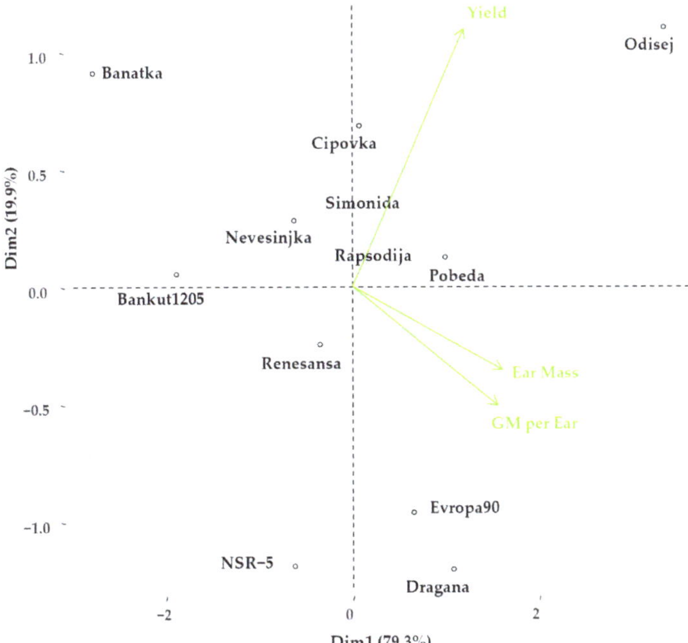

Figure 6. Principal component analysis for ear mass (EarMass), grain mass per ear (GM per Ear) and grain yield (Yield) for examined genotypes grown under 3S, during microclimate conditions A, B and C.

4. Discussion

4.1. Point Distribution of Genotype and Environmental Share: Markers of Genotype Stability

An assessment of the stability of different wheat genotypes, in terms of their grain yield and grain yield components, provides valuable information about genotype adaptation to specific environments, such as environments with increased soil salinity [25,37,38]. The results of this three-year experiment showed the extremely complex nature of the tested traits of phenotypic variation, which were considered components of the wheat grain yield. Besides the high level of monitored phenotypic characteristics in the tested cultivars, the stability of their reaction was also observed. However, when discussing stable reaction, i.e., the level of genotype × environment interaction, it is necessary to emphasize that this interaction was observed in two ways. A low level of genotype × environment interaction was observed and evaluated favorably, with the highest mean value of the observed trait, which indicated stability. At the same time, the nature of this interaction at high level was also assessed in cases where a marked change in rank occurred ("cross over" interaction). This was because the experiment was set up in the conditions of control and two levels of solonetz repair, so in some cases the high level of genotype × environment interaction indicated a favorable reaction of the cultivar to land reclamation measures. Therefore, in addition to accepting a stable reaction at average trait values higher than the average of the experiment, some cases of so-called unstable reaction were evaluated as favorable, if that reaction meant that the cultivar reacted violently to land reclamation measures by increasing the mean value of the observed trait. In that way, cultivars with low interaction and stable reaction to variation of agro-ecological conditions were separated. This research confirms that the local population, Banatka, and the ancient variety, Bankut 1205, have successfully adapted to 3S, which accords with Gharib, et al. [39], who found that local populations and old varieties of wheat could be a useful genetic resource for increasing genetic variability and specific adaptation to 3S conditions, as well as suitable parental material in breeding programs.

Still, a priori is not dismissed without detailed analysis, as with any reaction that would be assessed as unstable and unfavorable in normal growing conditions. The aspiration follows, from the above, to obtain the most complete and realistic variation scores, as well as the source of the experiment variation, especially on the level and the nature of genotype × environment interaction. This would describe the total or average stability of the genotype, so it is of utmost importance to analyze the complex nature of genotype × environment interaction in more details. Over the last few decades, several studies have been carried out to understand plant biology in response to 3S, with a major emphasis on genetic and other hereditary components [9,12,29]. Based on the outcome of these studies, several approaches are being followed, to enhance plants' ability to tolerate salt stress while still maintaining reasonable levels of crop yields [40,41]. This research indicates the importance of studying the correlations between yield components and wheat grain yield in 3S, which is in accordance with Matković Stojšin, et al. [42].

The complex nature of genotype × environment interaction was reflected in the large number of interaction axes, i.e., main axes from the main components analysis, which were applied in the AMMI model for more detailed analysis of genotype × environment interaction. The larger number of main interaction axes, which proved to be statistically significant, meant that the variance of the genotype × environment interaction was influenced by several agronomically significant and explanatory sources of variation. Due to this, the tested yield components showed a genotype × environment interaction that could not be explained only by the first interaction axis (IPCA$_1$).

4.2. The External Environment: Sculptor of Genetic Expression of Examined Traits

The ear mass showed a great dependence on the action of environmental factors, as the genetic share in the phenotypic variation was not high, nor the calculated heritability. This result was expected, given the distinct quantitative basis of this wheat yield component—for the expression of which, polygene groups are responsible—but also its dependence on grain mass per ear and number of grains per ear. The genotypes had the lowest averages of ear mass in microclimate condition A, regardless of treatment. Although this season was characterized by certain deviations in terms of climatic parameters, it was still observed as a whole, and had more favorable climate conditions. This resulted in a reduction of the differences in genotype response to the measures of solonetz repair for the ear mass. However, the average values of ear mass had a tendency to increase from control to treatment, with 50 t × ha^{-1} of phosphogypsum. The effect of reclamation measures was absent in the third vegetation season, because the highest average value of ear mass was obtained on the control variant of solonetz ($\bar{x} = 1.8$ g). Thus, during the second vegetation season, the highest average value of this trait was achieved ($\bar{x} = 1.9$ g on treatment with 50 t × ha^{-1} of phosphogypsum). This indicated that the high average value of the ear mass in such conditions could only be achieved due to the favorable reaction of genotypes to the measures of solonetz repair. Similar results were obtained by Ljubičić, et al. [43], who found that the basic source of ear mass variation is reclamation measure.

Ear mass stability analysis showed that there were differences between the genotypes, which were quantified by a statistically highly significant value of mean squares of genotype × environment interaction. However, the distribution points of the genotypes, and considerable agro-ecological environments that were more scattered by abscissa than by ordinate on the biplot, could lead to the conclusion that variation of the additive part of the total variance was more pronounced than the multivariate for the ear mass. This resulted in a stable reaction of almost all genotypes for the ear mass. Measures to repair solonetz contributed to the increase in stability and the decrease in interaction values. The highest genotype × environment interaction was observed in triticale cultivar Odisej, so this genotype was characterized as the least stable, which is consistent with the results of Purchase, et al. [44].

The grain mass per ear was characterized by low heritability and significant phenotypic variation, depending on the variation of environmental factors. The lower genotype × environment interaction of this trait was a good basis for selection for stable wheat yield. By analyzing the grain mass per ear, a high variability of this wheat yield component could be noticed, in relation to the examined microclimate conditions and different phosphogypsum treatments. The grain mass per ear showed complex genotype × environment interaction, whereby it was possible to group genotypes according to stability. The highest average values of grain mass per ear ($\bar{x} = 1.5$ g on treatment with 50 t $\times$ ha^{-1} of phosphogypsum) were achieved by genotypes in the season when the most unfavorable environmental conditions prevailed, especially in stages crucial for ear formation and grain filling. Differences in the mean values of the examined trait during different vegetation seasons were in accordance with the previous results, where the speed and completeness of the flowering, pollination and fertilization processes were determined by environmental conditions, above all by temperatures and humidity [26,45]. Microclimate condition A did not favor the development of grain mass per ear, as with the ear mass, just as the solonetz repair measures did not have a significant impact. However, the AMMI analysis determined that this season, in all three variants of the experiment, was the most favorable for achieving a stable reaction for grain mass per ear. In the same location, seed yield was tested, and its expression, controlled with multiple genes and traits, suffered decrease, compared to non-stressed conditions [26]. The great influence of the external environment on the grain mass per ear, and the number of grains per ear, has been confirmed by the research of [43,46], as has the high difference between the phenotypic and genotypic coefficient of variation, especially for grain mass per ear, indicating the greater influence of the environment. Dimitrijević, et al. [47], Knežević, et al. [16], Knežević, et al. [17], and Matković Stojšin, et al. [48], found a high share of the environment in the variation of grain mass per ear. Similar results were obtained by Matković Stojšin, et al. [42], for ear mass.

For both examined traits, values of genetic variance were almost equal to values of environmental variance. The consequence of this phenomenon was low heritability of these yield components, which could indicate that environmental factors could be attributed as epigenetics. Testing genotypes under abiotic stress is critical, because it allows researchers to analyze their first response to a particular situation, i.e., to primary stress [49]. Assessment of a genotype's tolerance to abiotic stress, in situ, is of great importance for forming a realistic picture because, in addition to 3S, all other abiotic factors also affect the plant [42,50–52].

When genotypes are tested at different phases of development, determining the response to primary stress makes sense. Better epigenetic mechanisms, which allow them to recall stress, are defined by varieties that respond best to primary stress with the least harm in the early stages of development, and then give the same response to the source of abiotic stress in the later stages of development. Because genotypes are not formed for specific growing circumstances, this response is not encoded in their genetic code, so it is hypothesized that DNA methylation and/or histone modification are responsible for this reaction, which is the focus of new research. In this research, phenotypic evaluation of genotypes was carried out at various phases of development, in order to show that genotypes with the best assessment of primary stress had the greatest mean values of the attributes studied in the whole physiological maturity phase.

Through three separate stages of plant cultivation, the memory of stress at the vertical level was examined in this experiment. This brings up the prospect of answering the question: "Can a seed remember?"—that is to say—"Can epigenetic factors be fixed in the genome?"—which will be a good parent for developing new genetic variability specific for growing under stressful conditions.

5. Conclusions

In experiments where the interaction of climate conditions, soil conditions and appropriate treatments is examined, it is important to consider the genetic potential of selected cultivars and factors that interfere. This is particularly important if the environmental conditions are the sources of abiotic stress as well, because this gives a clearer picture of the potential breeding usability of the cultivars. In this investigation, the stability of wheat genotypes, grown in the stressful conditions of solonetz soil type and steppe microclimate, was assessed. By comparing the results that analyzed genotypes had on the control treatment (without application of phosphogypsum) with the results that genotypes achieved on the treatments with phosphogypsum, potential gene donors for the creation of new genetic variability were selected from the existing germplasm.

Regardless of treatment, the genotypes in microclimate A had the lowest averages of ear mass. The genotype Odisej, on solonetz reclaimed by $50\,t \times ha^{-1}$ of phosphogypsum, in microclimate B, recorded the greatest ear mass values. Agro-ecological factors significantly controlled how ear mass was expressed. Genotypes Banatka, Renesansa, Rapsodija and Simonida demonstrated the most stable reactions. The effect of the applied amelioration measure depended on the meteorological conditions of the growing season. On solonetz reclaimed by $25\,t \times ha^{-1}$ and $50\,t \times ha^{-1}$ of phosphogypsum, in microclimate C, the genotypes showed the highest stability. It was found that the local population, Banatka, and the ancient variety, Bankut 1205, had successfully adapted to 3S, and that they could be good parents for new wheat cultivars for growing under those conditions. In all of the experiment's versions, microclimate B produced the greatest average values of grain mass per ear and the lowest average values of the coefficient of variation. In the microclimate A, solonetz reclaimed by $25\,t \times ha^{-1}$ of phosphogypsum came out as the ideal setting for obtaining a stable reaction of the genotypes. The most stable genotypes were Rapsodija and Renesansa. In less favorable circumstances, under 3S, genotype Simonida produced one of the most steady reactions for grain mass per ear.

The results obtained by this research may be important for the further process of creating stable wheat genotypes, both for 3S conditions and abiotic stress in general.

Author Contributions: Conceptualization, Methodology, Software, Validation, Investigation, Resources, Writing—original draft, Writing—review & editing, Visualization, Funding acquisition, B.B.; Validation, Visualization, Funding acquisition, V.M.; Methodology, Validation, Visualization, Funding acquisition, S.P.; Software, Validation, Visualization, Funding acquisition, M.M.-S.; Validation, Data curation, Visualization, Funding acquisition, Đ.K.; Validation, Visualization, Funding acquisition, S.V.; Validation, Visualization, Funding acquisition, K.M.; Validation, Visualization, Funding acquisition, B.K., Validation, Visualization, Funding acquisition, D.B. (Dušana Banjac); Validation, Visualization, Funding acquisition, S.J.; Validation, Visualization, Funding acquisition, D.B. (Danilo Begić), Validation, Visualization, Funding acquisition, R.Š. All authors have read and agreed to the published version of the manuscript.

Funding: This research received no external funding.

Institutional Review Board Statement: Not applicable.

Informed Consent Statement: Not applicable.

Data Availability Statement: Not applicable.

Acknowledgments: Ministry of Education, Science and Technological Development, Serbia, 451-03-68/2022-14/200117.

Conflicts of Interest: The authors declare no conflict of interest.

References

1. Shiferaw, B.; Smale, M.; Braun, H.J.; Duveiller, E.; Reynolds, M.; Mauricho, G. Crops that feed the world 10. Past scuccesses and future challenges to the role played by wheat in global food security. *Food Secur.* **2013**, *5*, 291–317. [CrossRef]
2. Knežević, D.; Zečcević, V.; Dukić, N.; Dodig, D. Genetic and phenotypic variability of grain mass per spike of winter wheat genotypes (*Triticum aestivum* L.). *Kragujev. J. Sci.* **2008**, *30*, 131–136.

3. Li, T.; Deng, G.; Tang, Y.; Su, Y.; Wang, J.; Cheng, J. Identification and Validation of a Novel Locus Controlling Spikelet Number in Bread Wheat (*Triticum aestivum* L.). *Front. Plant Sci.* **2021**, *12*, 611106. [CrossRef] [PubMed]
4. Popović, V.; Ljubičić, N.; Kostić, M.; Radulović, M.; Blagojević, D.; Ugrenović, V.; Popović, D.; Ivošević, B. Genotype × Environment Interaction for Wheat Yield Traits Suitable for Selection in Different Seed Priming Conditions. *Plants* **2020**, *9*, 1804. [CrossRef]
5. Sheoran, S.; Jaiswal, S.; Raghav, N.; Sharma, R.; Sabhyata; Gaur, A.; Jaisri, J.; Gitanjali, T.; Singh, S.; Sharma, P.; et al. Genome-Wide Association Study and Post-genome-Wide Association Study Analysis for Spike Fertility and Yield Related Traits in Bread Wheat. *Front. Plant Sci.* **2022**, *12*, 820761. [CrossRef]
6. Delvet, A. Crop Production and Yield Limiting Factors. *MAS J. Appl. Sci.* **2021**, *6*, 325–349.
7. Tóth, G.; Montanarella, L.; Stolbovoy, V.; Máté, F.; Bódis, K.; Jones, A.; Panagos, P.; Van, M.; Liedekerke, M. *Soils of the European Union*; Office for Official publications of the European Communities: Luxembourg, 2008; p. 85.
8. Bai, R.; Zhang, Z.; Hu, Y.; Fan, M.; Schmidhalter, U. Improving the salt tolerance of Chinese spring wheat through an evaluation of genotype genetic variation. *Aust. J. Crop Sci.* **2011**, *5*, 1173–1178.
9. Borzouei, A.; Kafi, M.; Akbari-Ghogdi, E.; Mousavi-Shalmani, M. Long term salinity stress in relation to lipid peroxidation, superoxide dismutase activity and proline content of saltsensitive and salt-tolerant wheat cultivars. *Chil. J. Agric. Res.* **2012**, *72*, 476–482. [CrossRef]
10. Liu, X.; Chen, D.; Yang, T.; Huang, F.; Fu, S.; Li, L. Changes in soil labile and recalcitrant carbon pools after land-use change in a semi-arid agro-pastoral ecotone in Central Asia. *Ecol. Indic.* **2020**, *110*, 105925. [CrossRef]
11. Sairam, R.K.; Rao, K.V.; Srivastava, G.C. Differential response of wheat genotypes to long term salinity stress in relation to oxidative stress, antioxidant activity and osmolyte concentration. *Plant Sci.* **2002**, *163*, 1037–1046. [CrossRef]
12. Mladenov, V.; Fotopoulos, V.; Kaiserli, E.; Karalija, E.; Maury, S.; Baranek, M.; Segal, N.; Testillano, P.S.; Vassileva, V.; Pinto, G.; et al. Deciphering the Epigenetic Alphabet Involved in Transgenerational Stress Memory in Crops. *Int. J. Mol. Sci.* **2021**, *22*, 7118. [CrossRef] [PubMed]
13. Ullah, K.; Khan, S.J.; Muhammad, S.; Irfaq, M.; Muhammad, T. Genotypic and phenotypic variability, heritability and genetic diversity for yield components in bread wheat (*Triticum aestivum* L.) germplasm. *Afr. J. Agric. Res.* **2011**, *6*, 5204–5207.
14. Azarbad, H.; Tremblay, J.; Giard-Laliberté, C.; Bainard, L.D.; Yergeau, E. Four decades of soil water stress history together with host genotype constrain the response of the wheat microbiome to soil moisture. *FEMS Microbiol Ecol.* **2020**, *96*, fiaa098. [CrossRef] [PubMed]
15. Petrović, S.; Dimitrijević, M.; Kraljević-Balalić, M. Stabilnost mase klasa divergentnih genotipova pšenice [Stability of spike weight of divergent wheat genotypes]. *Letop. Naučnih Rad. Poljopr. Fak.* **2001**, *25*, 32–39.
16. Knežević, D.; Zečević, V.; Kondić, D.; Marković, S.; Šekularac, A. Genetic and phenotypic variability of grain mass per spike in wheat under different dose of nitrogen nutrition. *Tur. J. Agric. Nat. Sci.* **2014**, *1*, 805–810.
17. Knežević, D.; Radosavac, A.; Zelenika, M. Variability of grain weight per spike of wheat grown in different ecological conditions. *Acta Agric. Serb.* **2015**, *20*, 85–95. [CrossRef]
18. Jiang, T.; Dou, Z.; Liu, J.; Gao, Y.; Malone, R.W.; Chen, S.; Feng, H.; Yu, Q.; Xue, G.; He, J. Simulating the Influences of Soil Water Stress on Leaf Expansion and Senescence of Winter Wheat. *Agric. For. Meteorol.* **2020**, *291*, 108061. [CrossRef]
19. Arzani, A.; Ashraf, M. Smart engineering of genetic resources for enhanced salinity tolerance in crop plants. *Crit. Rev. Plant Sci.* **2016**, *35*, 146–189. [CrossRef]
20. Mohammadi, R.; Armion, M.; Zadhasan, E.; Ahamdi, M.M.; Amir, A. The use of AMMI model for interpreting genotype × environment interaction in durum wheat. *Exp. Agric.* **2018**, *54*, 670–683. [CrossRef]
21. Verma, A.; Singh, G.P. AMMI with BLUP analysis for stability assessment of wheat genotypes under multi locations timely sown trials in Central Zone of India. *J. Agric. Sc. Food Technol.* **2021**, *7*, 118–124.
22. Sime, B.; Tesfaye, S.M. Stability performance of bread wheat (*Triticum aestivum* L.) genotype for yield and yield components in Oromia, Ethiopia. *J. Agric. Res. Dev.* **2021**, *12*, 625.
23. Zobel, R.W.; Wright, M.J.; Gauch, H.G. Statistical analysis of a yield trial. *Agron. J.* **1988**, *80*, 388–393. [CrossRef]
24. Gauch, H.G. A simple protocol for AMMI analysis of yield trials. *Crop. Sci.* **2013**, *53*, 1860. [CrossRef]
25. Petrović, S.; Dimitrijevic, M.; Belić, M.; Banjac, B.; Bošković, J.; Zečević, V.; Pejić, B. The variation of yield components in wheat (*Triticum aestivum* L.) in response to stressful growing conditions of alkaline soil. *Genetika* **2010**, *42*, 545–555. [CrossRef]
26. Banjac, B.; Mladenov, V.; Dimitrijević, M.; Petrović, S.; Bočanski, J. Genotype x environment interactions and phenotypic stability for wheat grown in stressful conditions. *Genetika* **2014**, *46*, 799–806. [CrossRef]
27. Neisse, A.C.; Kirch, J.L.; Hongyu, K. AMMI and GGE Biplot for genotype × environment interaction: A medoid–based hierarchical cluster analysis approach for high–dimensional data. *Biom. Lett.* **2018**, *55*, 97–121. [CrossRef]
28. Hongyu, K. Adaptability, stability and genotype by environment interaction using the AMMI model for multienvironmental trials. *Biodiversity* **2018**, *17*, 10–21.
29. Dimitrijević, M.; Petrović, S.; Belić, M.; Mladenov, N.; Banjac, B.; Vukosavljev, M.; Hristov, N. Utjecaj limitirajućih uvjeta solonjeca na variranje uroda krušne pšenice [The influence of solonetz soil limited growth conditions on bread wheat yield]. In Proceedings of the 45th Croatian & 5th International Symposium on Agriculture, Opatija, Croatia, 15–19 February 2010; Faculty of Agriculture, University of Josip Juraj Strossmayer: Osijek, Croatia, 2010; pp. 394–398.

30. Petrović, S.; Dimitrijević, M.; Kraljević-Balalić, M.; Crnobarac, J.; Lalić, B.; Arsenić, I. Uticaj genotipova i spoljne sredine na komponente prinosa novosadskih sorti pšenice [The effect of genotype and the environment on yield components in Novi Sad wheat varieties]. *Zb. Rad. Naučnog Inst. Za Ratar. I Povrtarsvo* **2005**, *41*, 199–206.
31. Zadoks, J.C.; Chang, T.T.; Konzak, C.F. A decimal code for the growth stage of cereals. *Weed Res.* **1974**, *14*, 415–421. [CrossRef]
32. *GENSTAT 9th Edition*, Trial Version; VSN International Ltd.: Indore, India, 2009.
33. Comstock, R.R.; Robinson, H.F. Genetic parameters, their estimation and significance. In Proceedings of the 6th International Grassland Congress, State College, PA, USA, 17–23 August 1952; National Publishing Company: Washington, DC, USA, 1952; Volume 1, pp. 248–291.
34. Singh, R.K.; Chaudhary, B.D. *Biometrical Methods in Quantitative Genetic Analysis*; Kalyani Publishers: New Delhi/Ludhiana, India, 1985; pp. 39–78.
35. Falconer, D.S. *Introduction to Quantitative Genetics*, 3rd ed.; Longman Scientific and Technical: New York, NY, USA, 1989; p. 438.
36. *R Project for Statistical Computing*, Version 4.2.0 (2022-04-22 ucrt); R Foundation for Statistical Computing: Vienna, Austria, 2022. Available online: https://www.R-project.org/ (accessed on 3 July 2022).
37. Banjac, B.; Dimitrijević, M.; Petrović, S.; Mladenov, V.; Banjac, D.; Kiprovski, B. Antioxidant variability of eheat genotypes under salinity stress. *Genetika* **2020**, *52*, 1145–1160. [CrossRef]
38. Matković Stojšin, M.; Petrović, S.; Dimitrijević, M.; Šućur Elez, J.; Malenčić, Đ.; Zečević, V.; Banjac, B.; Knežević, D. Effect of salinity stress on antioxidant activitz and grain yield of different wheat genotypes. *Turk. J. Field Crops* **2022**, *27*, 33–40. [CrossRef]
39. Gharib, M.; Qabil, N.; Salem, A.; Ali, M.; Awaad, H.; Mansour, E. Characterization of wheat landraces and commercial cultivars based on morpho-phenological and agronomic traits. *Cereal Res. Commun.* **2020**, *49*, 149–159. [CrossRef]
40. Saradadevi, G.P.; Das, D.; Mangrauthia, S.K.; Mohapatra, S.; Chikkaputtaiah, C.; Roorkiwal, M.; Solanki, M.; Sundaram, R.M.; Chirravuri, N.N.; Sakhare, A.S.; et al. Genetic, Epigenetic, Genomic and Microbial Approaches to Enhance Salt Tolerance of Plants: A Comprehensive Review. *Biology* **2021**, *10*, 1255. [CrossRef] [PubMed]
41. Denčić, S.; Kastori, R.; Kobiljski, B.; Duggan, B. Evaluation of grain yield and its components in wheat cultivars and landraces under near optimal and drought conditions. *Euphytica* **2000**, *113*, 43–52. [CrossRef]
42. Matković Stojšin, M.; Petrović, S.; Banjac, B.; Zečević, V.; Roljević Nikolić, S.; Majstorović, H.; Đorđević, R.; Knežević, D. Assessment of genotype stress tolerance as an effective way to sustain wheat production under salinity stress conditions. *Sustainability* **2022**, *14*, 6973. [CrossRef]
43. Ljubičić, N.; Popović, V.; Ćirić, V.; Kostić, M.; Ivošević, B.; Popivić, D.; Pandžić, M.; El Musafah, S.; Janković, S. Multivariate Interaction Analaysis of Winter Wheat Grown in Environmen of Limited Soil Conditions. *Plants* **2021**, *10*, 604. [CrossRef]
44. Purchase, J.L.; Hatting, H.; van Deventer, C.S. Genotype × environment interaction of winter wheat (*Triticum aestivum* L.) in South Africa: II. Stability analysis of yield performance. *S. Afr. J. Plant Soil* **2000**, *17*, 101–107. [CrossRef]
45. Hagos, H.G.; Abay, F. AMMI and GGE biplot analysis of bread wheat genotypes in the northern part of Ethiopia. *J. Plant Breed. Genet.* **2013**, *1*, 12–18.
46. Zhao, C.; Liu, B.; Piao, S.; Wang, X.; Lobell, D.B.; Huang, Y.; Huang, M.; Yao, Y.; Bassu, S.; Ciais, P.; et al. Temperature increase reduces global yields of major crops in four independent estimates. *Proc. Natl. Acad. Sci. USA* **2017**, *114*, 9326–9331. [CrossRef]
47. Dimitrijević, M.; Petrović, S.; Banjac, B. Wheat breeding in abiotic stress conditions of solonetz. *Genetika* **2012**, *44*, 91–100. [CrossRef]
48. Matković Stojšin, M.; Zečević, V.; Petrović, S.; Dimitrijević, M.; Mićanović, D.; Banjac, B.; Knežević, D. Variability, correlation, path analysis and stepwise regression for yield components of different wheat genotypes. *Genetika* **2018**, *50*, 817–828. [CrossRef]
49. Kakoulidou, I.; Avramidou, E.V.; Baranek, M.; Brunel-Muguet, S.; Farrona, S.; Johannes, F.; Kaiserli, E.; Lieberman-Lazarovich, M.; Martinelli, F.; Mladenov, V.; et al. Epigenetics for crops improvement of global change. *Biology* **2021**, *10*, 766. [CrossRef]
50. El-Hendawy, S.E.; Hassan, W.M.; Al-Suhaibani, N.A.; Refay, Y.; Abdella, K.A. Comparative performance of multivariable agro-physiological parameters for detecting salt tolerance of wheat cultivars under simulated saline field growing conditions. *Front. Plant Sci.* **2017**, *8*, 435. [CrossRef]
51. El-Hendawy, S.; Al-Suhaibani, N.; Mubushar, M.; Tahir, M.U.; Refay, Y.; Tola, E. Potential use of hyperspectral reflectance as a high-throughput nondestructive phenotypic tool for assessing salt tolerance in advanced spring wheat lines under field conditions. *Plants* **2021**, *10*, 2512. [CrossRef]
52. Mansour, E.; Moustafa, E.S.; Desoky, E.S.M.; Ali, M.; Yasin, M.A.; Attia, A.; Alsuhaibani, N.; Tahir, M.U.; El-Hendawy, S. Multidimensional evaluation for detecting salt tolerance of bread wheat genotypes under actual saline field growing conditions. *Plants* **2020**, *9*, 1324. [CrossRef]

 *sustainability*

Article

Improved and Highly Efficient *Agrobacterium rhizogenes*-Mediated Genetic Transformation Protocol: Efficient Tools for Functional Analysis of Root-Specific Resistance Genes for *Solanum lycopersicum* cv. Micro-Tom

Máté Tóth [1], Zoltán Gábor Tóth [1], Sándor Fekete [2], Zoltán Szabó [1] and Zoltán Tóth [1,*]

1   Applied Plant Genomics Group, Institute of Genetics and Biotechnology, Hungarian University of Agriculture and Life Sciences, Páter Károly Str. 1, 2100 Gödöllő, Hungary; toth.mate@uni-mate.hu (M.T.); toth.zoltan.gabor@uni-mate.hu (Z.G.T.); szabo.zoltan.gen@uni-mate.hu (Z.S.)
2   GTIG, Török Ignác Secondary School of Gödöllő, Petőfi Sándor Str. 12–14, 2100 Gödöllő, Hungary; feketes@gmail.com
*   Correspondence: toth.zoltan.gen@uni-mate.hu

Abstract: Gene *function* analysis, molecular breeding, and the introduction of new traits in crop plants all require the development of a high-performance genetic transformation system. In numerous crops, including tomatoes, *Agrobacterium*-mediated genetic transformation is the preferred method. As one of our ongoing research efforts, we are in the process of mapping a broad-spectrum nematode resistance gene (*Me1*) in pepper. We work to transform tomato plants with candidate genes to confer resistance to nematodes in *Solanaceae* members. The transformation technology development is designed to produce a reproducible, rapid, and highly effective *Agrobacterium*-mediated genetic transformation system of Micro-Tom. In our system, a transformation efficiency of over 90% was achieved. The entire procedure, starting from the germination of seeds to the establishment of transformed plants in soil, was completed in 53 days. We confirmed the presence of the NeoR/KanR and DsRed genes in the transformed roots by polymerase chain reaction. The hairy root plants were infected with nematodes, and after 3 months, the presence of DsRed and NeoR/KanR genes was detected in the transformant roots to confirm the long-term effectiveness of the method. The presented study may facilitate root-related research and exploration of root–pathogen interactions.

Keywords: *Agrobacterium*-mediated transformation; functional genomics; *Solanum lycopersicum* L.; Micro-Tom; DsRed fluorescence; *Agrobacterium rhizogenes*

Citation: Tóth, M.; Tóth, Z.G.; Fekete, S.; Szabó, Z.; Tóth, Z. Improved and Highly Efficient *Agrobacterium rhizogenes*-Mediated Genetic Transformation Protocol: Efficient Tools for Functional Analysis of Root-Specific Resistance Genes for *Solanum lycopersicum* cv. Micro-Tom. *Sustainability* **2022**, *14*, 6525. https://doi.org/10.3390/su14116525

Academic Editor: Balázs Varga

Received: 12 April 2022
Accepted: 25 May 2022
Published: 26 May 2022

Publisher's Note: MDPI stays neutral with regard to jurisdictional claims in published maps and institutional affiliations.

1. Introduction

With an estimated yield of 186 million tons in 2020 [1], the tomato (*Solanum lycopersicum* L.) is a widely cultivated vegetable crop of global significance. The tomato is important for scientific research as a model plant of the *Solanaceae* family, which is often used as an experimental plant for fruit-related studies. For this reason, tomato may offer new opportunities for investigating biological mechanisms that are not possible in other popular model plants such as *Arabidopsis*. In recent years, the use of tomato as a model crop has increased due to its readily available resources, such as high-quality assembled reference genomes [2,3], as well as the research efforts to improve the efficiency of *Agrobacterium*-mediated transformation and the recovery time of transgenic lines [4–7]. The tomato genome was published in 2012 [8], and as a result, tomatoes have become an increasingly favored research material over the years.

Among the many known tomato varieties, the "Micro-Tom" has become a widely used model plant. Micro-Tom is a tomato mutant with a short life cycle that allows harvesting of the ripe fruit within 70–90 days after sowing. Seedlings grow with ease under ordinary fluorescent lighting. It can tolerate close planting, it is easily transformed, and it has the same hereditary information as the common tomato except for two main genes (dwarf gene and miniature gene). Thus, it is most useful for studying the gene functions of model plants such as *Arabidopsis* [9,10]. Due to the rapid development of genome editing technologies, the availability of efficient transformation methods is especially important. When root biology research is the primary focus, *Agrobacterium rhizogenes*-mediated transformation provides a quick and reliable alternative method to stable transformation. *A. rhizogenes* contains root loci (rol) genes, which, upon infection, trigger the formation of genetically transformed roots (hairy root phenotype), referring to the striking abundance of adventitious roots present [11,12]. Composite plants with healthy, untransformed shoots and transgenic hairy roots can be obtained as a result of *A. rhizogenes* transformation. Protocols were proposed for the production of composite plants in order to efficiently identify root–pathogen resistance genes in plants. Furthermore, several molecular studies were carried out in vitro with tomato hairy root crops, e.g., Fusarium tolerance [13], transgenic nematode resistance [14,15], salt stress [16], and rhizobial symbiosis study [17].

Almost 40 years ago, McCormick et al. [18] published the first report of *Agrobacterium*-mediated tomato (*S. lycopersicum*) transformation. Transformations of various species and explant types, such as leaves, cotyledons, and hypocotyls, have been reported in recent years [19–21]. In recent years, great progress has been made in the development of efficient *Agrobacterium*-mediated transformation protocols for Micro-Tom [22–24]. In most published papers, the transformation rates of *A. rhizogenes*-mediated Micro-Tom transformation were around 40–60%. In the majority of cases, no essential details (medium components, transplantation frequency, and infection method) were provided on the transformation protocol, which would be essential for the reproducibility of Micro-Tom's transformation procedures. It was also unclear whether the transformation frequency was determined by the number of independent transgenic events or the number of transgenic plants.

In this paper, we present a high-throughput transformation system in detail developed in *S. lycopersicum* cv. Micro-Tom, using radicles and hypocotyls as starting explants and applying *A. rhizogenes* ARqua1 in order to solve the problem of long transformation periods and low-efficiency transformation rates in tomato plant. We set out to evaluate two different inoculation methods and compare their effectiveness. In an effort to make it easier to control the transformation process and to distinguish between non-transformed and transgenic roots, DsRed was used as a visual marker that produced red fluorescence, and its use as a marker was tested for the reliable identification and selection of transformed hairy roots of tomato during in vitro culture. One of the main aims of the present study is to show that the expression of transgenes persists in the transformant roots for several months [25]. We intended to demonstrate that this alternative transformation method by means of nematode infection is suitable for long-term studies since, in the majority of cases, 2–3 months are required for the development of the corresponding infection phenotype in *Meloidogyne* infection. Our final objective was to establish a highly efficient, reliable, and reproducible *A. rhizogenes* transformation system for *S. lycopersicum* cv. Micro-Tom to validate and describe candidate genes (*Me1*) of pepper (*Capsicum annuum*, PI201234) [26] that could confer resistance against root-knot nematode species (*Meloidogyne* spp.).

We hypothesized that *A. rhizogenes* is suitable for transforming *S. lycopersicum* cv. Micro-Tom with obtaining high efficiency. Furthermore based on our preliminary experiments in *C. annuum* [27], we expected that *rhizogenes* infection by puncturing the hypocotyl will show the greatest efficiency in the case of tomato.

2. Materials and Methods

2.1. Plant Material

The seeds of tomato (*S. lycopersicum*) cv. Micro-Tom were soaked in water for 3 h, surface sterilized with filtered 10% solution of $CaClO_2$ and a drop of Tween20 for 20 min, and rinsed three times with sterile distilled water. MS20 (Murashige and Skoog supplied with 20 g sucrose) media were used to germinate the sterilized seeds. A total of 360 seeds were germinated this way in 3 separate rounds, each round with 120 seeds.

2.2. Agrobacterium Strain and Vector

We used the *A. rhizogenes* strain ARqua1 harboring our desired plasmid vector for the transformation. For the experiments, we used the plasmid pKWGFS7-RR (VIB-Ugent Center for Plant Systems Biology modified by Á. Domonkos, Gödöllő, Hungary) [28] containing an aminoglycoside adenylyltransferase gene for the bacterial selection, an aminoglycoside phosphotransferase gene for the plant selection, and a DsRed reporter gene. The control line was transformed with the same plasmid pKGWFS7 [29] construction but without the DsRed reporter.

2.3. Agrobacterium Culture Preparation and Infection Medium

For the infection procedure to be carried out, *Agrobacterium* strains were cultured in plastic Petri dishes at 28 °C on solid Luria Broth (LB) medium containing 50 mg/L spectinomycin and 50 mg/L rifampicin for 48 h.

2.4. Preparation of Explants, Agrobacterium Infection and Co-Cultivation

The plant roots were cut horizontally at the desired parts (radicles and hypocotyl) using a sterile sharp scalpel. In this study, we used 3 transformation methods to compare their transformation efficiency and studied a total of 40 plants/method (30 + 10 control) in each round: (1) coating the wound surface of radicles in bacterial mass; (2) coating the wound surface hypocotyl in bacterial mass; (3) puncturing the hypocotyl with tungsten needle dipped in bacterial mass. For Methods (1) and (2), we crossed the wounded surface over the bacterial mass in the Petri dishes. For Method (3) we gently dipped the tungsten needle in the bacterial mass, then punctured the hypocotyl vertically numerous times. The co-cultivation with the bacteria was carried out on MS20 media. To observe the effect of co-cultivation length, the first round of germinated seeds (120 plantlets) was co-cultivated for 6 days, while the other two rounds (240 plantlets) were co-cultivated for 4 days. The media were poured obliquely into sterile glass cultivating tubes 24 mm in diameter and 15 cm in height in order to be able to observe the plants individually during the experiment. Figures 1 and 2 highlight the key steps in the transformation process.

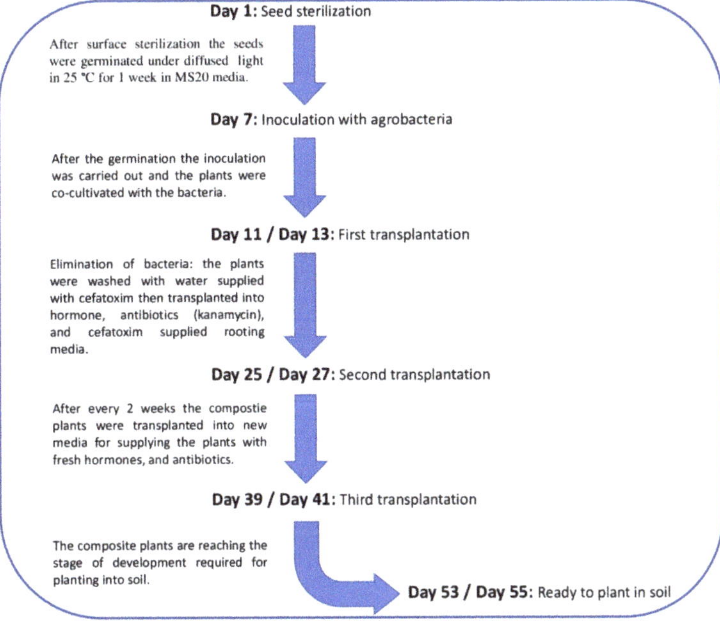

Figure 1. Schematic representation of the timeline for the production of composite tomato plants.

Figure 2. Essential steps in the production of tomato composite plants. Seed germination (**a**) and seeds placed in sterilized glass storage jars with MS20 media (**b**). Vigorous plants were removed from the MS20 medium (**c**). *A. rhizogenes* bacterial culture and observed transformation methods (**d–f**). Resulting vigorous composite tomato plantlets growing in sterile glass cultivating tubes 24 mm in diameter and 15 cm in height (**g,h**).

2.5. Tissue Culture and Transplantation

The plants were washed in sterile distilled water containing 500 mg/L cefotaxime after co-cultivation and then transferred to a rooting medium containing 400 mg/L cefotaxime, 60 mg/L kanamycin, and 0.5 mg/L IAA, to eliminate the bacteria and apply selection pressure to the roots. The plants were transferred every 2 weeks to a new rooting medium to supply them with fresh hormones, nutrients, and antibiotics.

2.6. Fluorescence Observation and Imaging

The red fluorescence in the seedling roots was observed using a Leica MZ10 F stereomicroscope (Leica, Wetzlar, Germany) with an external Leica EL6000 light source for enhanced fluorescence imaging, visualized using the manufacturer's own computer software (Leica Application Suite). The images were processed using Adobe Photoshop software ver. 23.2.2 (Adobe Inc., San Jose, CA, USA).

2.7. DNA Extraction and PCR Verification

Total DNA was isolated from fresh root tissue (100 mg/sample) using a Zenogene DNA purification kit (Zenon Bio, Szeged, Hungary) according to the manufacturer's instructions. The PCR reactions were carried out in 12 µL reaction mixtures containing 1.2 µL of 10× PCR buffer, 0.3 U Taq polymerase, 0.2 mM dNTP, 0.5 µM of each primer, and 10 ng template DNA. Thermal cycling conditions consisted of one cycle of initial denaturation for 4 min at 94 °C, followed by 30 cycles of 94 °C for 60 s, 58 °C for 60 s, 72 °C for 60 s, with a final extension step of 7 min at 72 °C. The amplified products were electrophoresed on a 1.0% (*w/v*) agarose gel stained with 0.5 mg/L ethidium bromide solution and visualized on a UV transilluminator. pKWGFS7-RR vector-specific NeoR/KanR primers (F-5′-GTTCTTTTTGTCAAGACCGACCT-3′; R-5′-CTCTTCAGCAATATCACGGGTAG-3′) and DsRed primers (F-5′-TGCCCTGCGCGCTCCTCCA-3′; R-5′-CTACAGGAACAGGTGGTGG-3′) were used for PCR identification of positive plants.

2.8. Determination of Agrobacterium Infection and Transformation Efficiency

DsRed fluorescence and PCR screening of NeoR/KanR and DsRed genes from roots of four-week-old in vitro putative T0 plants were used to calculate transformation efficiency. The statistical calculation is based on the following equation: the transformation efficiency (%) = (Number of positive transgenic plants/number of infected explants) × 100.

2.9. Nematode Susceptibility Test

The culture of *Meloidogyne incognita* was maintained in the susceptible tomato line. Before infection with nematodes, the two-month-old plantlets were transplanted individually into 250 mL plastic pots filled with steam-sterilized sandy soil and grown to the fourth/fifth true leaf stage. The tomato plants were inoculated with egg masses and J2 juveniles collected from infected roots and soil. Plants were grown in a 25 °C growth chamber, and after 8–10 weeks, the roots were evaluated, and transgenes in the root samples were tested.

3. Results

3.1. Rhizogenes Inoculation Method

The infiltration method and the length of co-cultivation play a key role in achieving a highly efficient *Agrobacterium* transformation. To identify the most appropriate method for tomato transformation, we applied three transformation methods to compare their transformation efficiency. *S. lycopersicum* plants were cut horizontally at the desired parts (radicles and hypocotyl) using a sterile sharp scalpel and infected by coating the wound surface in bacterial mass or puncturing the hypocotyl with a tungsten needle dipped in bacterial mass. The typical phenotype of the infection after one week was characterized by cell proliferation that could be observed at the site of infection and on the stem, as well as by brown spots, which were probably the signs of local cell death. The transformed roots typically grew from the cell proliferation zone. As in our previous experiments in peppers, both normal and hairy roots grew on the plant, and the formation of transformant roots did not depend on the development of the hairy root phenotype. Since the newly formed roots were highly fragile and easily damaged, during the transplantation, extra care had to be taken. The formation of the first roots started on the rooting medium after two weeks of cultivation. The plants infected with the tungsten needle (Method (3)) were the fastest in rooting while the coating of hypocotyl with bacterial mass (Method (2)) showed the slowest rooting (usually 4 weeks).

3.2. Elimination of A. rhizogenes

Upon each transplantation, we washed the plants with sterile water supplemented with 500 mg/L cefotaxime. The length of co-cultivation greatly influenced the survival rates of plants. After 6 days of co-cultivation, the bacteria typically overgrew the medium and were impossible to eliminate during transplantation despite the washing step. Because of the overgrown bacteria, we observed an 85% mortality rate among the plants. Method (1) had the highest survival rates with 11 living plants, while Method (3) only had 1, and Method (2) produced no surviving plants. The results showed that the plants infected with Method (2) were the least tolerant to the overgrowing bacteria, while plants infected with Method (1) showed significantly higher survival rates. Therefore, the infection with Method (1) showed the highest tolerance against the aggressive growth of bacteria. The dead plants were characterized by whitening (a typical symptom of kanamycin, most likely because the plants failed to develop roots due to the high amount of bacteria present in the rooting area), lack of cell proliferation, stem watering, and bacterial mass that penetrated the medium even after multiple elimination attempts. The surviving plants successfully developed roots and the elimination of the bacteria was successful. After 4 days of co-cultivation, the bacteria were eliminated after the first transplantation; therefore, we could not observe the high mortality rates that occurred during the 6-day-long co-cultivation period (Table 1). These results supported the finding that the length of co-cultivation played a key role in the survival rate of plants and the success of the *rhizogenes* transformation of *S. lycopersicum* cv. Micro-Tom. The 4-day-long co-cultivation turned out to be the most efficient way of infection in regard to transformation and survival rates.

Table 1. Mortality rate of different organs inoculated with binary vector inserted in the *A. rhizogenes* strain ARqua1 after 6 days of co-cultivation. Plants transformed with pKWGFS7-RR and pKWGFS7 (mock) are represented together in this table.

Organ Explant	Inoculation Method	Total No. of Transformed Plant	No. of Explants Surviving	No. of DsRed-Positive Roots	Mortality Rate (%)
Radicles	Coating—method (1)	40	11	11	73
Hypocotyl	Coating—method (2)	40	0	0	100
Hypocotyl	Puncturing—method (3)	40	1	1	98

3.3. Agrobacterium Transformation of Explants

Hypocotyl explants of the Micro-Tom genotypes were inoculated with pKWGFS7-RR by using the puncturing and coating inoculation method. Two weeks after the first transplantation approximately 60% of the puncture-inoculated plants succeeded in developing intact roots, while only 20% of the coating-inoculated plants had roots during the same period (Figure 3). Four weeks after the inoculation, this number was 90% in the case of Method (3) and 80% for Method (2). During this time, the puncture-inoculated plants rapidly developed multiple roots from shared nodes, which appeared to be the cell proliferation zones on the surface of the hypocotyl. Concerning transformation, the punctured plants showed higher efficiency, with 95% of the inoculated plants developing transformant roots, while in the case of Method (2), this number was 92%. Six weeks after the inoculation, the puncture-inoculated plants showed a substantially more extensive root system compared to the other methods. The hypocotyl-coating-inoculated plants showed the highest sensitivity to the overgrowing bacteria of the three inoculation methods. Our results suggested that if the study focused on the *rhizogenes* transformation of hypocotyl explants, the puncturing method offered a more efficient way to generate composite plants.

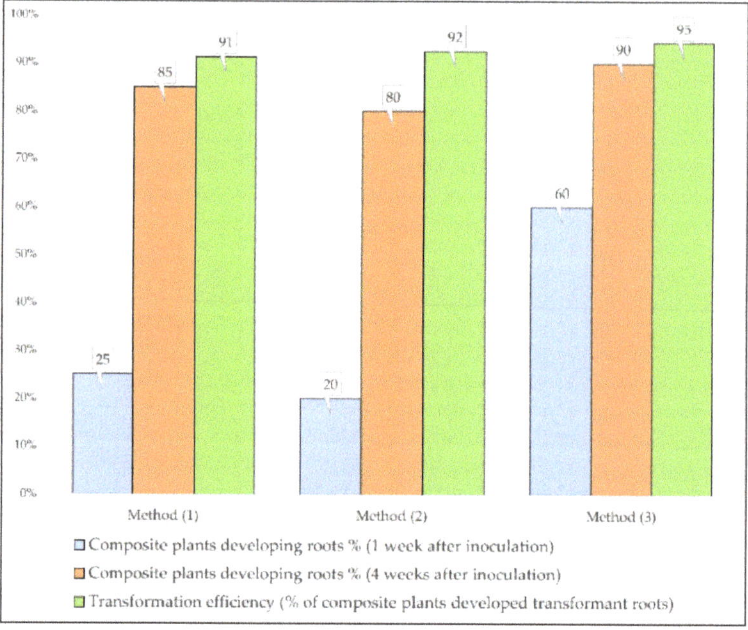

Figure 3. The effectiveness of different methods for the production of tomato composite plants and the development of roots on the composite plants transformed by different methods.

Radicle explants from Micro-Tom genotypes were inoculated with pKWGFS7-RR by using the coating infection method. The growing of roots was similar to the hypocotyl explants transformed by the coating of the wound surface. Two weeks after the first transplantation, only 25% of the plants succeeded in root generation, but 4 weeks after the first transplantation, 85% of all the radicle-transformed plants were able to grow intact roots. The transformation efficiency was similar to the methods involving the puncturing of hypocotyl, with 91% of the transformed plants growing transformant roots (Table 2).

Table 2. Transformation efficiency of different organs inoculated with binary vector inserted in the *A. rhizogenes* strain ARqua1 (pKWGFS7-RR) and mock inoculated (pKWGFS7). The table represents only the 4-day-long co-cultivation experiments.

Organ Explant	Inoculation Method	No. of Experiments	Inoculation with *A. rhizogenes* Strain ARqua1 (pKWGFS7-RR)			Mock Inoculation—*A. rhizogenes* Strain ARqua1 (pKWGFS7)			Transformation Efficiency [1] (%)
			Total No. of Transformed Plant	No. of Explants That Showed Hairy Roots	No. of DsRed-Positive Roots	Total No. of Transformed Plant	No. of Explants That Showed Hairy Roots	No. of DsRed-Positive Roots	
Radicles	Coating—method (1)	2×	60	57	55	20	18	0	91
Hypocotyl	Coating—method (2)	2×	60	58	56	20	18	0	92
Hypocotyl	Puncturing—method (3)	2×	60	58	57	20	19	0	95

[1] The transformation efficiency was calculated on the basis of the number of plants in 4 days of co-cultivation.

In comparison with the composite plants of Method (3), 6 weeks after the first transplantation, 99% of the plants with both transformation methods were ready to be transplanted into soil, but the radicle-transformed samples showed a less extensive root system compared to the plants obtained by Method (3). The radicle-transformed plants showed the highest tolerance to the aggressively overgrowing bacteria of all three methods; therefore, using radicle explants for the transformation can be advantageous if the plant type used is sensitive to the growing bacteria.

3.4. DsRed Allows for Quick and Easy Identification of Positive Plants

We decided to use the DsRed gene (pKWGFS7-RR) driven by the Ubiquitin promoter as a molecular marker to identify transformed hairy roots and to quantify transformation efficiency. The resultant red fluorescence in transgenic tissue could be observed through a fluorescence microscope because DsRed could be expressed in all tissues at every developmental stage. As a result, the red fluorescence could be used to screen positive roots at various stages of genetic transformation and quickly distinguish between positive and negative roots. Furthermore, DsRed provides an opportunity to eliminate non-transformant roots during the experiment. The wavelengths of excitation and emission for DsRed range from 554 to 563 nm for excitation and from 582 to 592 nm for emission. As we expected, strong red fluorescence appeared in the positive roots, while the transgene-negative roots showed almost no signal or weak autofluorescence (Figure 4).

We showed that the applied vector pKWGFS7-RR was highly suitable for *A. rhizogenes* transformation in the case of *S. lycopersicum* cv. Micro-Tom. In the roots, which showed a higher level of expression of the DsRed gene than the average, the presence of the red fluorescent protein was visible under normal light in the form of pale pink colorations on the roots of the composite plants.

Figure 4. Selection of transformed hairy roots expressing the DsRed gene in tomato plants. The roots were visualized with a stereomicroscope under bright light (lines **a,c,e**) and fluorescence (lines **b,d,f**). Control roots were visualized with stereomicroscope under bright light (line **g**) and under fluorescence (line **h**). Each bar represents 8 mm.

3.5. Confirmation of Genetic Co-Transformation of the Roots by PCR

To test whether the desired target DNA fragments were integrated into hairy roots that emitted red fluorescence and showed kanamycin resistance, all transgenic roots were selected by PCR amplification. DNA was isolated from the roots of all independent Micro-Tom plantlets and a non-transformed Micro-Tom plant, and by using DsRed and NeoR/KanR-specific primers, PCR analysis was performed. In line with our expectations, the DsRed and NeoR/KanR amplification products appeared as 678 bp and 558 bp long PCR products, respectively (Figure 5). PCR amplifications demonstrated the simultaneous presence of the NeoR/KanR and DsRed genes in the roots that were previously evaluated as DsRed positive under the stereomicroscope. In contrast, only the NeoR/KanR gene was amplified in the DNA sample deriving from control hairy roots. In the DsRed-negative roots (based on microscopy), none of the desired fragments appeared in the agarose gel. The results support the fact that upon transformation, the whole T-DNA sequence is integrated into the roots.

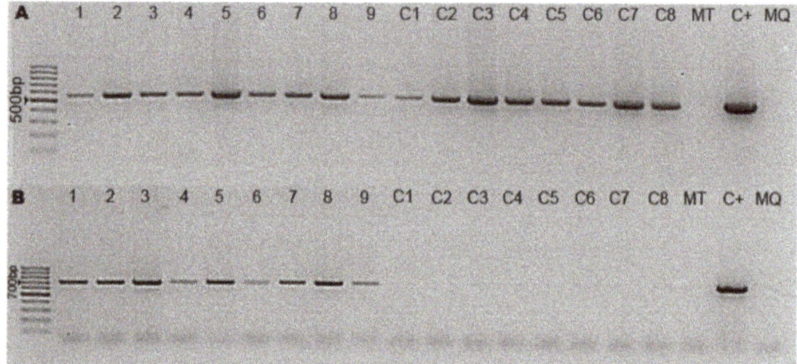

Figure 5. PCR screening for NeoR/KanR and Dsred genes. (**A**) PCR results of NeoR/KanR primers: Samples 1–9 represent transformant plants which were selected based on microscopy; Samples C1-C8 represent the control transformant line without the DsRed reporter; MT is a non-transformant Micro-Tom sample; C+ is the vector-positive control; MQ is a MilliQ water sample. (**B**) PCR results of DsRed-specific primers with the same samples in the same order.

3.6. Long-Term Evaluation of Transformed Roots after Nematode Infection and Analysis of the Presence of Transgenes

To confirm that the expression of transgenes persists for several months in transformed roots, we evaluated plants infected with *Meloidogyne* after 4 months. Several months after infection, transformed roots infected with *M. incognita* developed a variety of gall symptoms. We proved that the nematode could complete its life cycle in transformed roots. We observed the presence of DsRed in the hairy roots of each infected plant through a fluorescence microscope. The study confirmed that the strong red fluorescence signal was detected in the roots of all adult plants without exception (Figure 6). It can be concluded that the method we have developed is suitable for testing the long-term function of root-specific genes in *S. lycopersicum* cv. Micro-Tom plants.

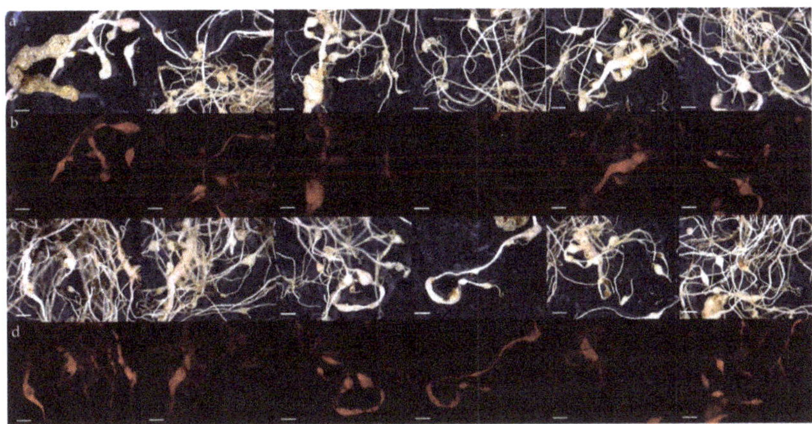

Figure 6. Selection of transformed hairy roots expressing the DsRed gene in tomato plants infected with *Meloidogyne incognita* after 4 months. The roots of adult plants were visualized with a stereomicroscope under bright light (line **a,c**) and under fluorescence (line **b,d**). Rounded or irregular galls are clearly observed on the roots as a phenotype of typical *Meloidogyne* spp. infection. Each bar represents 8 mm.

4. Discussion

In comparison with *A. tumefaciens*-mediated transformation, *A. rhizogenes*-mediated transformation has a range of benefits [30], including the ability to induce genetic transformation in hairy roots deriving from explants quickly and efficiently. Furthermore, individual transformed clones can be identified by using selection marker genes, which enable this technique to be applied in studying gene functions [31], plant transformation events [32], and secondary metabolism [33]. In tomato, the cotyledon, hypocotyl, and leaf discs [18] were used as materials for transformation [34]. The most common inoculation method used for the production of composite plants involves stab inoculation [35]. Recently, several authors reported that the growth rates of some species could substantially be improved when *A. rhizogenes* was inoculated directly onto the radicles or hypocotyls of freshly sliced seedlings [36,37]. Similar to other studies, hypocotyl explants produced more hairy roots than leaf discs or cotyledon segments in tomato when inoculated with *A. rhizogenes* strains [31].

In the present study, we used radicles and hypocotyls for inoculation with *A. rhizogenes* strain ARqua1, and 120 hairy root clones per method were obtained after 7 weeks of cultivation (Figure 1). The DsRed gene was also used as a selection marker, with its expression pattern monitored at various developmental stages, followed by the confirmation of its integration through PCR analysis. The number of re-generate explants was increased by establishing a durable and efficient transformation method. Using roots transformed by

A. rhizogenes made the functional validation of candidate genes involved in pathogen–plant root interactions possible [38,39]. Since most of the studies only presented short-term investigations, it was necessary to confirm that the plants transformed could maintain the expression level of the gene of interest for several months until the infection phenotype of *Meloidogyne* species appeared. An exception is the study in which Kátia et al. [25] verified by PCR and enzyme assays that the nptII gene remained active for several months in *S. lycopersicum* Mill. cv. Ailsa Craig explants co-cultivated with *A. rhizogenes* strain R1601. It was also an important goal to investigate whether DsRed could possibly be used to screen transformant roots from the early stages of the experiment because a number of tissues could provide strong autofluoroescent signals at certain wavelengths.

In brief, our results verified that DsRed could be used as a visual screening marker to reduce the difficulties and improve the efficiency of the screening process for positive transgenic plants at different developmental stages and that the red fluorescent signals remained visible throughout the whole cycle of root-knot nematode infection. We established a method for producing *A. rhizogenes*-dependent transformed roots from *S. lycopersicum* cv. Micro-Tom that was efficient, convenient, and fast. Three important factors were investigated in our experiments: first, the part of the plant to be transformed; second, the difference between inoculation methods; and third, the quality and efficiency of the transformation method. We found that even though both radicle and hypocotyl segments could be infected by the same ARqua1, the highest transformation efficiency was obtained with hypocotyls of Micro-Tom infected by Method (3) and the lowest with radicles of Micro-Tom infected by Method (1). Our results showed that the puncturing inoculation method proved to be highly efficient in producing composite plants harboring our genes of interest in the roots. Presumably, the puncturing process allowed the bacteria to penetrate deeper into the stem, while the effect of small individual wounds promoted enhanced rooting on the wound sites. In the past, numerous tools were used for puncturing such as sterile toothpicks, stainless steel needles, and sterile scalpels. The tungsten needle we used might be effective since, due to its thin and smooth surface, the needle was able to inflict minimal damage, making it easier for the plant to survive.

In this experiment, hairy roots with red fluorescence were observed in the transformed radicles and hypocotyls using a stereo fluorescence microscope and UV light. Over 90% of the hairy roots showing strong DsRed expression were derived from hypocotyls and radicles. Seven to ten days after the infection, a large number of adventitious roots appeared on the surface of each hypocotyl, which took place much faster than in the radicle segments, which took 15–20 days to develop the roots. In addition, our results showed that the hypocotyl was the most suitable material for *A. rhizogenes*-mediated transformation of tomato and that the hypocotyl explants of *S. lycopersicum* cv. Micro-Tom and the *A. rhizogenes* strain ARqua1 were an appropriate combination for tomato transformation. The microscopy results showed the possibility of extended visual screening with DsRed for *Meloidogyne* studies, as well as the opportunity to efficiently screen the positive transformation events in the early stages of root formation. Similarly, the PCR validation results confirmed the presence of DsRed in the plants previously evaluated as DsRed positive under the microscope.

The improved *A. rhizogenes*-mediated transformation method had numerous benefits. Firstly, DsRed served as an observable marker for identifying transformed plants, making the process easier and faster and shortening the transformed plant selection period. Furthermore, the highest transformation efficiency was found to be 95% in tomato plants.

The co-transformation efficiency (number of transformed explants/number of infected explants) of *S. lycopersicum* cv. Micro-Tom varied from 90% to 95% according to the type of organ (hypocotyl or radicles) and the transformation method used (Figure 3). In a previous study with *S. lycopersicum* cv. Moneymaker, Plágaro et al. [31] achieved a similar high transformation efficiency of almost 90% using *A. rhizogenes* strain MSU440. The transformation method presented in this study provided a fast and reliable means of studying root-specific genes, promoters, and host–pathogen interactions. The *A. rhizogenes-*

mediated transformation can be used in species that appear to be recalcitrant to plant regeneration or *A. tumefaciens* transformation such as *Capsicum* species [40], since the method only involves plant rooting.

Several studies reported the possibility of generating stable transformant plants (whole-plant level) by regenerating *rhizogenes*-transformed root tissues, although the publications mostly focused on ornamental plants [41–43]. In the future, we would like to explore this novel approach to generating a high number of transformant plants from transformed root tissues since, in theory, this procedure excludes the generation of mosaic transformation. This method can also aid research work in the field of mycorrhizal studies, as well as be applied in the creation of transgenic plants producing secondary metabolites.

5. Conclusions

In this study, we developed a novel and highly efficient protocol for *A. rhizogenes*-mediated transformation of *S. lycopersicum* Micro-Tom cultivars. Furthermore, using the fluorescent molecular marker DsRed, the selection of transformed tomato hairy roots could be effectively performed. Hairy root culture has a number of advantages, including intensive growth, repeatability, and vegetative propagation, which allows the transformed roots to be permanently maintained. This method improves the ability to study root-specific genes and could be used for molecular studies of root–pathogen interactions as well.

As a result of composite plants with well-developed roots that were acclimatized, transformed rootstocks were produced in 2 months, which could be directly used for nematode resistance validation and functional testing. This method had the significant advantage of allowing functional analysis studies on root genes at the whole-plant level. In comparison, in most cases, it takes 1 year to produce well-developed transformed seedlings from primary explants, such as leaf cuttings, using *A. tumefaciens*-mediated transformation.

The presented study can assist root-specific experiments by offering an effective in vitro protocol to introduce genes into plant roots and enabling the gene products to be screened and localized when the reporter is used in a fusion-protein construction. The method we present can facilitate the advancement of root-related research in the future.

Author Contributions: Conceptualization, M.T. and Z.T.; Data curation, M.T. and Z.T.; Formal analysis, M.T., Z.G.T., Z.S. and Z.T.; Investigation, M.T. and Z.T.; Methodology, M.T. and Z.T.; Project administration, Z.T.; Resources, Z.T.; Supervision, S.F. and Z.T.; Validation, M.T., Z.G.T. and Z.T.; Visualization, Z.T.; Writing—original draft, M.T. and Z.T.; Writing—review & editing, M.T., Z.S. and Z.T. All authors have read and agreed to the published version of the manuscript.

Funding: This research received no external funding.

Data Availability Statement: Not applicable.

Conflicts of Interest: The authors declare no conflict of interest.

References

1. Food and Agriculture Organization (FAO). FAOSTAT. 2020. Available online: http://www.fao.org/faostat/en/#home (accessed on 15 April 2020).
2. Hosmani, P.S.; Flores-Gonzalez, M.; van de Geest, H.; Maumus, F.; Bakker, L.V.; Schijlen, E.; van Haarst, J.; Cordewener, J.; Sanchez-Perez, G.; Peters, S.; et al. An improved de novo assembly and annotation of the tomato reference genome using single-molecule sequencing, Hi-C proximity ligation and optical maps. *bioRxiv* **2019**. [CrossRef]
3. Su, X.; Wang, B.; Geng, X.; Du, Y.; Yang, Q.; Liang, B.; Meng, G.; Gao, Q.; Yang, W.; Zhu, Y.; et al. A high-continuity and annotated tomato reference genome. *BMC Genom.* **2021**, *22*, 898. [CrossRef] [PubMed]
4. Sun, H.-J.; Uchii, S.; Watanabe, S.; Ezura, H. A Highly Efficient Transformation Protocol for Micro-Tom, a Model Cultivar for Tomato Functional Genomics. *Plant Cell Physiol.* **2006**, *47*, 426–431. [CrossRef] [PubMed]
5. Cruz-Mendívil, A.; Rivera-López, J.; Germán-Báez, L.J.; Lopez-Meyer, M.; Hernández-Verdugo, S.; López-Valenzuela, J.A.; Reyes-Moreno, C.; Valdez-Ortiz, A. A Simple and Efficient Protocol for Plant Regeneration and Genetic Transformation of Tomato cv. Micro-Tom from Leaf Explants. *HortScience* **2011**, *46*, 1655–1660. [CrossRef]
6. Wu, Z.; Sun, S.; Wang, F.; Guo, D. Establishment of Regeneration and Transformation System of *Lycopersicon esculentum* MicroTom. *Br. Biotechnol. J. Int.* **2011**, *1*, 53–60. [CrossRef]

7. Van Eck, J.; Keen, P.; Tjahjadi, M. *Agrobacterium* tumefaciens-Mediated Transformation of Tomato. *Methods Mol. Biol.* **2019**, *1864*, 225–234. [CrossRef]
8. Tomato Genome Consortium. The tomato genome sequence provides insights into fleshy fruit evolution. *Nature* **2012**, *485*, 635–641. [CrossRef]
9. Sun, W.; Dunning, F.M.; Pfund, C.; Weingarten, R.; Bent, A.F. Within-species flagellin polymorphism in *Xanthomonas campestris* pv campestris and its impact on elicitation of *Arabidopsis* FLAGELLIN SENSING2–dependent defenses. *Plant Cell* **2006**, *18*, 764–779. [CrossRef]
10. Chopra, R.; Johnson, E.B.; Daniels, E.; McGinn, M.; Dorn, K.M.; Esfahanian, M.; Folstad, N.; Amundson, K.; Altendorf, K.; Betts, K.; et al. Translational genomics using Arabidopsis as a model enables the characterization of pennycress genes through forward and reverse genetics. *Plant J.* **2018**, *96*, 1093–1105. [CrossRef]
11. Van Altvorst, A.C.; Bino, R.J.; Van Dijk, A.J.; Lamers AM, J.; Lindhout, W.H.; Van der Mark, F.; Dons, J.J.M. Effects of the introduction of *Agrobacterium rhizogenes* rol genes on tomato plant and flower development. *Plant Sci.* **1992**, *83*, 77–85. [CrossRef]
12. Bettini, P.P.; Marvasi, M.; Fani, F.; Lazzara, L.; Cosi, E.; Melani, L.; Mauro, M.L. *Agrobacterium rhizogenes* rolB gene affects photosynthesis and chlorophyll content in transgenic tomato (*Solanum lycopersicum* L.) plants. *J. Plant Physiol.* **2016**, *204*, 27–35. [CrossRef] [PubMed]
13. Xue, R.F.; Wu, X.B.; Wang, Y.; Zhuang, Y.; Chen, J.; Wu, J.; Ge, W.; Wang, L.; Wang, S.; Blair, M.W. Hairy root transgene expression analysis of a secretory peroxidase (PvPOX1) from common bean infected by Fusarium wilt. *Plant Sci.* **2017**, *260*, 1–7. [CrossRef] [PubMed]
14. Plovie, E.; De Buck, S.; Goeleven, E.; Tanghe, M.; Vercauteren, I.; Gheysen, G. Hairy roots to test for transgenic nematode resistance: Think twice. *Nematology* **2003**, *5*, 831–841. [CrossRef]
15. Wiśniewska, A.; Dabrowska-Bronk, J.; Szafrański, K.; Fudali, S.; Święcicka, M.; Czarny, M.; Wilkowska, A.; Morgiewicz, K.; Matusiak, J.; Sobczak, M.; et al. Analysis of tomato gene promoters activated in syncytia induced in tomato and potato hairy roots by Globodera rostochiensis. *Transgenic Res.* **2013**, *22*, 557–569. [CrossRef] [PubMed]
16. Talano, M.A.; Agostini, E.; Medina, M.I.; Forchetti, D.; Milrad, S.; Tigier, H.A. Tomato (*Lycopersicon esculentum* cv. Pera) Hairy Root Cultures: Characterization and Changes in Peroxidase Activity under NaCl Treatment. *Vitr. Cell. Dev. Biol. Plant* **2003**, *39*, 354–359. [CrossRef]
17. Rudaya, E.S.; Dolgikh, E.A. Production and analysis of composite tomato plants *Solanum lycopersicum* L. carrying pea genes encoding the receptors to rhizobial signal molecules. *Agric. Biol.* **2021**, *56*, 465–474. [CrossRef]
18. McCormick, S.; Niedermeyer, J.; Fry, J.; Barnason, A.; Horsch, R.; Fraley, R. Leaf disc transformation of cultivated tomato (*L. esculentum*) using Agrobacterium tumefaciens. *Plant Cell Rep.* **1986**, *5*, 81–84. [CrossRef]
19. Moghaieb, R.E.; Saneoka, H.; Fujita, K. Shoot regeneration from GUS-transformed tomato (*Lycopersicon esculentum*) hairy root. *Cell. Mol. Biol. Lett.* **2004**, *9*, 439–450.
20. Karmakar, S.; Molla, K.A.; Gayen, D.; Karmakar, A.; Das, K.; Sarkar, S.N.; Datta, K.; Datta, S.K. Development of a rapid and highly efficient *Agrobacterium*-mediated transformation system for pigeon pea [*Cajanus cajan* (L.) Millsp]. *GM Crop. Food* **2019**, *10*, 115–138. [CrossRef]
21. Xiao, Y.; Zhang, S.; Liu, Y.; Chen, Y.; Zhai, R.; Yang, C.; Wang, Z.; Ma, F.; Xu, L. Efficient *Agrobacterium*-mediated genetic transformation using cotyledons, hypocotyls and roots of 'Duli' (*Pyrus betulifolia* Bunge). *Sci. Hortic.* **2022**, *296*, 110906. [CrossRef]
22. Dan, Y.; Yan, H.; Munyikwa, T.; Dong, J.; Zhang, Y.; Armstrong, C.L. MicroTom—A high-throughput model transformation system for functional genomics. *Plant Cell Rep.* **2006**, *25*, 432–441. [CrossRef] [PubMed]
23. Peres, L.E.P.; Morgante, P.G.; Vecchi, C.; Kraus, J.E.; van Sluys, M.-A. Shoot regeneration capacity from roots and transgenic hairy roots of tomato cultivars and wild related species. *Plant Cell Tissue Organ Cult.* **2001**, *65*, 37–44. [CrossRef]
24. Shikata, M.; Ezura, H. Micro-Tom Tomato as an Alternative Plant Model System: Mutant Collection and Efficient Transformation. *Methods Mol Biol.* **2016**, *1363*, 47–55. [CrossRef]
25. Joäo KH, L.; Brown, T.A. Long-term stability of root cultures of tomato transformed with *Agrobacterium rhizogenes* R1601. *J. Exp. Bot.* **1994**, *45*, 641–647. [CrossRef]
26. Tóth, Z.; Szabó, Z.; Földi, T.; Szabadi, N.; Hajnik, L.; Jeney, A.; Kiss, G.; Kaló, P. Genetic mapping and identification of the Me1 gene conferring resistance to root-knot nematodes in pepper (*Capsicum annuum* L.). In Proceedings of the XVIth EUCARPIA, Kecskemét, Hungary, 12–14 September 2016; pp. 542–545.
27. Mate, T.; Zoltan, S.; Zoltan, T. Alternative method for the transformation of *Capsicum* species. *J. Plant Sci. Phytopathol.* **2021**, *5*, 1–3. [CrossRef]
28. Horváth, B.; Domonkos, Á.; Kereszt, A.; Szűcs, A.; Ábrahám, E.; Ayaydin, F.; Kaló, P. Loss of the nodule-specific cysteine rich peptide, NCR169, abolishes symbiotic nitrogen fixation in the Medicago truncatula dnf7 mutant. *Proc. Natl. Acad. Sci. USA* **2015**, *112*, 15232–15237. [CrossRef]
29. Domonkos, A.; Kovács, S.; Gombár, A.; Kiss, E.; Horváth, B.; Kováts, G.Z.; Farkas, A.; Tóth, M.T.; Ayaydin, F.; Bóka, K.; et al. NAD1 Controls Defense-Like Responses in *Medicago truncatula* Symbiotic Nitrogen Fixing Nodules Following Rhizobial Colonization in a BacA-Independent Manner. *Genes* **2017**, *8*, 387. [CrossRef]
30. Sharma, P.K.; Goud, V.V.; Yamamoto, Y.; Sahoo, L. Efficient *Agrobacterium tumefaciens*-mediated stable genetic transformation of green microalgae, *Chlorella sorokiniana*. *3 Biotech* **2021**, *11*, 196. [CrossRef]

31. Ho-Plágaro, T.; Huertas, R.; Tamayo-Navarrete, M.I.; Ocampo, J.A.; García-Garrido, J.M. An improved method for *Agrobacterium rhizogenes*-mediated transformation of tomato suitable for the study of arbuscular mycorrhizal symbiosis. *Plant Methods* **2018**, *14*, 34. [CrossRef]

32. Aggarwal, P.R.; Nag, P.; Choudhary, P.; Chakraborty, N.; Chakraborty, S. Genotype-independent *Agrobacterium rhizogenes*-mediated root transformation of chickpea: A rapid and efficient method for reverse genetics studies. *Plant Methods* **2018**, *14*, 55. [CrossRef]

33. Cui, M.-L.; Liu, C.; Piao, C.-L.; Liu, C.-L. A Stable *Agrobacterium rhizogenes*-Mediated Transformation of Cotton (*Gossypium hirsutum* L.) and Plant Regeneration from Transformed Hairy Root via Embryogenesis. *Front. Plant Sci.* **2020**, *11*, 604255. [CrossRef]

34. Shahin, E.A.; Sukhapinda, K.; Simpson, R.B.; Spivey, R. Transformation of cultivated tomato by a binary vector in *Agrobacterium rhizogenes*: Transgenic plants with normal phenotypes harbor binary vector T-DNA, but no Ri-plasmid T-DNA. *Theor. Appl. Genet.* **1986**, *72*, 770–777. [CrossRef] [PubMed]

35. Quandt, H.J.; Pühler, A.; Broer, I.N.G.E. Transgenic root nodules of *Vicia hirsuta*: A fast and efficient system for the study of gene expression in indeterminate-type nodules. *Mol. Plant Microbe Interact.* **1993**, *6*, 699–706. [CrossRef]

36. Boisson-Dernier, A.; Chabaud, M.; Garcia, F.; Bécard, G.; Rosenberg, C.; Barker, D.G. *Agrobacterium rhizogenes*-transformed roots of *Medicago truncatula* for the study of nitrogen-fixing and endomycorrhizal symbiotic associations. *Mol. Plant Microbe Interact.* **2001**, *14*, 695–700. [CrossRef] [PubMed]

37. Tomilov, A.; Tomilova, N.; Yoder, J.I. *Agrobacterium tumefaciens* and *Agrobacterium rhizogenes* transformed roots of the parasitic plant *Triphysaria versicolor* retain parasitic competence. *Planta* **2007**, *225*, 1059–1071. [CrossRef]

38. Kifle, S.; Shao, M.; Jung, C.; Cai, D. An improved transformation protocol for studying gene expression in hairy roots of sugar beet (*Beta vulgaris* L.). *Plant Cell Rep.* **1999**, *18*, 514–519. [CrossRef]

39. Hwang, C.-F.; Bhakta, A.V.; Truesdell, G.M.; Pudlo, W.M.; Williamson, V.M. Evidence for a Role of the N Terminus and Leucine-Rich Repeat Region of the *Mi* Gene Product in Regulation of Localized Cell Death. *Plant Cell* **2000**, *12*, 1319–1329. [CrossRef]

40. Li, D.; Zhao, K.; Xie, B.; Zhang, B.; Luo, K. Establishment of a highly efficient transformation system for pepper (*Capsicum annuum* L.). *Plant Cell Rep.* **2003**, *21*, 785–788. [CrossRef]

41. Bhat, S.R.; Chitralekha, P.; Chandel, K.P.S. Regeneration of plants from long-term root culture of lime, *Citrus aurantifolia* (Christm.) Swing. *Plant Cell Tissue Organ Cult.* **1992**, *29*, 19–25. [CrossRef]

42. Giri, A.; Narasu, M.L. Transgenic hairy roots: Recent trends and applications. *Biotechnol. Adv.* **2000**, *18*, 1–22. [CrossRef]

43. Choi, P.S.; Kim, Y.D.; Choi, K.M.; Chung, H.J.; Choi, D.W.; Liu, J.R. Plant regeneration from hairy-root cultures transformed by infection with *Agrobacterium rhizogenes* in *Catharanthus roseus*. *Plant Cell Rep.* **2004**, *22*, 828–831. [CrossRef] [PubMed]

sustainability

MDPI

Article

General Defense Response under Biotic Stress and Its Genetics at Pepper (*Capsicum annuum* L.)

János Szarka [1], Zoltán Timár [2], Regina Hári [2], Gábor Palotás [2,*] and Balázs Péterfi [1]

[1] Pirospaprika Ltd., 1222 Budapest, Hungary; janos.szarka.dr@gmail.com (J.S.); bpeterfi67@gmail.com (B.P.)
[2] Univer Product Plc., 6000 Kecskemét, Hungary; zoltan.timar@univer.hu (Z.T.); regina.hari@univer.hu (R.H.)
* Correspondence: gabor.palotas@univer.hu

Abstract: Since the beginning of resistance breeding, protection of plants against pathogens has relied on specific resistance genes encoding rapid tissue death. Our work has demonstrated in different host–pathogen relationships that plants can defend themselves against pathogens by cell growth and cell division. We first demonstrated this general defence response (GDR) in plants by identifying the *gds* gene in pepper. Subsequently, the existence of a genetic system for tissue defence became apparent and we set the goal to analyse it. The *gdr 1 + 2* genes, which operate the complete GDR system, protect plant tissues from pathogens in a direcessive homozygous state in both host and non-host relationships. The inheritance pattern of the two genes follows a 12:3:1 cleavage of the dominant epistasis. With the knowledge of the *gds* and *gdr 1 + 2* genes, the role of tissue-preserving (GDR) and tissue-destructive (HR) pathways in disease development and their relationship was determined. The genes encoding the general defence response have a low stimulus threshold and are not tissue-destructive and pathogen-specific. They are able to fulfil the role of the plant immune system by providing a general response to various specific stresses. This broad-spectrum general defence system is the most effective in the plant kingdom.

Keywords: pepper; general defence response; tissue retention; hypersensitivity response; resistance breeding

Citation: Szarka, J.; Timár, Z.; Hári, R.; Palotás, G.; Péterfi, B. General Defense Response under Biotic Stress and Its Genetics at Pepper (*Capsicum annuum* L.). *Sustainability* 2022, 14, 6458. https://doi.org/10.3390/su14116458

Academic Editor: Balázs Varga

Received: 9 March 2022
Accepted: 20 May 2022
Published: 25 May 2022

Publisher's Note: MDPI stays neutral with regard to jurisdictional claims in published maps and institutional affiliations.

1. Introduction

Since the beginning of our half-century of work on resistance breeding, experiments reported in publications of interest to us have been repeated to help understanding. We also took photographs of pathologies described in other authors' publications but not shown in pictures. With these and with photographs of our own work, we aim to make the knowledge of the symptoms in this speciality more visible and clearer.

The century-old history of plant resistance breeding has been based on the rapid destruction of the attacked tissue of the host plant, the hypersensitive response (specific HR).

In the relationship between cereals and *Puccinia graminis*, Ward [1] and Stakman [2], in the interaction between *Nicotiana glutinosa* and TMV, Holmes [3,4], in the host–pathogen relationship between *Capsicum annum* and *Xanthomonas vesicatoria*, Cook and Stall [5] described the local lesion in the *Bs1* gene, and Cook and Guevara [6] described the local lesion in the *Bs2* gene (Figure 1).

In addition to the study of HR induced by specific resistance genes in the host–pathogen relationship, the observation of the non-host–pathogen relationship has also started. The leaves of the *Nicotiana tabacum* L. *White Burley* plant were infiltrated with inoculum concentration of 10^5, 10^6, 10^7, 10^8 cells/mL of *Pseudomonas tabaci* being in a host–pathogenic relationship with tobacco, and of *Pseudomonas* species as non-host–pathogenic plant pathogen. Inoculum concentration of 10^6, 10^7, 10^8 cells/mL of bacteria being in a non-host–pathogenic relationship with tobacco induced rapid tissue destruction in 24–48 h for all *Pseudomonas* species. The 10^5 inoculum concentration induced partial tissue destruction only in *P. tabaci*, whereas in the other pathogenic *Pseudomonas* species, small local lesions were formed in the area of infected leaf spots. The authors found that phytopathogenic *Pseudomonas* species can also multiply in the intercellular ducts of non-host plants, inducing (nonspecific) HR, and that this is considered a naturally occurring phenomenon, but the authors overlooked the fact that this phenomenon can only occur if the pathogen has entered the intercellular ducts in very large quantities. A very important aspect of the experiment, but not interpreted by the authors, is that the dilution series of *Pseudomonas* pathogen inoculum, the reaction type switch, was in all cases triggered by increasing the inoculum concentration from 10^5 cells/mL to 10^6 cells/mL, thus setting the limit of the overall tissue retention capacity of *White Burley* [7].

Figure 1. Local lesions in wheat-*Puccinia graminis* (**A**)—susceptible, (**B**)—resistant, *Nicotiana glutinosa*-TMV (**C**), *Capsicum annuum* (*Bs2* gene)-*Xanthomonas vesicatoria* (**D**).

The relationship between susceptibility and HR in host–pathogen and non-host–pathogen relationships was studied by Szarka et al. [8]. *Nicotiana tabacum* cv. *Pallagi* leaves were infiltrated with $10^1–10^9$ cells/mL inoculum of *P. tabaci* (Figure 2A), and *P. phaseolicola* (Figure 2B) bacteria (Figure 2), irrespective of the host–pathogen or non-host–pathogen relationship.

Figure 2. Reaction exhibited by *Nicotiana tabacum* cv. *Pallagi* after infiltration with 10^7, 10^8 and 10^9 cells/mL concentrations of the bacteria *Pseudomonas tabaci* (**A**) and *Pseudomonas phaseolicola* (**B**).

The 10^9 cells/mL inoculum induced nonspecific HR in both bacterial species in the same way. The 10^8 cells/mL inoculum of *P. tabaci* induced the typical pathological symptoms of the disease on its host plant, tobacco, and 10^1–10^7 cell/mL inoculum of *P. tabaci* and 10^1–10^8 cells/mL inoculum of *P. phaseolicola* induced at most a slight bulging towards the front surface of the leaf and a chlorotic spot with indeterminate margins on infiltrated tissue sections. In other words, the tissue retention capacity of the *Pallagi* tobacco variety was able to tolerate stress caused by a bacterial suspension of *P. tabaci* at a concentration of 10^7 cell/mL and of *P. phaseolicola* at a concentration of 10^8 cell/mL. From this observation, conclusions were drawn on the tissue retention capacity of tobacco [8].

In evaluating the results, the term General Defence Reaction (GDR) was used to describe the tissue retention capacity of the plant in host–pathogen and non-host–pathogen relationships, i.e., the specific HR counterpoint, and its interpretation was described in 1995 [9].

The 10^8 cells/mL inoculum of *Pallagi* and *P. tabaci* formed a host–pathogen relationship in which a delicate equilibrium was established between the stress of the pathogen and the tissue retention capacity of the host plant, which was upset by the stress effect of the pathogen's increased proliferation. The plant tissue began to die, and this balance of power was eventually manifested in the susceptible disease. The *P. tabaci* suspension at concentration of 10^7 cells/mL showed the dominance of the GDR of the plant, while the rapid tissue death caused by the suspension at concentration of 10^9 cells/mL showed the destruction of the GDR. Since susceptibility as a physiological state is intermediate between the two, it was likely that GDR levels play a dominant role in the susceptibility of plants to pathogens [8].

To study susceptibility status and specific resistance gene activation, pepper cell lines with similar GDR levels, containing susceptible and *Bs2* resistance genes, were tested using a non-host–pathogen relationship, using *X. vesicatoria inocula* at concentrations of 10^1–10^9 cells/mL [8]. Leaf spots of susceptible and *Bs2* resistance gene-containing peppers infiltrated with suspensions of *X. vesicatoria* at concentrations of 10^1–10^7 cells/mL were at most chlorotic. The inoculum at concentration of 10^8 cells/mL produced a greasy spot on the susceptible cell line and a maroon discolouration on the leaf containing the *Bs2* gene, which did not dry out and caused resistance symptoms. Since pepper cell lines with similar tissue retention were tested, it was concluded that the *Bs2* gene is only activated when one of the stress levels represented by the *X. vesicatoria* dilution series breaks through the plant's general defence system. On the basis of this observation, it is likely that the general defence response is activated first in the defence process, and that its insufficiency leads to a state of susceptibility, which induces a physiological disturbance that triggers the specific resistance gene, which, thus, only becomes active after the development of the disease.

Contrary to the "resistance = HR" theory, the "destruction is not resistance" position could not be proven until a gene encoding a resistant response without tissue death was found.

The non-hypersensitive, nonspecific recessive gene found in cell line PI 163,192 of *Capsicum Annuum* became applicable for resistance identification in 1995. Based on the characteristics it encodes, it is a proposed marker of monogenic recessive trait: general defence system—*gds* [9]. In contrast to the *Bs2* gene (Figure 3), the response regulated by the *gds* gene (Figure 4) is manifested by cell growth, cell division, tissue compaction, and bulging of the infected leaf plate.

Unlike *X. vesicatoria* (Figure 5A) being in a host–pathogen relationship with pepper, *P. phaseolicola* (Figure 5B) and *X. Phaseoli* (Figure 5C) being in a non-host relationship, and the saprophytic *P. Fluorescens* (Figure 5D) bacteria, the *gds* gene responds with the same tissue lesions.

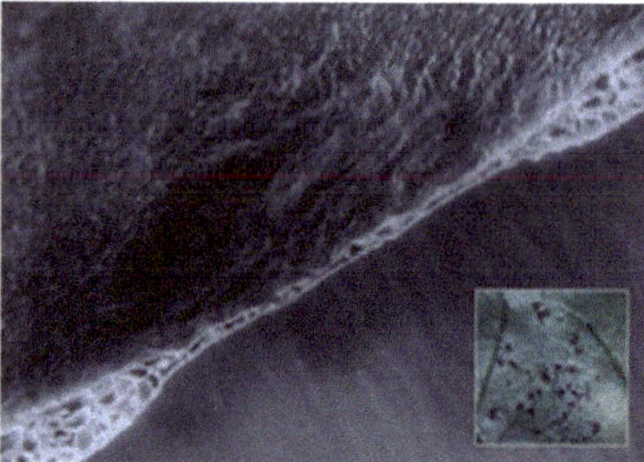

Figure 3. Pathology and section of a pepper leaf infected by *Xanthomonas vesicatoria*, containing the HR-inducing *Bs2* gene [8].

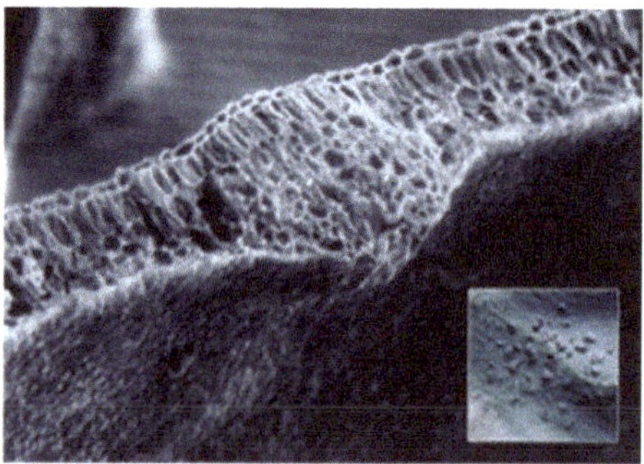

Figure 4. Pathology and section of a pepper leaf infected with *Xanthomonas vesicatoria*, containing the *gds* gene encoding tissue retention [10].

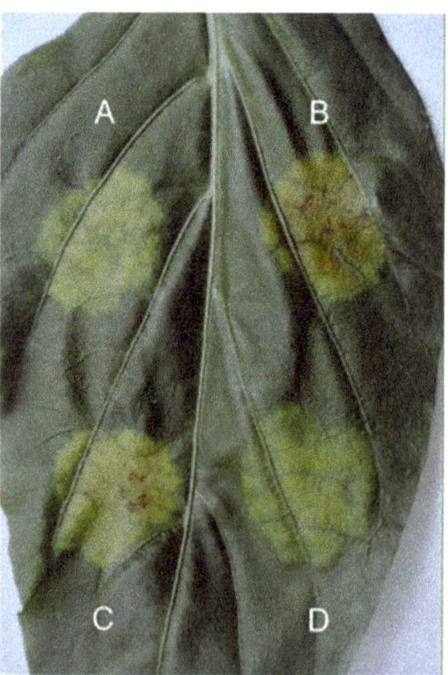

Figure 5. Tissue retention response of pepper leaves containing the *gds* gene to inoculation with a suspension of *Xanthomonas vesicatoria* (**A**), *Pseudomonas phaseolicola* (**B**), *Xanthomonas phaseoli* (**C**), and *Pseudomonas fluorescens* (**D**) at concentration of 10^8 cells/mL.

Thus, in addition to being non-hypersensitive, *gds* is also a non-specific gene, as it gives a generalised response (GDR) to specific stresses induced by different pathogenic species in peppers containing the *gds* gene.

The gene that protects pepper by producing a tissue retention defence response was published in 2002 [11]. The gene was given the name *bs5* and another recessive tissue retention gene, *bs6*, was also described at the same time.

Since the *bs5* gene was found in the same PI 163,192 cell line in which Szarka and Csilléry were the first to find and describe a tissue retention recessive gene in 1995 [9], the authors performed allele testing of *gds* and the *bs5* gene [12]. Based on the susceptibility and tissue retention pathogenicity of *X. vesicatoria*, the heritability tests confirmed that the two genes were identical. Pepper containing the *bs6* gene (ECW60) was shown to be susceptible in testing, which is presumably why it is not used in breeding today, unlike the *gds* (*bs5*) gene.

Another type of tissue retention response other than *gds* has also been reported [9]. In this experiment, pepper cell lines carrying the *Bs2* and *Bs3* genes were infected with *X. vesicatoria* and the infected leaves were detached from the plants and dried 24 h after inoculation (Figure 6). The leaves of the *Bs3* plant showed tissue damage indicative of HR already at the time of detachment, while the leaves of the *Bs2* plant showed only a slight purple discolouration indicating the location of the infected spots. For the *Bs3* gene, the dried infected tissue (Figure 6A) was only half as thick as the uninfected tissue, whereas for the *Bs2* gene, in addition to the slight purple discolouration, the infected tissue was twice as thick (Figure 6B) as the uninfected tissue.

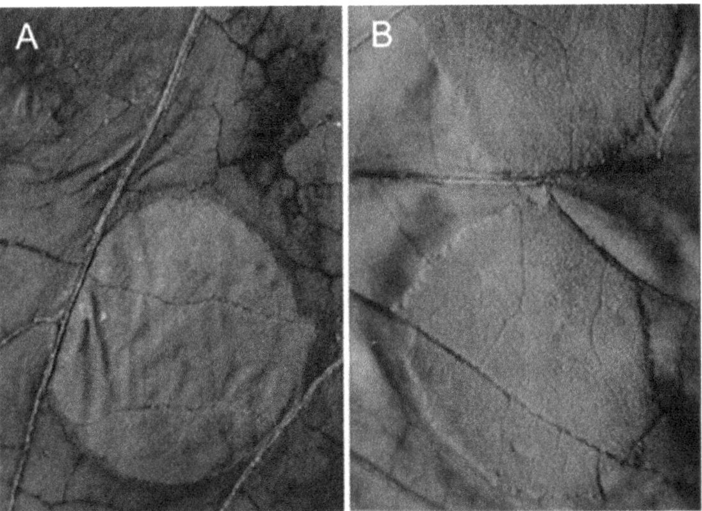

Figure 6. Tissue structural changes in leaves of pepper plants containing the *Bs3* (**A**) and *Bs2* (**B**) genes, detached at 24 h after infection with *Xanthomonas vesicatoria* and dried.

A similar tissue retention property was observed in the non-host–pathogenic relationship between pepper containing HR genes (*Bs2*, *Bs3*) and the bean pathogen *X. Phaseoli*.

The GDR phenomenon has also been described in several host–pathogen relationships where the host plant did not contain a known resistance gene and nonetheless developed tissue retention pathology suggestive of non-susceptibility (Figure 7). Such factors include: *Cucumis sativus–Pseudoperonospora cubensis* (Figure 7A,B), *Phaseolus vulgaris–Pseudomonas phaseolicola* (Figure 7C) [13].

Figure 7. General Defence Reaction, GDR in Cucumber—*Pseudoperonospora cubensis*, (**A**)—adaxial leaf surface, (**B**)—abaxial leaf surface, and bean—*Pseudomonas phaseolicola* (**C**) host–pathogen relationship.

Hydrogen peroxide plays an essential role in the development of specific HR determined by specific resistance genes. H_2O_2 in plant-microbe or host–pathogen interactions, depending on its amount, either enhances or destroys host plant cells. The production of H_2O_2 by susceptible plants containing the *Bs2* and *gds* genes, upon the effect of *X. vesicatoria*, was as follows [10]. The author investigated the amount of H_2O_2 by infiltration of the entire leaf surface of the plants for 10 h after infection. In the susceptible host–pathogen relationship, the pathogen enters the plant without causing a stress effect, so no significant H_2O_2 change occurs. The pathogen only becomes detectable to the plant during its accumulation. In response to H_2O_2 as a signalling molecule for the initiation of death in the susceptible phase, the *Bs2* gene is activated, causing a specific HR characterised by H_2O_2 "burst". In the experiment, the H_2O_2 "burst" occurred within 30 min and after 8 h it was reduced to the control level, while the infected tissues were destroyed. H_2O_2 levels remained constant in plants containing the *gds* gene, which encodes a strong tissue retention capacity.

H_2O_2 induces lignin synthesis and cross-linking between phenolic compounds and cellular wall proteins. This results in increased resistance to the enzymes degrading the cellular wall [14]. The formation of cross-links between cellular wall proteins is very rapid, occurring within 2–5 min after a stimulus [15]. Non-host resistance is primarily based on general responses linked to the cellular wall. These include thickening, lignification of the cellular wall, accumulation of phenols, flavonoids, which are highly localized responses expressed at the point of pathogen entry [16].

The aim of our work is to describe the functional defects of specific resistance genes used in resistance breeding and to elucidate their causes. Furthermore, we will describe the pathophysiology of a previously unknown defence response without tissue destruction, its genetic analysis, and its application in resistance breeding.

2. Materials and Methods

The experiments were carried out at the Kecskemét site of Univer Product Plc. between 2017 and 2021.

A significant part of our observations on plant material for variety production were made in the framework of resistance breeding work. The plants were planted in the soil of the growing house, i.e., we worked with plants with strong roots and vigorous growth. The plants were inoculated before flowering, in the vegetative stage, which is characterised by strong growth. The leaves selected for this purpose were at 70–80% of their development. Suspensions of different concentrations of *X. vesicatoria* and *P. phaseolicola* bacteria prepared from 48 h culture were used for inoculation. To form lesions, inoculum was applied to the abaxial surface of the leaf by brushing, mimicking the natural infection process. The primary criterion of evaluation in this case was the size of the lesion. For evaluation by tissue retention, intercellular ducts were flooded by injection. This method is less dependent on the environment and allows differentiation based on the quality of the infected tissue. The assessment of symptoms was performed on days 7 and 14, with occasional continuous evaluation.

For the genetic analysis, two infected leaves of a plant, directly above each other, were inoculated (Figure 8A,B).

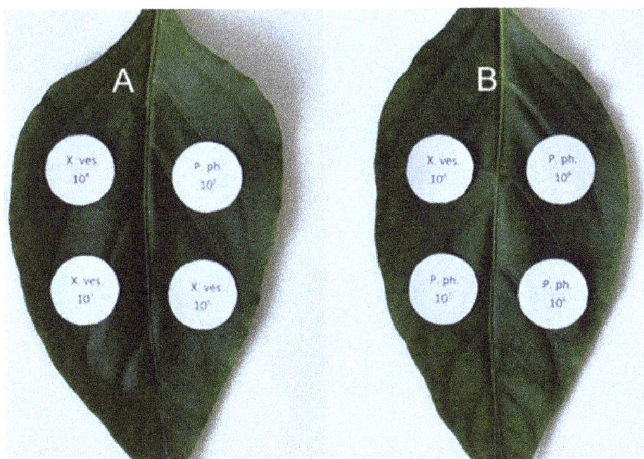

Figure 8. Location of tissue spots on leaves infected with a suspension of *Xanthomonas vesicatoria* (**A**) and *Pseudomonas phaseolicola* (**B**) at concentrations of 10^8, 10^7, 10^6 cells/mL for genetic analysis of the general defence response of pepper.

In both leaves, the upper left spot shows the inoculation of 10^8 cells/mL of *X. vesicatoria* bacteria in a host–pathogenic relationship with pepper, and the upper right spot shows the inoculation of 10^8 cells/mL of *P. phaseolicola* bacteria in a non-host–pathogenic relationship with pepper. The lower two spots on the upper leaf show the result of inoculation by the bacterium *X. pesicatoria* at concentrations of 10^7 (left spot) and 10^6 (right spot) cells/mL. The lower two spots on the lower leaf show the spots resulting from the inoculation of *P. phaseolicola* at concentrations of 10^7 (left spot) and 10^6 (right spot) cells/mL. Both leaves were tested with a 10^8 cells/mL suspension of the two pathogens to account for possible differences due to the age of the leaves.

The reaction of the leaves was classified into pathogen groups according to whether the inoculation resulted in a green, tissue-retaining spot or a drying, necrotic spot (Figures 9–11), as required for genetic analysis.

Figure 12A shows the design of experiments to investigate the rate of response of genes controlling specific and general defence. This involved infecting a spot with 5 mm of diameter with one pathogen (a) and then, after absorption of the watery patch, re-infecting it with the other pathogen (b), knowing the location of the first spot (a).

Figure 12B illustrates the inoculation process to study the function of specific and general defence responses in damaged cells. This involved inducing a convex tissue spot infiltrated by mechanical pressure with an object that consequently damaged the cells to different degrees (c) and inoculating it with a bacterial suspension (d).

Of the hot peppers used in the experiments, *Unihot* is a commercial variety, the others are breeding lines (Table 1).

Figure 9. Pathogen symptom clusters induced by the inoculum of *Xanthomonas vesicatoria* (**A**) and *Pseudomonas phaseolicola* (**B**) at 10^8 cells/mL and their dilution series (10^7, 10^6 cells/mL). *X. vesicatoria*: Z-Z-Z-Z, *P. vhaseolicola*: Z-Z-Z-Z.

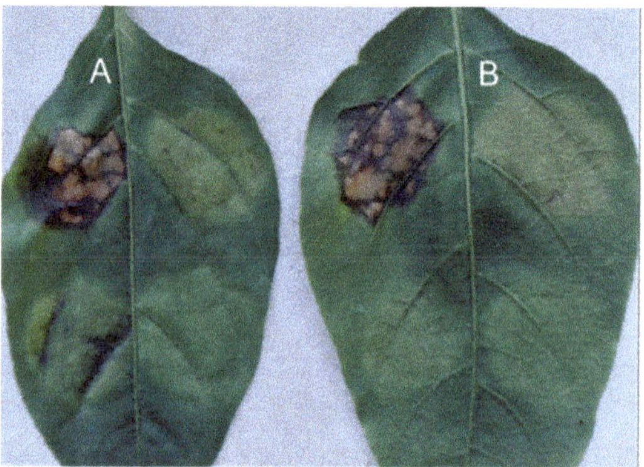

Figure 10. Pathogen symptom clusters induced by the inoculum of *Xanthomonas vesicatoria* (**A**) and *Pseudomonas phaseolicola* (**B**) at concentration of 10^8 cells/mL and their dilution series (10^7, 10^6 cells/mL). *X. vesicatoria*: L-Z-L$_X$-Z, *P. phaseolicola*: L-Z-Z-Z.

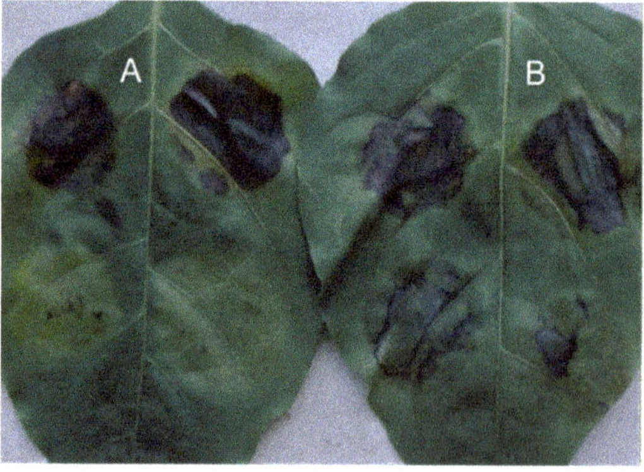

Figure 11. Pathogen symptom clusters induced by the inoculum of *Xanthomonas vesicatoria* (**A**) and *Pseudomonas phaseolicola* (**B**) at concentration of 10^8 cells/mL and their dilution series (10^7, 10^6 cells/mL). *X. vesicatoria*: P-P-Z-Z, *P. phaseolicola*: P-P-P-P.

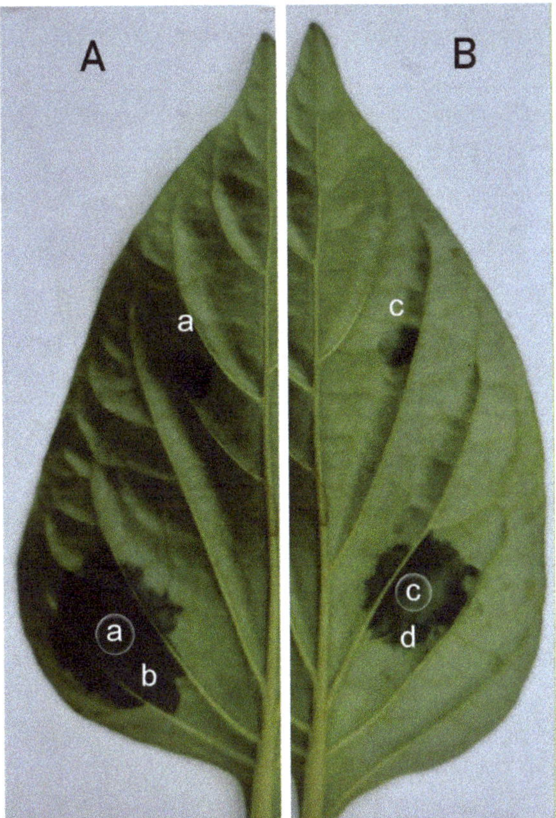

Figure 12. Pre-infection with a suspension of *Xanthomonas vesicatoria* in a spot with 5 mm of diameter (**Aa**), followed by super-infection with a suspension of *Pseudomonas phaseolicola* (**Ab**). Mechanical pressure-induced infiltrated tissue spot (**Bc**), followed by super-infection with a suspension of *Xanthomonas vesicatoria* (**Bd**).

Table 1. Reactions of the hot pepper lines used in hybridizations to inoculation with *X. vesicatoria* at concentration 10^8 cells/mL.

Pepper Materials Used	Reaction Type			
	S (P_X) [1]	L_f [2]	L_x [3]	X [4]
Unihot	+			
L330		+		
L1786			+	
L1710/2				+
L1713/2				+
L1715/3				+
L1716/4				+

[1] S (P_X): necrotic spot (susceptible); [2] L_F: rapidly dying slightly purple spot (*Bs2* gene); [3] L_x: green spot with purple vessels (*Bs2* gene); [4] X: uniform green tissue-retaining spots after inoculation with *X. vesicatoria* at concentration of 10^8 cells/mL and *P. phaseolicola* at 10^8 cells/mL.

The cleavages observed in F2 were subjected to Chi^2 test regression analysis. Our null hypothesis is that the observed genetic cleavage rates correspond to a given theoretical genetic cleavage.

There is no established terminology yet for plant traits and pathological processes related to tissue retention. Hence, we see the need for an interpretation of the terms and concepts we use.

Susceptibility (S): in the case of a host-pathogen relationship, plant tissue death characteristic of the pathogen. The rate of the process depends on the plant's capacity to retain tissue.

Hypersensitive response (HR): rapid tissue death in response to biotic stresses.

- Specific hypersensitive response (sHR): in a host–pathogen relationship, the destruction of the attacked plant tissue by a specific resistance gene.
- Nonspecific hypersensitive reaction (aHR): rapid tissue death in a non-host relationship due to stress induced by a pathogen, or rapid tissue death in a host–pathogen relationship without a specific resistance agent due to over-infection.

Tissue retention capacity: the genetically determined trait that protects plant tissues from destruction by biotic and abiotic stresses.

General defence response (GDR): a defence process under biotic stresses, in host–pathogen and non-host–pathogen relationships, based on the plant's tissue retention capacity to exclude pathogens in a general response to specific stresses.

- Weak/strong GDR: the degree of tissue retention manifested in pathogenicity without knowledge of the genotype.
- Complete GDR: the ability of a plant containing both genes (*gdr 1 + 2*) to retain tissue.
- Incomplete GDR: the degree of tissue retention in a plant containing either or neither of the *gdr 1* and *2* genes, under stress expressed as inoculum concentration.

3. Results

3.1. Insufficient Function of Specific HR

The observed disturbances in the function of specific HR genes used in resistance breeding under cultivation conditions were investigated in different host–pathogen relationships.

In tobacco-*TMV* and pepper-*ToMV-Ob* host–pathogen relationships, leaf viability functions are reduced by senescence and excision, respectively, and pathogens previously blocked in lesions are released. Despite the presence of resistance genes, the virus spreads unhindered in tissues, as indicated by tissue lesions characteristic of the pathogen (Figure 13A–C).

The function of the *Bs2* gene, which protects pepper against *X. vesicatoria*, is inhibited in cool, rainy weather. The bacterium escapes from the lesions already formed and infiltrates, showing the susceptibility of the flooded tissues. When weather conditions become more favourable (warming, reduction in humidity), the tissues invaded by the bacteria start to get purple. Leaf-drop as a self-destruction associated with the delayed onset of the *Bs2* gene function is an additional burden for the plant.

If plant life processes are weakened or inhibited due to senescence or adverse environmental factors, specific HR genes are unable to prevent the spread of the pathogen in the leaf, but only kill damaged cells already affected by the pathogen in a delayed, downstream manner. This characteristic results in their inadequacy and insufficiency in resistance inhibition.

Under conditions favourable to the plant, specific HR genes prevent the systemisation of pathogens in the plant by forming lesions. The smaller the lesion, the better the resistance efficacy, i.e., the less tissue loss. In turn, the pathogen releases and spreads further from the lesions formed on the leaves of plants whose life processes are inhibited. Based on this observation, it is clear that other plant properties besides specific HR genes are involved in the formation of lesions. This could be the plant's ability to retain tissue, which is manifested as a general defence response (GDR) under biotic stresses. A strong correlation was found between small lesion diameter and high levels of GDR detected by infiltration.

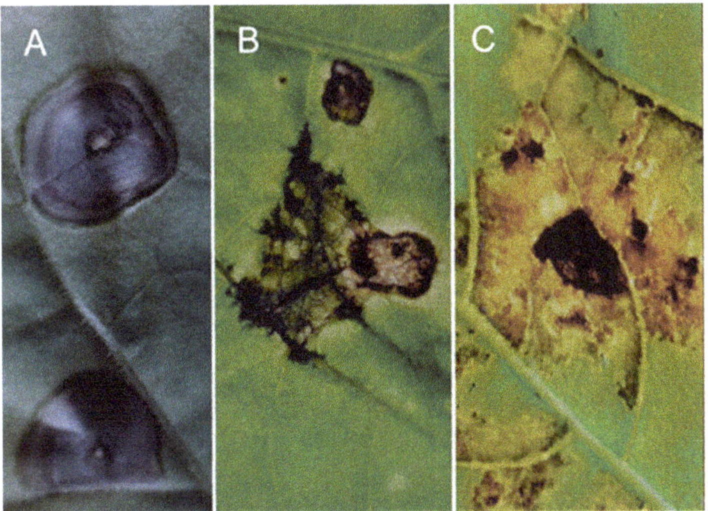

Figure 13. Local lesions of tobacco (**A**) and pepper (**B,C**) leaves as a consequence of senescence, excision, reduced life cycle, and unfavourable environmental conditions, leading to the release of the pathogen (*TMV, ToMV-Ob, X. vesicatoria*) and its spread in the plant.

3.2. Specific HR and GDR

The previous association, together with the fact that in a host–pathogen relationship, significant tissue thickening in the presence of a specific resistance gene (*Bs2*) inhibited rapid tissue death within 24 h, justified a detailed investigation of GDR (Figure 6).

We chose the *Bs2* gene for our work because the purple discoloration showing its function is a good indicator for pathological genetic studies. *Bs2*, a so-called specific resistance gene capable of destroying pepper tissue, induces different pathogenic symptoms that also accurately reflect the GDR levels of the plants (Figure 14).

The manifestations of increasingly higher GDR levels are shown in Figure 14A–C. Figure 14C also shows that in the presence of high levels of GDR, the specific HR gene could only be activated along the vessels and in the vessels and induce partial tissue death. These levels of tissue retention induced by *X. vesicatoria* at a concentration of 10^8 cells/mL were also confirmed by testing with a suspension of *P. phaseolicola* at the same concentration. The pathologies of the *Bs2* gene with different genetic structures in terms of tissue retention were designated L_F (Figure 14A), LL (Figure 14B), and L_X (Figure 14C).

3.3. Inheritance of GDR

3.3.1. Finding Plants X

For the genetic work, we selected plants marked L_X (Figure 15A), which were hybridized to the susceptible (S) variety *Unihot* (Figure 15B). Individuals of the F_1 generation were infected with *X. vesicatoria* inoculum at a concentration of 10^8 cells/mL, resulting in purple and desiccated uniform L_L disease symptoms (Figure 15C).

For accurate genetic analysis, the F_2 generation was also tested with inoculum at concentrations of 10^6, 10^7, 10^8 cells/mL and a stress band of appropriate width was established to assess the spectrum of tissue retention.

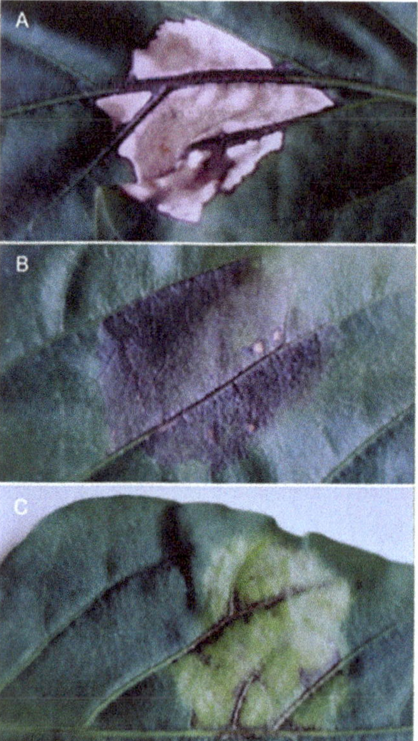

Figure 14. Pathogenicity induced by *Xathonomas vesicatoria* inoculum at concentration of 10^8 cells/mL in pepper lines carrying the *Bs2* gene. (**A**) L_F—rapid tissue death (HR), (**B**) L_L—purple-coloured tissue, (**C**) L_X—tissue retention (GDR), vessels getting purple.

Figure 15. Response of *Bs2* gene-containing peppers with L_x pathotype (**A**) and susceptible (S) *Unihot* variant (**B**) and F_1 generation (**C**) to *Xanthomonas vesicatoria*.

Two directly overlapping infected leaves of each plant were inoculated (Figure 8). This allowed us to characterise the tissue retention capacity of the plants by pathogen symptom clusters (Figures 9–11). Based on these experiments, individual plants of the F_2 generation were infected with *X. vesicatoria* inoculum at concentration of 10^8 cells/mL and *P. phaseolicola* inoculum at concentration of 10^8 cells/mL. Individual plants with both infected spots green in colour and a thickened leaf plate were selected from the fission population. These two characteristics always occurred together. The resulting plants with a high level of tissue retention were marked X. The non-X marked individuals of the fission population were not yet divided into phenotypic categories and were treated as a single whole (Σ Non-Z_x + Z_p) (Table 2).

Table 2. Hybridization of L_x and susceptible (S) lines (2017/1).

Hybridization	Phenotype Categories				
	S (P_x) [1]	L_L [2]	L_X [3]	Z_x [4] + Z_p [5] (X)	Σ Non-Z_x + Z_p
P1 L1786			+		
P2 Unihot	+				
F1		+			
F2				33	451

[1] S (P_x): a necrotic susceptible spot formed by inoculation of *X. vesicatoria* at a concentration of 10^8 cells/mL. [2] L_L: uniform purple spot formed by inoculation of *X. vesicatoria* at a concentration of 10^8 cells/mL. [3] L_X: green patch with purple vessels formed by inoculation of *X. vesicatoria* at a concentration of 10^8 cells/mL. [4] Z_X: a green tissue-retaining spot formed by inoculation of *X. vesicatoria* at a concentration of 10^8 cells/mL. [5] Z_P: a green tissue-retaining spot formed by inoculation of *P. phaseolicola* at a concentration of 10^8 cells/mL.

The results of this experiment, despite its shortcomings, suggested that it is a dihybrid (two-gene) inheritance system in which two recessive genes must be homozygous for the Z_x + Z_p symptom, i.e., the complete tissue retention GDR system, to manifest. This can occur in 9:3:3:1, 9:6:1, 12:3:1, and 15:1 dihybrid cleavages, and based on the available data, we started with the 15:1 possible cleavage as a working hypothesis, which became our null hypothesis and on which we performed the Chi^2 test regression analysis: $\Sigma (Y_i - X_i)^2/X_i = (451 - 454)^2/454 + (33 - 30)^2/30 = 0.0198 + 0.3 = 0.3198 < 3.84$, which shows that the calculated Chi^2 value is lower than the value in the contingency table for the given degree of freedom (1), and, therefore, the null hypothesis for the 15: 1 cleavage was accepted, i.e., the observed cleavage may correspond to the trait inherited by the two recessive genes (double recessive homozygotes) at $p = 0.05$ significance level.

X (Z_x + Z_p) plants selected from the F2 generation were strictly self-fertilized and their progeny were tested with inoculum of *X. vesicatoria* at concentration of 10^8 cells/mL and *P. phaseolicola* at a concentration of 10^8 cells/mL. It was found that the progeny of the X lines highlighted in generation F2 in generation F3 also all responded to infection with both bacteria with green, non-necrotic spots, so the lines did not show any symptomatic cleavage. However, it was still necessary to exclude the possibility that the X-labelled plants also contain the *Bs2* gene in a repressed, hypostatic state. For this, X-labelled individual plants were hybridized with a variety (*Unihot*) susceptible to the *X. vesicatoria* pathogen. The F1 generation gave susceptible symptoms when tested with inoculum of *X. vesicatoria* at a concentration of 10^8 cells/mL, thus, making it clear that the *Bs2* gene is not present in the X plants even in a repressed state. This demonstrates that *gdr* genes can act without the presence of the *Bs2* gene and are, therefore, not helper, intensifier, or modifier genes, which can only act in the presence of resistance genes.

3.3.2. Hybridization of Plants X with Line L330

In further studies, the pepper line L330 was used as a hybridization partner for the X lines. This line contains the *Bs2* gene and develops a very fast, 24 h, drying out and dying HR under infection by *X. vesicatoria* (Figure 16(A1)) and *P. phaseolicola* (Figure 16(A2)), and its symptom type belongs to the L_F category.

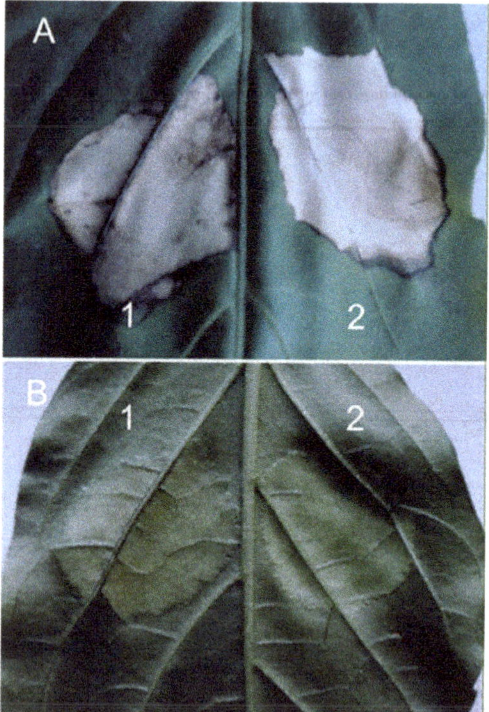

Figure 16. Pathological history of pepper line L330 (**A**) containing the *Bs2* gene and line X–(**B**) with strong tissue retention capacity induced by inocula of *Xanthomonas vesicatoria* (**1**) and *Phseudomonas phaseoicola* (**2**) at a concentration of 10^8 cells/mL, used for genetic analysis of GDR.

Because of the extremely rapid tissue death, it was assumed that none of the tissue-retaining *gdr* genes were present and, therefore, it is suitable for determining the inheritance of *gdr* genes. The F_1 generation of hybridization between the X lines and the L330 line showed signs of purple discoloration and partial tissue death suggestive of *Bs2* in all hybridization, which we identified as the L_L symptom type.

Individual plants of the F_2 generation were infected with inoculum of *X. vesicatoria* at concentration of 10^8 cells/mL and inoculum of *P. phaseolicola* at a concentration of 10^8 cells/mL, and individual plants of the fission population were phenotyped. This made it clear that, from a genetic point of view, the symptoms can be classified into three distinct phenotype categories, which at first sight assume 12:3:1 cleavage with the null hypothesis being that this distribution exists. Our assumption was subjected to Chi^2 test regression analysis.

The contingency table shows a value of 5.99 at a probability level of $p = 0.05$ at degree of freedom 2, the calculated Chi^2 value is less than this, so in all cases, the null hypothesis that the empirical frequencies are consistent with 12:3:1 cleavage of the dominant epistasis is accepted (Table 3). The genes carried by the direcessive homozygotes were named *gdr 1 + 2* genes and identified as manifestations of the complete GDR system. The *gdr 1 + 2* genes protect the plant tissue from pathogen destruction in both host and non-host relationships. The cleavages also show that one of the *gdr* genes can eradicate *P. phaseolicola* on its own (phenotype category $P_X + Z_P$) but can only defend against *X. vesicatoria* attack in combination with the other *gdr* gene in the form of a tissue retention green spot (phenotype category $Z_x + Z_p$).

Table 3. F2 population of X × L330 hybridizations.

Hybridization	Empirical Frequency			Theoretical Frequency 12:3:1			Chi ^{2}	Level of Significance
	P_X [1] + P_P [2]	P_X + Z_P [3]	Z_x [4] + Z_p	P_X + P_P	P_X + Z_P	Z_x + Z_p		
L1710/2 × L330	640	131	42	610	152	51	5.964	$p = 0.05$
L1713/2 × L330	260	51	23	251	63	21	2.799	$p = 0.05$
L1715/3 × L330	142	22	11	132	33	11	4.516	$p = 0.05$
L1716/4 × L330	368	91	20	359	90	30	5.034	$p = 0.05$

[1] P_x: necrotic spot formed by inoculation of *X. vesicatoria* at a concentration of 10^8 cells/mL. [2] P_P: necrotic spot formed by inoculation of *P. phaseolicola* at a concentration of 10^8 cells/mL. [3] Z_P: a green tissue-retaining spot formed by inoculation of *P. phaseolicola* at a concentration of 10^8 cells/mL. [4] Z_X: a green tissue-retaining spot formed by inoculation of *X. vesicatoria* at a concentration of 10^8 cells/mL.

3.4. Morphological and Histological Characterisation of GDR and Comparison with Other Resistance Genes

3.4.1. GDR and Bs2 Gene

Recognition and identification of GDR helps to interpret pathophysiology based on microscopic observation (Figure 17).

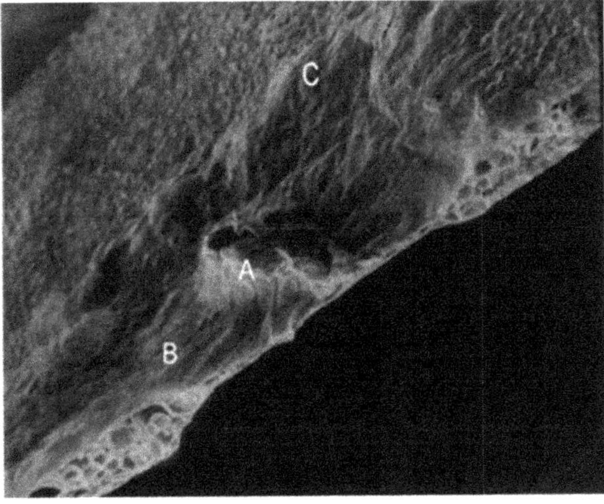

Figure 17. Lesion induced by *Xanthomonas vesicatoria* on a pepper line containing the *Bs2* gene with an incomplete GDR system.

In the presence of incomplete GDR, the tissue compaction at the infection point (Figure 17A) is not sufficient to block the pathogen. Therefore, the *Bs2* gene, which senses pathogen-induced physiological disturbances (susceptibility state), is activated and induces specific HR in the tissues surrounding the infection site with tissue compaction (Figure 17B). However, the HR-induced stress activates distant tissues, resulting in a cell enlargement-mediated tissue compaction barrier around the necrotic tissue spot (Figure 17C).

With the knowledge of the GDR, macroscopic modelling of the processes observed in the lesions was also performed. The design of the experiment is shown in Figure 12A. Susceptible pepper leaves were infiltrated with a suspension of *X. vesicatoria* at concentration of 10^8 cells/mL in a spot of 5 mm in diameter. After 10 min, when the inoculum squeezed into the intercellular passages was absorbed, it was superinfected with a suspension of *P. phaseolicola* at concentration of 10^6 cells/mL. In this way, we activated the incomplete GDR in a non-host relationship (Figure 18).

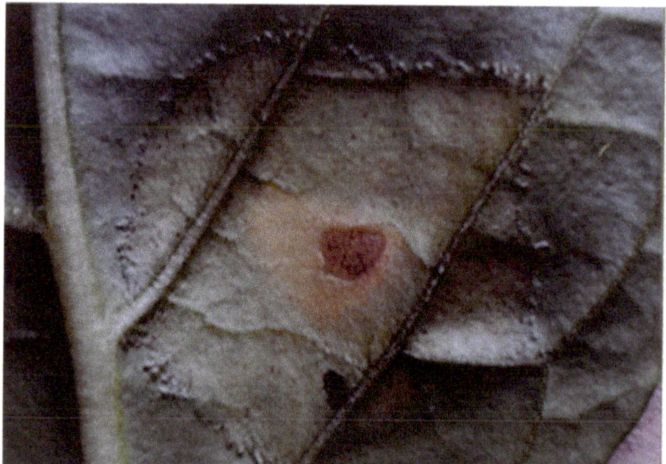

Figure 18. Leaves of susceptible peppers with incomplete GDR infiltrated with *Xanthomonas vesicatoria* (at a concentration of 10^8 cells/mL) in a spot of 5 mm, followed by super-infection with a suspension of *Pseudomonas phaseolicola* (at a concentration of 10^6 cells/mL), the border of which is marked by scarring.

The super-infection also resulted in further drift of *X. vesicatoria* bacteria in the intercellular passages, as indicated by yellowing of the area. Due to dilution and the activation of the GDR, this area of tissue has not dried out as the 5 mm spot in the centre. Scar tissue at the border of the spot infiltrated by *P. phaseolicola* indicates that the distal part of the leaf plate has been involved in the exclusion of the pathogen (Figure 19).

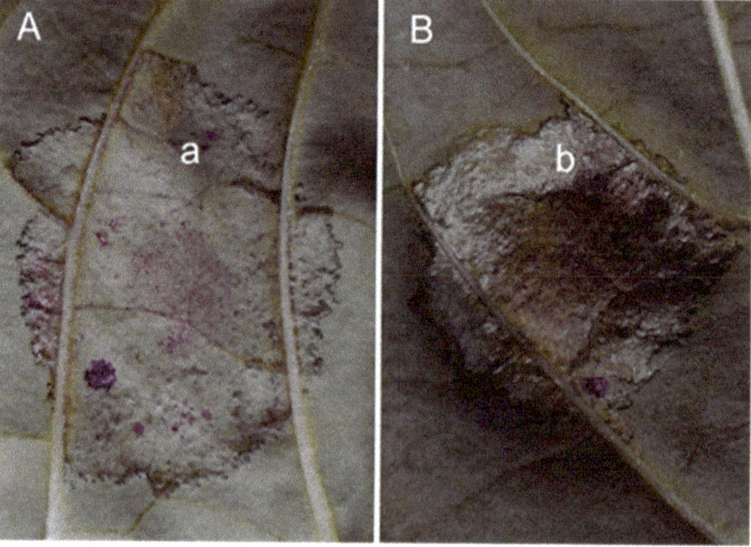

Figure 19. Tissue changes after infiltration of pepper lines having the *Bs2* gene and incomplete (**A**) and complete (**B**) GDR system (*gdr 1 + 2 genes*) with *Xanthomonas vesicatoria* bacteria (at a concentration of 10^8 cells/mL) in a spot of 5 mm and subsequent super-infection with a suspension of *Pseudomonas phaseolicola* at concentrations of 10^6 (**a**), 10^8 (**b**) cells/mL.

As described above, testing of pepper lines having incomplete (Figure 19A) and complete (Figure 19B) GDR system (*gdr 1 + 2 genes*) carrying the *Bs2* gene was also performed.

Suspension of *P. phaseolicola* at concentration of 10^6 cells/mL was used for superinfection of plants with incomplete GDR, and at concentration of 10^8 cells/mL for plants with complete GDR.

In the presence of incomplete GDR (19/A), the *Bs2* gene was activated by *X. vesicatoria*, although its role in defence is questionable. Complete GDR (19/B) formed a solid, shiny tissue spot in a non-host relationship, with no discolouration indicative of *Bs2* gene activation.

3.4.2. GDR and *gds* Gene

The *gds* and *gdr 1 + 2 genes* are indistinguishable on the basis of pathological symptoms induced by natural infection, as they induce cell enlargement and cell division that are common to plants. However, when unnatural inoculation is used, the two tissue retention mechanisms show distinctive differences. The *gds* gene induces a pronounced convexity in the leaves of young plants due to the 'accelerated' growth of infected spots (Figure 20A), whereas the *gdr 1 + 2* gene leaves the leaf plate flat upon emergence of the pathogenic symptom (Figure 20B).

Figure 20. Leaf bulging induced by the *gds* gene in young pepper leaves (**A**) and the absence of bulging with *gdr 1 + 2* genes (**B**) in response to *Xanthomonas vesicatoria* infection.

In the case of complete infiltration of the leaf plate (Figure 21), the *gds* gene develops its characteristic convexity (Figure 21A), whereas the leaf plate containing the *gdr 1 + 2* genes remains flat (Figure 21B), the latter with purple vessels indicating that the *Bs2* gene is also present in this genome. In the presence of weak GDR, the leaf containing the *Bs2* gene responded with complete destruction (Figure 21C).

Differences in the structural changes of infected tissues between *gds* and *gdr 1 + 2* genes can also be observed (Figure 22).

The *gdr 1 + 2* genes responded to stress induced by *P. Phaseolicola* (at concentration of 10^8 cells/mL) by forming a shiny compacted tissue (Figure 22A). The *gds* gene in an incomplete GDR background induces strong cell division and proliferation in response to *X. Vesicatoria* infection (at concentration of 8 cells/mL) (Figure 22B). The *gds* gene-induced cell proliferation is strongest in spongy parenchymal tissue (Figure 23).

However, such cell proliferation only occurs in plants with weak GDR. A characteristic difference between the *gds* and *gdr 1 + 2* genes for tissue retention includes the change in leaf vessels in response to infection (Figure 24).

Figure 21. Response to whole-leaf infiltration with *Xanthomonas vesicatoria* bacteria in the presence of *gds* (**A**), *gdr 1 + 2* (**B**) and *Bs2* gene in an incomplete GDR background (**C**).

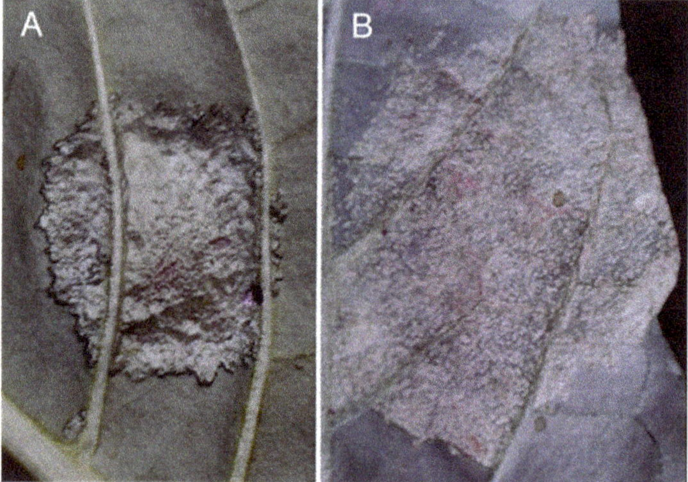

Figure 22. The *gdr 1 + 2* genes induced a shiny compacted tissue spot (**A**), and the *gds* gene resulted in uncontrolled cell division in an incomplete GDR background, upon *Xanthomonas vesicatoria*-induced stress (**B**) in pepper leaves.

Vessels passing through *X. vesicatoria*-infected spots rupture along the transport routes due to cell division induced by the *gds* gene (Figure 24A). Veins in infected spots of leaves containing *gdr 1 + 2* genes, as shown in Figure 19B, Figure 20B, and Figure 22A, show no change. If the *Bs2* gene is also present in the plant, the vessels may turn slightly purple (Figure 24B). (In susceptible plants, whitening of the vessels is the first indication of damage.)

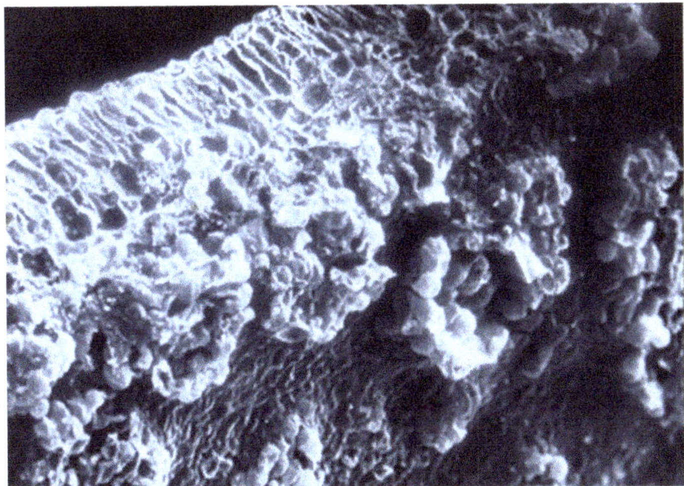

Figure 23. Cell proliferation in spongy parenchymal tissue of pepper leaves containing the *gds* gene, in response to *Xanthomonas esivcatoria* infection.

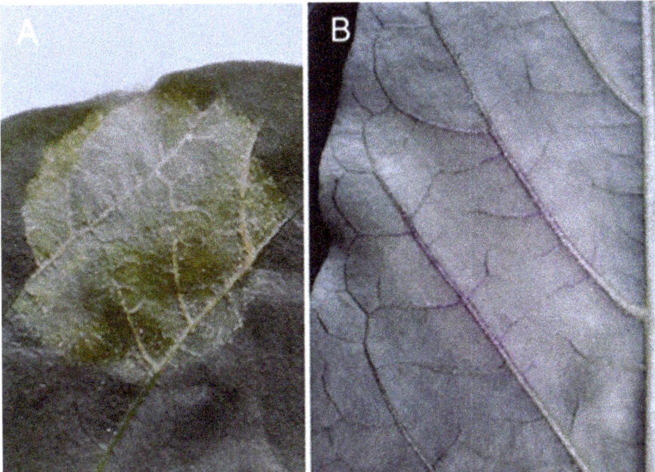

Figure 24. Vessel lesion caused by *Xanthomonas vesicatoria* on pepper vessels containing the *gds* gene (**A**) and mild purple discolouration of vessels of an infected spot also containing the *Bs2* gene, which responds in a manner characteristic of *gdr 1 + 2* genes (**B**).

The difference between the processes controlled by the *gds* and *gdr 1 + 2* genes is also reflected in the relationship to the *Bs2* gene. When a pepper line containing both the *gds* and *Bs2* genes in the homozygous state was infected with *X. vesicatoria*, a pathological symptom characteristic of the *gds* gene was observed, but the purple discolouration characteristic of the *Bs2* gene was not seen.

A pepper line homozygous for both the *gdr 1 + 2* and *Bs2* genes also shows lesions characteristic of the *gdr 1 + 2* gene, with the difference that the vessels may be slightly purple.

This may be due to, among other things, the compression on or damage of protruding vessels on the abaxial surface during infiltration.

In an incomplete GDR background, the vessels show strong purple discolouration (Figure 14C).

Mechanical damage to tissues has given rise to the idea of investigating the role of genes encoding tissue retention or tissue destruction in the pathogenesis.

In this experiment, a spot was infiltrated by applying strong mechanical pressure on the leaves of double homozygous pepper lines containing only the *gds* gene or both the *gds* and *Bs2* genes, and the affected spot and its environment were infiltrated at these points with a suspension of *vesicatoria* at concentration of 10^8 cells/mL. The inoculation procedure is shown in Figure 12B. In the case of the *gds* gene, the inoculated tissue spot responded in a manner showing susceptibility, while tissue retention characteristic of the *gds* gene was observed in the surrounding tissue (Figure 25A). The centre of the point of the double homozygous line injured by pressure dried out, but the surrounding ring was purple, indicative of the *Bs2* gene. The *Bs2* gene was, therefore, only activated in the damaged tissue. Surrounding tissues showed a tissue retention response of the *gds* gene (Figure 25B).

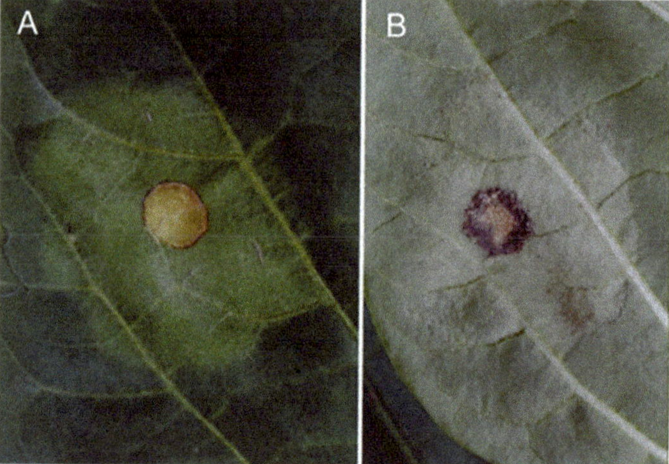

Figure 25. Healthy and pressure-stressed tissue of a pepper line containing the *gds* gene inoculated with *Xanthomonas vesicatoria* responding with tissue compaction, indicative of susceptibility (**A**) and the specific defence response of healthy and pressure-stressed tissue of a plant carrying *gds* and *Bs2* genes, respectively, indicative of the *Bs2* gene (**B**).

It is difficult to observe the phenotypic characteristics of the *gds* and *gdr 12* genes because their appearance in the same plant is uncertain due to their interactions. Importantly, in the presence of weak GDR, as with HR genes, the *gds* gene does not appropriately function.

Since all plants have some level of general defence response, function, and expression of neither the specific resistance gene (*Bs2*), nor the *gds* gene can be studied without the influence of GDR. However, the GDR can be tested on its own. This fact also points to its fundamental role in the plant's defence mechanism.

In the case of a weak GDR system, the *gds* gene does not function properly in environmental conditions unfavourable to the plant, whereas the complete GDR (*gdr 1 + 2*) gene provides sufficient protection even in this case (Figure 26).

The *gds* gene (Figure 26A) was unable to protect infected spots from the stress effect of either *X. vesicatoria* (Figure 26(A1)) or *P. phaseolicola* (Figure 26(A2)) inoculum at concentration of 10^8 cells/mL, resulting in chlorosis of the entire leaf and subsequent shedding. In the case of *gdr 1 + 2* genes (Figure 26B), neither *X. vesicatoria* (Figure 26(B1)) nor the bacterial suspension of *P. phaseolicola* (Figure 26(B2)), which induced a stronger stress due to exotoxin production, was able to cause tissue destruction and necrosis.

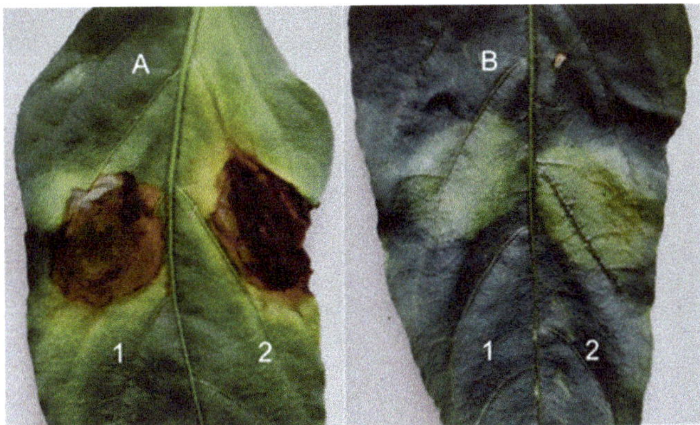

Figure 26. Protection provided by the *gds* gene (**A**) in a weak GDR structure and by the complete GDR (*gdr 1 + 2* genes) (**B**) under unfavourable environmental conditions against stress caused by *Xanthomonas vesicatoria* (**1**) and *Pseudomonas phaseolicola* (**2**) bacterial suspension at concentration of 10^8 cells/mL.

4. Discussion

4.1. Tissue Destruction and Tissue Retention in Resistance

The decades-long history of resistance breeding based on rapid tissue death has revealed several shortcomings. The findings hitherto considered as fundamental that a specific resistance gene prevents the spread of the pathogen in the plant by the development of specific HR, is only partially true.

4.1.1. Role of Specific Resistance Genes Encoding Hypersensitivity in Pathogenesis

A plant carrying a specific resistance gene suffers significant tissue loss due to the action of its tissue-destructive resistance gene when exposed to a pathogen that disrupts its life processes. This loss can be increased by breeding with a cluster of specific HR genes, which, in the event of an epidemic, may result in susceptible varieties producing more than those with more HR genes [17]. The strong dependence of HR gene function on environment also poses a high risk (Figure 13).

In our work we found that specific HR genes alone are not able to inhibit the pathogen even under optimal conditions. The general defence response (GDR), based on the tissue retention capacity of plants, plays a crucial role in inhibition. We have demonstrated that there is a strong correlation between small lesion diameter and high GDR. In general, this property determines the size of the lesions in the presence of specific HR genes, i.e., the effectiveness of the defence against the pathogen.

The practice in resistance breeding is to identify resistant plants with specific HR genes based rapid tissue death 24–48 h after inoculation. As a consequence, plants with poor GDR are the basis for resistance breeding work. Specific HR genes, on the other hand, are ineffective in weak GDR, as they attempt to prevent pathogen spread in a downstream way rather than in a preventive manner. This is a fundamental problem in their effectiveness.

4.1.2. Relationship between Susceptibility and the General Defence Reaction

The hypersensitivity of plants lacking specific resistance genes to host–pathogen and non-host–pathogen interactions was investigated in tobacco.

We determined the overall tissue resistance of the tested plants, which was characterized by stress levels expressed as inoculum concentrations. Since tissue retention was present in both host and non-host relationships, we recognized this phenomenon as the general defence response, GDR.

We further found that in the host–pathogen relationship, a state of susceptibility was formed at the boundary of the GDR and the switch to nonspecific HR (Figure 2). We concluded that tissue retention is a fundamental determinant of plant susceptibility to pathogens.

4.1.3. Interaction between the General Defense Reaction and the Specific Defense Reaction

The role of GDR, susceptibility, and HR in the pathogenesis was investigated in the case of pepper-*X. vesicatoria* relationship.

In our experiment, susceptible pepper lines containing the *Bs2* gene had similar levels of GDR. The susceptible host–pathogen relationship is characterized by the fact that the pathogen enters the plant without causing stress and only becomes detectable to the plant during its accumulation. In the experiment, this stress effect was induced by the concentration of 10^8 cells/mL and the susceptible pathogenic symptom developed. The *Bs2* gene of the pepper line was also activated at the same stress level expressed as the inoculum concentration and developed its characteristic pathology. Thus, when the GDR is exceeded, the specific resistance gene is activated only after the development of the susceptible physiological state, in a sort of downstream manner.

The level of tissue resistance, therefore, determines the initiation of the susceptible pathological process. For the plant, the pathogenesis is the result of the general tissue retention of the GDR and the tissue destruction of the specific HR. The resulting balance between the plant and the pathogen determines the size of the lesion and, in the case of infiltration, the degree of tissue retention.

Electron microscopy images of the lesions also show that, in addition to the tissue-destructive capacity of the plant to produce HR, another property, GDR, which is involved in tissue retention, also participates in the process. Starting from the point of infection, cells infected by the pathogen are destroyed by HR in a downstream manner until the GDR prevents further spread of the pathogen by forming a ring of tissue compaction (Figure 17). The diameter of the lesion, and, thus, the effectiveness of pathogen inhibition, is determined by the GDR.

Examining the role of the GDR-labelled trait in the formation of lesions, it was confirmed that GDR forced to function in a non-host relationship significantly inhibited the destruction of tissue formed in a susceptible relationship (Figure 18).

In the case of infection by infiltration, the *Bs2* gene used as an indicator clearly shows the difference in GDR levels between pepper lines (Figure 14). The pathology demonstrates that GDR plays a crucial role not only in determining susceptibility but also in the development of pathological symptoms caused by the *Bs2* gene. Incomplete GDR provides partial protection against *Bs2* gene-driven tissue destruction, whereas complete GDR (*gdr 1 + 2*) provides complete protection against *Bs2* gene-driven tissue destruction (Figure 19).

4.2. Genetic Analysis

In order to apply the general defence response in resistance breeding, we started to investigate the heritability of GDR, assuming that we are likely to be dealing with multiple gene traits. A line with a complete GDR system, which gave a tissue-retaining green spot after leaf infection with the inoculum of *X. vesicatoria* at concentration of 10^8 cells/mL and *P. phaseolicola* at a concentration of 10^8 cells/mL for both pathogens, was marked with X. This line was hybridized with a line containing the *Bs2* gene, which, on the contrary, responded to the same two pathogens with extremely rapid and vigorous tissue destruction, and is, therefore, presumed to lack any tissue retention gene. Three distinct phenotypic categories were found in the F2 generations. The empirical cleavages followed well the 12:3:1 phenotypic frequencies of the dominant epistasis, which was confirmed by the Chi^2 statistical regression test for all four hybridizations. It was found that the complete GDR system is only established when both *gdr* genes (*gdr 1 + 2*) are present in the homozygous state. Based on the observed symptoms and cleavages, it can be assumed that one of the *gdr* genes is able to form a tissue-retaining green spot in the homozygous state following infection with *P. phaseolicola* inoculum at a concentration of 10^8 cells/mL (Figure 10) and *X. vesicatoria* inoculum at a concentration of 10^8 cells/mL (Figure 10). However, both *gdr* genes must be in a homozygous state to form the tissue-retaining green spot (Figure 9). This clearly shows that the two *gdr* genes form one system in which one gene works independently and the other helps the other *gdr* gene to fight against *Xanthomonas* infection. We plan to explore these two processes in more detail in the future.

The F_1 generation of a hybridization performed by Csilléry et al., to study the genetic relationship between *Bs2* and *gds*, despite the inheritance pattern of the *Bs2* gene being found to be dominant, responded to infection with *X. vesicatoria* bacteria with green spots and purple vessels indicative of tissue retention, instead of purple tissue changes [18].

The pathological symptoms were correctly defined, but due to the incomplete knowledge of GDR at that time, the relationship between the two genes was not fully elucidated.

4.3. Phenotypical Deviation of Tissue Retaining Reactions

The known *gds* gene and the *gdr 1 + 2* genes, which also regulate tissue retention, are difficult to distinguish on the basis of their pathogenic symptoms, since both tissue retention mechanisms protect against microbial attack by inducing cell enlargement and cell division, which are fundamental to the plant. However, if excessive inoculation methods are used, the characteristics of the two traits can be distinguished (Figures 20–24).

4.3.1. Comparison of Pathogenic Symptoms Induced by *gdr 1 + 2* Genes and *gds* Gene on Pepper Leaves

In young leaves of plants containing the *gds* gene, the infected spot may bulge due to enlargement of columnar parenchyma cells (Figure 20A), whereas in more mature leaves, spongy parenchyma cells may respond to infection by meristematic cell proliferation (Figures 22B and 23). In *gdr 1 + 2* gene-directed defence, the leaf plate remains flat (Figures 20B and 21B), tissues may become shiny, and the leaf plate may thicken (Figure 22A). In both cases, infected spots on the leaf plate may be slightly chlorotic.

When the whole leaf is infiltrated (Figure 21), the *gds* gene is characterised by bulging (Figure 22A) and the *gdr 1 + 2* genes by flattening (Figure 21B), indicating a differential change in leaf tissue.

The vascular response induced for the *gds* gene is the rupture of vessels as a consequence of cell division. For the *gdr 1 + 2* genes, no visible change occurs in the vessels (Figure 20B), but if the *Bs2* gene is present in the plant, the vessels may turn slightly purple (Figure 21B).

Plants are protected against any microbial attack by the general defence response, as a primary defence front, by strengthening the cellular wall. Reactive oxygen species (ROS) play an important role in the immediate strengthening of the cell. Basic plant life processes, such as lignin synthesis for cellular wall formation and degradation processes, are also accompanied by specific amounts of H_2O_2 [14]. Since H_2O_2 is also released during the initiation of pathogen-induced tissue destruction [19–21]), it confuses the *Bs2* gene, which detects H_2O_2 as a signalling molecule and is triggered by it. The purple discolouration of vessels in the presence of the *gdr 1 + 2 gene* indicates excess *Bs2* gene activation.

4.3.2. Difference between the Processes Controlled by *gds* and *gdr 1 + 2* Genes in Relation to the *Bs2* Gene

If a pepper line homozygous for both the *gds* and *Bs2* genes is infected with *X. vesicatoria*, pathological symptoms specific for the *gds* gene are obtained. We have never observed purple discolouration characteristic of the *Bs2* gene. This is confirmed by testing the ECW 1–6 pepper lines containing the *Bs1* (sHR), *Bs2* (sHR), *Bs3* (sHR), *Bs4* (sHR), *bs5* (*gds*), and *bs6* (S) genes.

If, in addition to the *gdr 1 + 2* genes in the homozygous state, the *Bs2* gene is also part of the plant genome, then purple vessels may occur, but this is not inherent to the disease process, but is a mis-activation of the *Bs2* gene caused by excess H_2O_2 accompanying lignin synthesis in the transportation pathways. This was never observed for the *gds* gene.

When mechanical pressure was applied to a point on a leaf of a pepper line containing only the *gds* gene or double homozygous for the *gds* and *Bs2* genes, inoculated with the bacterium *X. vesicatoria*, and a disruption of life processes was induced, the first case showed susceptible disease, the second case showed purple lesions indicating the destruction by the *Bs2* gene. The *Bs2* gene only acted in a downstream manner upon detecting a disturbance in cell function (Figure 25). The unaffected tissues around the pressure points responded with a green tissue spot induced by *X. vesicatoria* bacterium, characteristic of the *gds* gene.

It applies for both *gds* and *gdr 1 + 2* genes that they only function in cells with complete integrity and that only intact tissues are capable of tissue retention responses, which are characterised by a low stimulus threshold and high reaction rate.

Our experiments demonstrate that specific resistance genes work most effectively in a background of strong general defence responses. So much so that in the case of a complete GDR (*gdr 1 + 2* genes), they remain in a latent, repressed state and, therefore, become redundant and represent only a genetic burden to the plant (Figure 19).

4.3.3. Application of the Tissue Retention *gds* and *gdr* 1 + 2 Genes in Resistance Breeding

The importance of specific HR genes in resistance breeding has been overestimated, as plant-microbe interactions are numerous, host–pathogen interactions are numerous, and host–pathogen interactions carrying specific resistance genes are few in number compared to these. The latter are not sufficient to protect different plant species against microorganisms.

The search for specific HR in resistance breeding goes back hundreds of years. Their widespread use has made resistance breeding a one-plane process, excluding the consideration of any other plant defence system in the selection work. The use of the known dominant resistance genes (*Bs1, Bs2, Bs3*) does not result in durable resistance because their use in field monocultures leads to the emergence of new pathogenic variants [22]. In contrast to specific HR responses involving different tissue destruction, for almost thirty years we have been investigating types of defence systems involving tissue retention, which are a different way for resistance breeding. We expressed this by not putting the tissue retention *gds* gene described in 1995 in the then next fourth position of the *Bacterial spot = Bs* gene sequence, because it would have narrowed the spectrum of action of the gene found. In fact, *gds* is not only a resistance gene against Xanthomonas, but also a non-pathogen specific gene with a broad spectrum of action (less sensitive to temperature). It is the first known resistance gene that protects pepper tissues and does not destroy them. Resistance defined by the *gds* gene is not associated with a hypersensitive response and no special effector or avirulence factor appears to be involved in the interaction of the host plant with the gene products in the development of resistance symptoms [23]. When the *gds* gene is used, the assimilation surface of the plant is not reduced during pathogen control.

In order to naturally incorporate the beneficial properties of genes conferring tissue retention into the breeding program, we needed to understand their physiological roles and interactions. In doing so, we looked at the role of tissue retention in manifestation of pathogenic symptoms in other plant species (Figure 7). Until then, the phenomenon that tissues of organs, including leaves, that have specialised from a meristematic state can regain their ability to divide in response to biotic stress, was unknown in plant pathology. This property has led to a significant increase in the efficiency of producing dihaploid plants from plants containing the *gds* gene and to initiatives to incorporate it into other plant species.

The discovery and practical application of the recessive resistance gene in pepper is undoubtedly a paradigm shift in resistance breeding. Following the description of the novel *gds* gene [9], only the *bs5* gene was reported [11], followed by the isolation of the *xcv-1* gene [24]. The equivalence of *gds = bs5 = xcv-1* has been demonstrated based on literature and test hybridization [12,25].

The last decade has witnessed a spread of practical applications of the *gds* gene. A clear sign of this spread is that since 2014, major breeding companies around the world have been acquiring the know-how to use the *bs5* gene from the 2Blades Foundation (Evanston, IL, USA) for their pepper breeding programmes (Figure 27) [26].

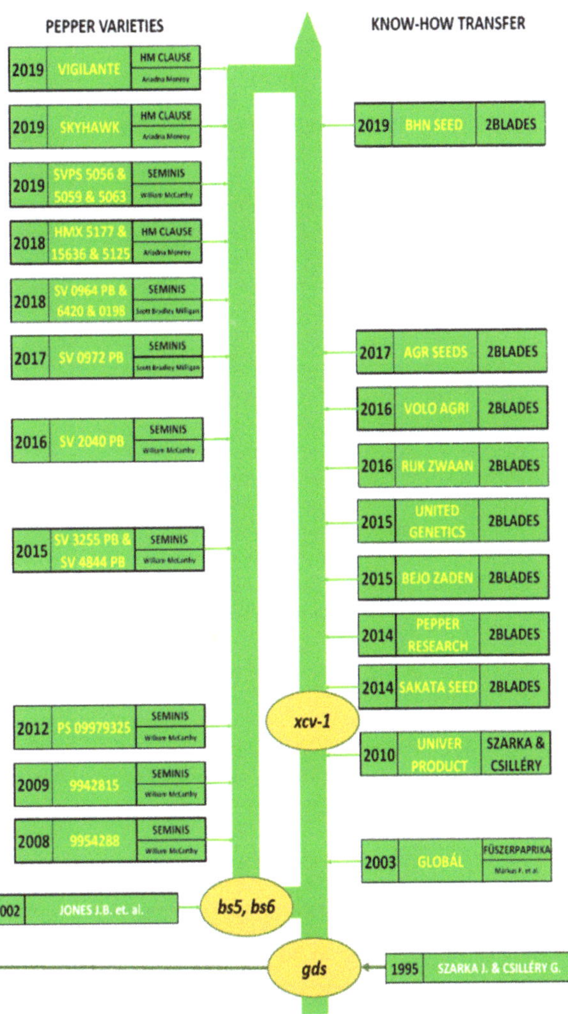

Figure 27. Twenty-five years of non-hypersensitive, non-specific recessive resistance in pepper [26].

In addition to pepper, the applicability of the *gds* (*xcv-1*) gene in other plant species has also been discussed. A patent application has been filed for the identification of the *xcv-1* gene by cloning based on gene mapping and for procedures to develop plants with resistance. The name *xcv-1* was derived from the renaming of *gds*. During the process, this right was also transferred from the Agricultural Biotechnology Research Centre (Hungary) to the 2Blades Foundation (USA) [24].

The complex trait system of tissue retention has been under continuous investigation since the discovery of *gds*. The result of this work includes the *gdr 1 + 2* gene and pathological experience with GDR function. The use of the *gds* gene in pepper breeding or its incorporation into other plant species carries the risk that the tissue retention gene may not function properly in the presence of incomplete GDR (Figure 26).

The *gds* is a genetic code that is activated by various stresses and has so far only been found in one pepper line. Its efficient functioning and success in breeding is attributable to the GDR background, as is the case for the HR gene.

As for the relationship between the genes encoding tissue retention, the *gds* gene is dependent on GDR for its function, while the *gdr 1 + 2* genes can function completely independently of *gds*.

GDR is the basis of plant defence. It has been experimentally demonstrated to reduce susceptibility-induced mortality, to protect against self-destruction encoded by specific HR genes, to promote *gds* gene function, to aid adaptation to abiotic stresses and resistance to biotic stresses, to have a low stimulus level, and a high and rapid responsiveness. All these factors make it suitable for the role of the plant immune system.

The classification of the multitude of plant–microbe interactions into host–pathogen and non-host–pathogen or compatible and incompatible opposition pairs makes it difficult to understand the natural processes that we want to express.

On pepper, *X. campestris pv. vesicatoria* (compatible host–pathogen relationship) develops a susceptibility pathogenesis, but in the presence of *gds*, *gdr 1 + 2* genes, pepper defends against infection by *X. campestris pv. vesicatoria* with the same tissue retention response as against infection by the bean pathogen *P. phaseolicola* (incompatible non-host–pathogen relationship with pepper).

Consequently, the genes encoding tissue retention have transformed a host–pathogen relationship into a non-host–pathogen relationship. This provides us with a form of defence that has evolved over the evolutionary history of plants with exceptional efficiency.

The role of the newly identified *gdr 1 + 2* genes in disease processes downstream of the *gds* gene may point to a new direction in resistance breeding.

Half a century of experience in studying hundreds of thousands of pathogens provoked and studied in various plant–microbe relationships has guided us from specificity to generalised protection.

We have found confirmation of this in the observations and ideas of the founder of stress theory, János Selye, in his book entitled In vivo. Such observations include the "vigorous mitotic cell proliferation" observed in animal experiments in response to biological stressors and the summarizing formulation that, "All observable biological effects produced by various stimuli are the sum of two components, the specific effect and the nonspecific response. The latter may even mask the specific effect!" [27].

5. Conclusions

The inadequacies of the history of resistance breeding and our present work demonstrate that specific HR genes alone cannot safely protect the plant because they only act in a downstream manner in cells attacked by the pathogen, inducing cell death.

A decisive role in the defence of plants containing HR genes is played by the plant's tissue retention capacity, which is expressed in the general defence response (GDR). GDR pre-emptively excludes the pathogen without tissue loss.

Specific HR genes do not protect but destroy the cells affected by pathogens. GDR genes protect the plant by strengthening the cells. The difference in the stimulus thresholds of the two responses also determines the order and effectiveness of the response. The non-pathogen specific GDR plays the role of the plant immune system due to its low stimulus threshold and high reaction rate.

Tissue retention capacity can be increased by breeding to a level where the GDR alone can provide adequate protection for the plant. In this case, the presence of specific HR genes is unnecessary; sometimes it is even a burden to the plant.

In our work on tissue retention in pepper, we found that the GDR system we studied is regulated by two recessive genes (*gdr 1 + 2*) that are completely independent of the HR-conferring *Bs2* gene and the previously identified tissue retention *gds* gene.

Studies on pepper demonstrate that plant resistance can be made safer without the incorporation of specific HR genes, relying solely on GDR.

An important point is that while the function of specific HR genes is dependent on the environment, GDR is as independent as possible from environmental factors. It functions under the extreme conditions that the plant can tolerate, acting as a plant immune system.

Host plant–pathogen contacts carrying specific HR genes are very rare in nature, while there are a lot of host–pathogen contacts and plant–microbe contacts are countless. This is possible because all plants have GDRs. Consequently, if we want to integrate resistance breeding into the order of nature, our work must be based on the tissue retention capacity of plants.

Author Contributions: Conceptualization, J.S. and B.P.; methodology, J.S., B.P. and Z.T.; validation, J.S. and B.P.; formal analysis, B.P.; investigation, J.S., Z.T. and R.H.; resources, J.S. and G.P.; data curation, B.P. and J.S.; writing—original draft preparation, J.S. and B.P.; writing—review and editing, Z.T. and G.P.; visualization, J.S.; supervision, Z.T. and G.P.; project administration, J.S. and R.H.; funding acquisition, G.P. and J.S. All authors have read and agreed to the published version of the manuscript.

Funding: This research received no external funding.

Institutional Review Board Statement: Not applicable.

Informed Consent Statement: Not applicable.

Data Availability Statement: Not applicable.

Conflicts of Interest: The authors declare no conflict of interest.

References

1. Ward, H.M. On the relations between Host and Parasite in the Bromes and their Brown Rust, *Puccinia dispersa* (Erikss.). *Ann. Bot.* **1902**, *16*, 233–315. [CrossRef]
2. Stakman, E.C. Relation between *Puccinia Graminis* and Plants Highly Resistant to Its Attack. *J. Agric. Res.* **1915**, *IV*, 193–200.
3. Holmes, F.O. Local Lesions in Tobacco Mosaic. *Bot. Gaz.* **1929**, *87*, 39–55. [CrossRef]
4. Holmes, F.O. Inheritance of resistance to tobacco mosaic disease in tobacco. *Phytopathology* **1938**, *28*, 553–561.
5. Cook, A.A.; Stall, R.E. Inheritance of resistance in pepper to bacterial spot. *Phytopathology* **1963**, *53*, 1060–1062.
6. Cook, A.A.; Guevara, Y.G. Hypersensitivity in *Capsicum chacoense* to race 1 of the bacterial spot pathogen of pepper. *Plant Dis.* **1984**, *68*, 329–330. [CrossRef]
7. Klement, Z.; Farkas, G.; Lovrekovich, L. Hypersensitive reaction induced by phytopathogenic bacteria in tobacco leaf. *Phytopathology* **1964**, *54*, 474–477.
8. Szarka, J.; Toldi, O.; Szarka, E.; Remenyik, J.; Csilléry, G. General defense reaction in the plant kingdom. *Acta Agron. Hung.* **2006**, *54*, 221–232. [CrossRef]
9. Szarka, J.; Csilléry, G. Defence systems against *Xanthomonas campestris pv. vesicatoria* in pepper. In Proceedings of the IXth Meeting on Genetics and Breeding on Capsicum and Eggplant, Budapest, Hungary, 21–25 August 1995.
10. Szarka, E. A növények általános védekezési rendszerének biokémiai és genetikai vizsgálata. Ph.D. Thesis, Corvinus University, Budapest, Hungary, 2008.
11. Jones, J.B.; Minsavage, G.V.; Roberts, P.D.; Johnson, R.R.; Kousik, C.S.; Subramanian, S.; Stall, R.E. A Non-Hypersensitive Resistance in Pepper to the Bacterial Spot Pathogen Is Associated with Two Recessive Genes. *Phytopathology* **2002**, *92*, 273–277. [CrossRef] [PubMed]
12. Timár, Z.; Palotás, G.; Csilléry, G.; Szarka, J. Study of recessive bacterial leaf spot resistance genes in *Capsicum annuum* L. In Proceedings of the 17th EUCARPIA Meeting on Genetics and Breeding of Capsicum and Eggplant, Avignon, France, 11–13 September 2019.
13. Szarka, J.; Csilléry, G. General defense system in the plant kingdom II. *Int. J. Hortic. Sci.* **2001**, *7*, 69–71. [CrossRef]
14. Brisson, L.F.; Tenhaken, R.; Lamb, C.J. Function of Oxidative Cross-linking of Cell Wall Structural Proteins in Plant Disease Resistance. *Plant Cell* **1994**, *6*, 1703–1712. [CrossRef] [PubMed]
15. Templeton, M.D.; Lamb, C.J. Elicitors and defence gene activation. *Plant Cell Environ.* **1988**, *11*, 395–401. [CrossRef]
16. Heath, M.C. Nonhost resistance and nonspecific plant defences. *Curr. Opin. Plant Biol.* **2000**, *3*, 315–319. [CrossRef]
17. Barabás, Z.; Matuz, J. A levélrozsda- és a lisztharmat-epidémia, illetve különféle rezisztenciatípusok befolyása őszibúza-genotípusok termésére. *Növénytermelés* **1983**, *32*, 193–198.
18. Csilléry, G.; Mitykó, J.; Szarka, E.; Szarka, J. Genetics of interactions between general and specific plant defense reactions. In Proceedings of the XIIIth Eucarpia Meeting on Genetics and Breeding of Capsicum and Eggplant, Warsaw, Poland, 5–7 September 2007.

19. Király, Z.; Barna, B.; Érsek, T. Hypersensitivity as a Consequence, Not the Cause, of Plant Resistance to Infection. *Nature* **1972**, *239*, 456–458. [CrossRef]
20. Goulden, M.G.; Baulcombe, D.C. Functionally Homologous Host Components Recognize Potato Virus X in *Gomphrena globosa* and Potato. *Plant Cell* **1993**, *5*, 921–930. [CrossRef] [PubMed]
21. Bendahmane, A.; Kanyuka, K.; Baulcombe, D.C. The Rx Gene from Potato Controls Separate Virus Resistance and Cell Death Responses. *Plant Cell* **1999**, *11*, 781–791. [CrossRef] [PubMed]
22. Dangl, J.L.; Horvath, D.M.; Staskawicz, B.J. Pivoting the Plant Immune System from Dissection to Deployment. *Science* **2013**, *341*, 746–751. [CrossRef] [PubMed]
23. Vallejos, C.E.; Jones, V.; Stall, R.E.; Jones, J.B.; Minsavage, G.V.; Schultz, D.C.; Rodrigues, R.; Olsen, L.E.; Mazourek, M. Characterization of two recessive genes controlling resistance to all races of bacterial spot in peppers. *Theor. Appl. Genet.* **2010**, *121*, 37–46. [CrossRef] [PubMed]
24. Kiss, G.B.; Szabó, Z.; Iliescu, C.E.; Balogh, M. Identification of a *Xanthomonas Euvesicatoria* Resistance Gene from Pepper (*Capsicum annuum*) and Method for Generating Plants with Resistance. World Intellectual Property Organization Patentscope Database. Available online: https://www.wipo.int/patentscope/en/ (accessed on 19 May 2016).
25. Palotás, G. 20 years of non-hypersensitive, non-specific, recessive resistance in pepper—Review. In Proceedings of the XVIth Capsicum and Eggplant Working Group Meeting, Kecskemét, Hungary, 12–14 September 2016.
26. Palotás, G.; Timár, Z. A paprika nem hiperszenzitív, nemspecifikus, recesszív rezisztenciájának 25 éve. In Proceedings of the XXVI Növénynemesítési Tudományos Napok, Szeged, Hungary, 4–5 March 2020.
27. Selye, J. *In Vivo—The Case for Supramolecular Biology*, 1st ed.; Liveright Publishing Corporation: New York, NY, USA, 1967.

sustainability

Article

Phenotypic and Molecular Characterization of Rice Genotypes' Tolerance to Cold Stress at the Seedling Stage

Nasira Akter [1], Partha Sarathi Biswas [2], Md. Abu Syed [2,*], Nasrin Akter Ivy [3], Amnah Mohammed Alsuhaibani [4], Ahmed Gaber [5] and Akbar Hossain [6,*]

[1] Plant Breeding Division, Bangladesh Agricultural Research Institute (BARI), Gazipur 1701, Bangladesh; nasirabsmrau@gmail.com

[2] Plant Breeding Division, Bangladesh Rice Research Institute (BRRI), Gazipur 1701, Bangladesh; psbiswasbrri@gmail.com

[3] Department of Genetics and Plant Breeding, Bangabandhu Sheikh Mujibur Rahman Agricultural University (BSMRAU), Gazipur 1701, Bangladesh; ivy.bsmrau@yahoo.com

[4] Department of Physical Sport Science, College of Education, Princess Nourah bint Abdulrahman University, P.O. Box 84428, Riyadh 11671, Saudi Arabia; amalsuhaibani@pnu.edu.sa

[5] Department of Biology, College of Science, Taif University, P.O. Box 11099, Taif 21944, Saudi Arabia; a.gaber@tu.edu.sa

[6] Department of Agronomy, Bangladesh Wheat and Maize Research Institute, Dinajpur 5200, Bangladesh

*   Correspondence: asyed.breeding@brri.gov.bd (M.A.S.); akbar.hossain@bwmri.gov.bd (A.H.)

Citation: Akter, N.; Biswas, P.S.; Syed, M.A.; Ivy, N.A.; Alsuhaibani, A.M.; Gaber, A.; Hossain, A. Phenotypic and Molecular Characterization of Rice Genotypes' Tolerance to Cold Stress at the Seedling Stage. *Sustainability* **2022**, *14*, 4871. https://doi.org/10.3390/su14094871

Academic Editor: Balázs Varga

Received: 20 March 2022
Accepted: 17 April 2022
Published: 19 April 2022

Publisher's Note: MDPI stays neutral with regard to jurisdictional claims in published maps and institutional affiliations.

Abstract: Rice plants are affected by low-temperature stress during germination, vegetative growth, and reproductive stages. Thirty-nine rice genotypes including 36 near-isogenic lines (NILs) of BRRI dhan29 were evaluated to investigate the level of cold tolerance under artificially induced low temperature at the seedling stage. Three cold-related traits, leaf discolouration (LD), survivability, and recovery rate, were measured to determine the level of cold tolerance. Highly significant variation among the genotypes was observed for LD, survivability, and recovery rate. Three NILs, IR90688-74-1-1-1-1-1, IR90688-81-1-1-1-1-1, and IR90688-103-1-1-1-1-1, showed tolerance in all three traits, while IR90688-118-1-1-1-1-1 showed cold tolerance with LD and recovery rate. IR90688-92-1-1-1-1-1, IR90688-125-1-1-1-1-1, IR90688-104-1-1-1-1-1, IR90688-124-1-1-1-1-P2, IR90688-15-1-1-1-1-1, and IR90688-27-1-1-1-1-1 showed significantly higher yield coupled with short growth duration and good grain quality. Genetic analysis with SSRs markers revealed that the high-yielding NILs were genetically 67% similar to BRRI dhan28 and possessed cold tolerance at the seedling stage. These cold-tolerant NILs could be used as potential resources to broaden the genetic base of the breeding germplasm to develop high-yielding cold-tolerant rice varieties.

Keywords: low-temperature stress; leaf discolouration; seedling stage; SSR markers; near-isogenic line; rice (*Oryza sativa* L.)

1. Introduction

Compared to other cereal crops, rice is more sensitive to low-temperature stress (LTS) as it has originated from tropical regions [1–4]. Cold stress negatively affects rice plants throughout various growth stages, from germination to maturity, and causes significant yield losses because of poor germination and seedling establishment, stunted growth pattern, non-vigorous plants, vast spikelet sterility, delay in flowering, and lower grain filling in the temperate, subtropical, and even in the tropical rice-growing regions [5–10]. Boro rice, which is grown from November to May in Bangladesh and some parts of Eastern India, is affected by LTS at both the seedling and reproductive stage. In the northern and north-eastern districts of Bangladesh, LTS at the seedling stage is a major problem for rice cultivation that seriously affects crop establishment. Seedling mortality occurs 10–90% of the time due to severe cold injury during transplanting in late December to early January [11]. Farmers are bound to replant the seedlings and, as a result, the cost of cultivation is increased. Sometimes, farmers delay the planting of the boro rice crops until the ambient temperature increases. Thus, they delay the next crop, which in turn results in the low total productivity of the farm. Moreover, it increases the risk of crop loss due to flash floods during the ripening stage in some low-lying areas of the north-eastern haors [4]. On the other hand, 100% spikelet sterility is observed in the short-duration rice varieties due to LTS at the booting stage of the rice crop during early to mid-February [12]. This problem could be resolved easily by developing cold-tolerant rice varieties. The success of a rice-breeding program depends on the presence of genetic variability among the parental lines used in it.

Genetic similarity/diversity can be assessed by using plant morphology, physiology, isozymes, storage protein profiles, DNA markers, etc. [13]. Among the different DNA markers, SSR markers have been extensively used as a powerful tool in variety protection [14], molecular diversity studies [15], quantitative trait loci (QTL) analysis, pedigree analysis, and marker-assisted breeding [16]. Morphological differences among near-isogenic lines (NILs) are very narrow [17]. The characterization of such a population is unrealistic if only based on morphological traits. Genetic analysis using molecular markers could reveal the difference between them at the DNA level. The molecular analysis of genes or QTLs underlying cold tolerance is the best approach, along with phenotypic cold tolerance screening. There are many reports available on QTL's underlying cold tolerance in the literature [18–26]; however, most of these QTL were detected from *japonica* cultivars.

A wide range of variations in cold tolerance exists among rice germplasms [19,27] and, overall, *Japonica* rice germplasms are more tolerant to cold stress than the *indica* rice germplasm [28–30]. To enhance cold tolerance in the *indica* rice, the commonly grown rice varieties in south and southeast Asian countries, it is imperative to use germplasms from temperate rice-growing countries as donor parents in the breeding program [18]. Crosses between *indica* and *japonica* rice usually show spikelet sterility due to cross incompatibility. However, marker-assisted backcrossing can recover recurrent parental genomes successfully. NILs, thus developed, could be used directly as a cold-tolerant variety after necessary evaluation or as the cold-tolerant parent as a donor to introgress into an elite *indica* background. In this study, we evaluated a set of NILs derived from a cross between a cold-tolerant *japonica* rice variety, Jinbubyeo and BRRI dhan29, a mega rice variety [31] of the boro ecosystem, to select potential lines for direct varietal release or use in the breeding program as the donor parents for cold tolerance.

2. Materials and Methods

2.1. Sources of Materials Used

Thirty-six BC$_3$F$_5$ near-isogenic lines (NILs) derived from a cross between a Korean *japonica* variety (Jinbubyeo) and a Bangladeshi *indica*-type high-yielding rice variety (BRRI dhan29), along with four check varieties (BR1, BR18, BRRI dhan28, and BRRI dhan29), were evaluated for seedling stage cold tolerance and agronomic performance. The same set of NILs, along with their parents and cold-tolerant and susceptible check varieties, were used in genotyping to explore the genetic variability/similarity that exists among them. The seeds of the NILs and the check varieties (BR1, BR18, BRRI dhan28, and BRRI dhan29) were collected from the International Rice Research Institute (IRRI) and the Bangladesh Rice Research Institute (BRRI), respectively.

2.2. Evaluation of Near Isogenic Lines (NILs) for Cold Tolerance at Seedling Stage

The cold screening was performed using a cold water tank in the cold screening laboratory of the Bangladesh Rice Research Institute (BRRI) following the protocol described in Khatun et al. [32] (Figure 1). Briefly, 10 pre-germinated seeds of each genotype were sown in one-row plots spaced at 3.0 cm in plastic trays of 60 cm × 30 cm × 2.5 cm in size. The trays were filled with gravel and crop residue-free granular and fertilized soil.

BR18 and BR1 were used as tolerant and susceptible check varieties, respectively, in each tray (Figure 1).

Figure 1. A pictorial view of the cold-screening nursery by using a cold water tank in the cold-screening laboratory of the Bangladesh Rice Research Institute (BRRI).

After seeding, a thin layer of fine granular soil was used to cover the germinating seeds. The seedlings were allowed to grow at ambient temperature. At the three-leaf stage (approximately 12–15 days after seeding depending on the ambient temperature), the plastic trays were placed in a cold water tank pre-set at 13 °C. Sensitivity to cold stress was recorded using arbitrary leaf discolouration (LD) scores (1 to 9), survivability, and recovery rate as described in Biswas et al. [4]. The LD score and survivability were recorded at seven days of cold treatment or when the susceptible check variety (BR1) died. After LD and survivability recording, the trays were placed at ambient temperature in a sunny place. The recovery rate was recorded 7 days after the removal of cold stress. The survivability and recovery rate were calculated as the percentage of the green plants to the total number of plants used in the screening activity following the formula given below:

$$\text{Survivability rate } (\%) = \frac{\text{No. of Green plants} \times 100}{\text{No. of plants treated}} \tag{1}$$

$$\text{Recovery rate } (\%) = \frac{\text{No. of recovered plants} \times 100}{\text{No. of plants treated}} \tag{2}$$

The analysis of variance (ANOVA) and mean comparison of the LD scores, survivability, and recovery rate were performed using the analytical software Statistix version 10. Spearman correlation coefficients between the traits were calculated using the Corplot 0.92 package in R 3.6.2.

2.3. Evaluation of Near Isogenic Lines (NILs) for Agronomic Performance

All 36 NILs along with four check varieties were grown in the field following augmented RCB design [33]. At 35 days old, seedlings were transplanted with a single seedling per hill at a spacing of 20 cm × 20 cm in 5.4 m^2 plots. The standard crop management protocol described in BRRI [34] was followed throughout the crop growth period. Observations on different morphological characters viz. plant height (PH), number of effective tillers per hill (ETPH), panicle length, number of fertile grains per panicle (FGPP), spikelet fertility (%), thousand-grain weight (g), growth duration (days), and grain yield per hill (g) were recorded at the appropriate growth stage of the rice plant following the descriptor for rice [35]. The data collected for all eight quantitative characteristics were subjected to analysis of variance for augmented block design according to the method suggested by Federer and Raghavarao [36]. The major descriptive statistics such as mean, range, and coefficient of variation were estimated following Panse and Sukhatme [37]. The plot means were analysed using standard statistical analysis suggested by Federer [33] and elaborated by Sharma [38]. The means of the genotypes were adjusted with the block effects, which were measured from the replicated check plots following the formula:

$$Rj = Bj - M \qquad (3)$$

where, Rj = effect of jth block, Bj = mean of all check-in the *j*th block, and M = grand mean of the check.

Principal component analysis (PCA) was performed using the R package FactoMineR and visualized in biplot using the ggplot2 package.

2.4. Genetic Analysis of Near-Isogenic Lines (NILs) of BRRI dhan29 Using SSR Markers

A total of 58 simple sequence repeat (SSR) markers (Table S1) randomly distributed over 12 chromosomes of rice were used in the genotyping of 42 genotypes, including 36 NILs and their parents (BRRI dhan29 and Jinbubyeo), one susceptible check variety (BR1), and two tolerant check varieties (BR18 and Hbj. B. VI). Genomic DNA was isolated from young leaf tissues following the modified mini-scale method as described by Syed et al. [39]. The polymerase chain reaction (PCR) and gel electrophoresis were performed as described by Syed et al. [40]. Briefly, PCR was performed in 10 µL reactions containing 2 µL DNA template (around 50 ng), 1 µL of 10X TB buffer (containing 0.2 µL dd water; 0.5 µL of 1 M Tris-HCl, pH 8.4; 0.2 µL of 5 M KCl; 0.1 µL of 15 mM MgCl$_2$; 0.00000001 g gelatin), 0.5 µL of 1 mMdNTPs, 0.5 µL each of 10 µM forward and reverse primers, 5.3 µL nano pure ddH$_2$O, and 0.2 µLTaq DNA polymerase (5 U/µL) using a thermal cycler. After initial denaturation for 2 min at 94 °C, each cycle comprised 30 s of denaturation at 94 °C, 30 s of annealing at 55 °C, and 30 s of extension at 72 °C with a final extension for 5 min at 72 °C at the end of 32 cycles. After completion of the reaction, the PCR product was preserved at a temperature of 4 °C. The amplified PCR products were separated in 6% polyacrylamide gel at 100 V for 1.5 to 2.5 h against 1Kb DNA marker. DNA bands were visualized with the exposure to ultraviolet light in a gel documentation system and saved in jpeg format. Allele scoring was performed considering the relative position of the bands in the gel images compared to the position of parental bands. Summary statistics, including the number of alleles per locus, major allele frequency, and polymorphic information content (PIC) values, were determined using Power Marker version 3.25 [41] following the formula proposed by Anderson et al. [42].

$$\text{PICi} = \sum_{j=1}^{n} \left(Pij \right)^2 \qquad (4)$$

where n is the number of marker alleles for marker i and Pij is the frequency of the *j*th allele for marker i.

Spearman's correlation coefficient was analyzed using the cor.test() function in R studio. The formula for calculating the Spearman correlation is as follows:

$$r_s = 1 - \frac{6 \sum di2}{n(n2 - 1)} \qquad (5)$$

where r_s = Spearman correlation coefficient; di = the difference in the ranks given to the two variables values for each item of the data; n = total number of observations.

The genetic distance was calculated using Nei distance [43]. The similarity matrix was calculated with the Simqual sub-program using the Shannon coefficient [44] and subjected to cluster analysis by the unweighted pair group method for the arithmetic mean (UPGMA), and a dendrogram was generated using the program NTSYS-pc [45].

3. Results

3.1. Evaluation of NILs for Cold Tolerance at Seedling Stage

The analysis of variance for LD score, survivability, and recovery rate showed highly significant variation ($p \leq 0.01$) among the genotypes (Table 1).

Table 1. Variance and mean performance of 39 genotypes in LD score, % survivability, and % recovery under artificial cold stress.

SL	Designation	LD Score			% Survivability			% Recovery		
		Mean	Diff. from BR1	Diff. from BR18	Mean	Diff. from BR1	Diff. from BR18	Mean	Diff. from BR1	Diff. from BR18
1	IR90688-5-1-1-1-1-1	2.3	−5.9 *	−2.7 *	67.60	28.9	−9.3	61.10	29.6	5.9
2	IR90688-13-1-1-1-1-p1	3.2	−5 *	−1.8 *	70.00	31.3	−6.9	57.40	25.9	2.2
3	IR90688-15-1-1-1-1-1	3.1	−5.1 *	−1.9 *	71.90	33.2	−5	69.20	37.7 *	14
4	IR90688-19-1-1-1-1-1	3.1	−5.1 *	−1.9 *	79.20	40.5 *	2.3	75.40	43.9 *	20.2
5	IR90688-20-1-1-1-1-1	2.8	−5.4 *	−2.2 *	86.30	47.6 *	9.4	84.40	52.9 *	29.2
6	IR90688-27-1-1-1-1-1	3.8	−4.4 *	−1.2	62.20	23.5	−14.7	60.00	28.5	4.8
7	IR90688-30-1-1-1-1-1	3.4	−4.8 *	−1.6 *	73.30	34.6	−3.6	63.30	31.8	8.1
8	IR90688-42-1-1-1-1-p1	3.3	−4.9 *	−1.7 *	78.30	39.6 *	1.4	69.40	37.9 *	14.2
9	IR90688-42-1-1-1-1-p2	4.1	−4.1 *	−0.9	68.10	29.4	−8.8	61.50	30	6.3
10	IR90688-43-1-1-1-1-P1	3.5	−4.7 *	−1.5 *	74.30	35.6	−2.6	66.70	35.2 *	11.5
11	IR90688-52-1-1-1-1-1	2.9	−5.3 *	−2.1 *	82.40	43.7 *	5.5	85.40	53.9 *	30.2
12	IR90688-54-1-1-1-1-1	3.9	−4.3 *	−1.1	76.70	38	−0.2	59.30	27.8	4.1
13	IR90688-56-1-1-1-1-1	3.8	−4.4 *	−1.2	63.10	24.4	−13.8	60.40	28.9	5.2
14	IR90688-62-1-1-1-1-1	4.2	−4 *	−0.8	52.70	14	−24.2	53.60	22.1	-1.6
15	IR90688-64-1-1-1-1-1	3.3	−4.9 *	−1.7 *	71.50	32.8	−5.4	74.30	42.8 *	19.1
16	IR90688-73-1-1-1-1-P1	3.0	−5.2 *	−2	71.70	33	−5.2	68.90	37.4 *	13.7
17	IR90688-74-1-1-1-1-1	2.8	−5.4 *	−2.2	95.00	56.3 *	18.1	92.60	61.1 *	37.4 *
18	IR90688-77-1-1-1-1-1	3.3	−4.9 *	−1.7 *	71.10	32.4	−5.8	58.70	27.2	3.5
19	IR90688-81-1-1-1-1-1	2.5	−5.7 *	−2.5 *	100.00	61.3 *	23.1	89.60	58.1 *	34.4 *
20	IR90688-82-1-1-1-1-1	3.0	−5.2 *	−2 *	86.80	48.1 *	9.9	71.50	40 *	16.3
21	IR90688-82-1-1-1-1-1	3.1	−5.1 *	−1.9 *	84.30	45.6 *	7.4	66.20	34.7 *	11
22	IR90688-92-1-1-1-1-1	3.4	−4.8 *	−1.6 *	80.60	41.9 *	3.7	65.00	33.5 *	9.8
23	IR90688-94-1-1-1-1-1	3.1	−5.1 *	−1.9 *	90.90	52.2 *	14	86.70	55.2 *	31.5
24	IR90688-95-1-1-1-1-1	3.0	−5.2 *	−2 *	81.90	43.2 *	5	76.70	45.2 *	21.5
25	IR90688-96-1-1-1-1-1	4.1	−4.1 *	−0.9	59.00	20.3	−17.9	53.30	21.8	−1.9
26	IR90688-103-1-1-1-1-1	2.2	−6 *	−2.8 *	98.30	59.6 *	21.4	92.00	60.5 *	36.8 *
27	IR90688-105-1-1-1-1-1	2.3	−5.9 *	−2.7 *	90.00	51.3 *	13.1	83.90	52.4 *	28.7
28	IR90688-106-1-1-1-1-1	3.0	−5.2 *	−2 *	83.30	44.6 *	6.4	71.50	40 *	16.3
29	IR90688-108-1-1-1-1-1	3.3	−4.9 *	−1.7 *	75.00	36.3 *	−1.9	73.30	41.8 *	18.1
30	IR90688-109-1-1-1-1-1	3.1	−5.1 *	−1.9 *	85.00	46.3 *	8.1	76.90	45.4 *	21.7
31	IR90688-114-1-1-1-1-1	3.0	−5.2 *	−2 *	73.30	34.6	−3.6	76.50	45 *	21.3
32	IR90688-118-1-1-1-1-1	2.4	−5.8 *	−2.6 *	80.00	41.3 *	3.1	90.00	58.5 *	34.8 *
33	IR90688-120-1-1-1-1-1	5.3	−2.9 *	0.3	53.30	14.6	−23.6	43.30	11.8	−11.9
34	IR90688-124-1-1-1-1-P2	5.0	−3.2 *	0	54.50	15.8	−22.4	42.00	10.5	−13.2
35	IR90688-125-1-1-1-1-1	2.9	−5.3 *	−2.1 *	90.30	51.6 *	13.4	82.00	50.5 *	26.8
36	BR1	8.2	0	3.2 *	38.70	0	−38.2	31.50	0	−23.7
37	BR18	5.0	−3.2 *	0	76.90	38.2 *	0	55.20	23.7	0
38	BRRI dhan28	7.6	−0.6	2.6 *	64.70	26	−12.2	47.10	15.6	−8.1
39	BRRI dhan29	6.9	−1.3	1.9 *	58.70	20	−18.2	58.00	26.5	2.8
	Variance	3.43 **			526.69 **			604.45 *		
	Mean	3.49			76.18			69.07		
	CV	15.17			19.20			23.72		
	Mean	3.49			76.18			69.07		
	LSD (0.05)	1.4			37.70			32.20		

* and ** indicate significant difference from BR18 at 5% and 1% level of probability, respectively.

3.1.1. LD Score

LD scores among the genotypes varied from 2.2 to 8.2, with an average of 3.49 and CV of 15.17 (Table 1). The highest LD value was observed with the cold-susceptible check variety BR1 (8.2), which was statistically similar to those of IR90688-120-1-1-1-1-1, IR90688-124-1-1-1-1-P2, and BRRI dhan28 at the 5% level of significance. On the other hand, IR90688-103-1-1-1-1-1, IR90688-105-1-1-1-1-1, IR90688-5-1-1-1-1-1, IR90688-105-1-1-1-1-1, IR90688-118-1-1-1-1-1, IR90688-81-1-1-1-1-1, IR90688-20-1-1-1-1-1, and IR90688-74-1-1-1-1-1 showed significantly lower LD values than that of BR18, which showed a moderate level of cold tolerance (LD score of 5.0). In addition, IR90688-62-1-1-1-1-1, IR90688-42-1-1-1-1-p2, IR90688-96-1-1-1-1-1, IR90688-54-1-1-1-1-1, and BRRI dhan29 showed LD values ranging from 3.9 to 4.5, indicating a moderate level of cold tolerance at the seedling stage.

3.1.2. Survivability Rate

The survivability rate varied widely among the genotypes, ranging from 38.70% to 100% with 19.20% CV. Out of 39 genotypes, 31 showed more than 60% survivability, while susceptible variety BR1 showed only 38.7% survivability (Table 1). IR90688-81-1-1-1-1-1 showed the highest level of survivability (100%), followed by IR90688-103-1-1-1-1-1 (98.3%), IR90688-74-1-1-1-1-1 (95.0%), and IR90688-94-1-1-1-1-1 (90.9%), which were, in fact, not significantly different from the cold-tolerant check variety BR18 (76.9%), but were significantly different from the susceptible check variety BR1.

3.1.3. Recovery Rate

The recovery rate also showed a wide range of variations among the genotypes, starting from 31.5% to 92.6% with an average value of 69.07 (Table 1). Although the tolerant and susceptible check varieties showed no significant difference in recovery rate, 23 NILs showed significantly higher recovery than BR1, and 4 (IR90688-74-1-1-1-1-1, IR90688-103-1-1-1-1-1, IR90688-118-1-1-1-1-1, and IR90688-81-1-1-1-1-1) showed significantly higher recovery than even the cold-tolerant check variety BR18.

3.1.4. Correlation among LD Score, Survivability, and Recovery Rate

The Spearman correlation among the cold-responsive traits measured in this study showed that the LD score, survivability, and recovery rate were significantly correlated to each other (Figure 2).

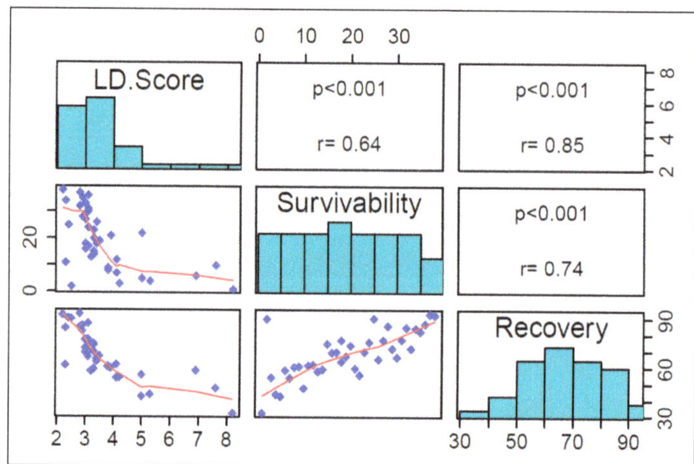

Figure 2. Spearman correlation between LD score, survivability, and recovery rate. The upper pan describes the correction coefficients and their significance level, the lower pan shows the directions of correlation, and the diagonal pan shows the distributions of the trait value across the genotypes.

The highest correlation coefficient was observed between LD score and recovery rate, and the lowest correlation was between LD score and survival rate. However, both were in a negative or reverse direction. The survival rate and recovery rate were positively correlated. The principal component analysis also showed that LD score was negatively associated with survivability and recovery rate (Figure 3).

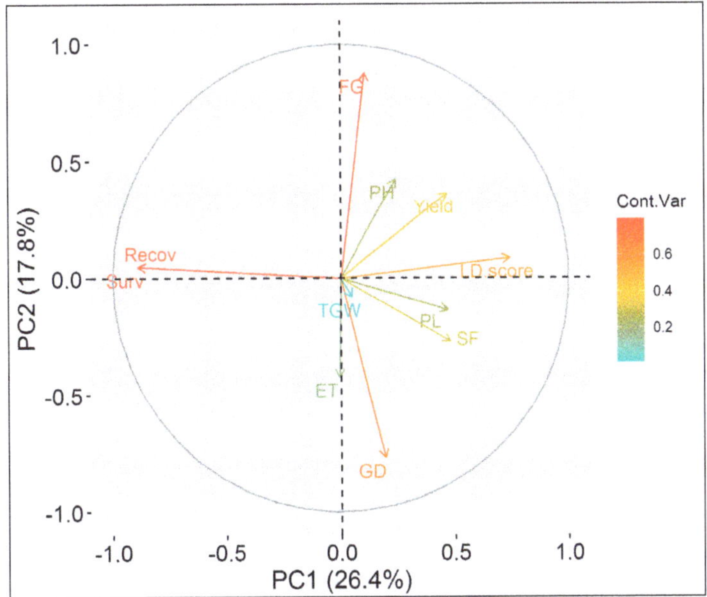

Figure 3. PCA plot showing an association between agronomic and cold-related seedling traits. The traits representing the same quadrant and in the same direction are positively associated, and the traits in the opposite direction are negatively associated. The angular distance between traits denotes the strength of association.

3.2. Evaluation of NILs for Agronomic Performance

The NILs and the check varieties showed distinct genetic variation, particularly in grain yield per plant, the number of fertile grains per panicle, and % spikelet fertility. Days to flowering, plant height, and panicle length were also variable. Fertile grain per panicle had the widest range, followed by spikelet fertility. IR90688-92-1-1-1-1-1 had the highest value and IR90688-52-1-1-1-1-1 had the lowest value of FGPP. The ranges of panicle length, spikelet fertility, and thousand-grain weight were smaller compared to other characteristics. The maximum coefficient of variation was found in the case of fertile grain (12.7%) followed by grain yield per plant (11.43%).

The mean values of the NILs varied significantly from the check varieties for the characters measured in this study (Table 2). The check varieties also differed from each other. There were many NILs that exceeded the mean values of the four high-yielding check varieties. Eleven NILs showed a significantly higher yield than the susceptible check varieties BR1 and BRRI dhan28. Out of these 11 NILs, IR90688-92-1-1-1-1-1, IR90688-125-1-1-1-1-1, IR90688-104-1-1-1-1-1, IR90688-124-1-1-1-1-1-P2, IR90688-15-1-1-1-1-1, and IR90688-27-1-1-1-1-1 had a significantly shorter growth duration than BR1, while only IR90688-92-1-1-1-1-1 and IR90688-125-1-1-1-1-1 were on par with BRRI dhan28 in terms of days to maturity.

Table 2. Adjusted mean of yield and yield-contributing traits of Jinbubyeo-BRRI dhan29 NILs.

SN	Genotype	GD (Days)	PH (cm)	PL (cm)	ET	FG	% SF	TGW	Yield (g/plant)
1	IR90688-5-1-1-1-1-1	143.7	94.1	22.7	17.7	61.6	38.3	22.7	11.5
2	IR90688-13-1-1-1-1-P1	150.7	98.2	22.4	14.7	71.4	97.4	18.5	27.7
3	IR90688-15-1-1-1-1-1	146.7	95.2	22.4	14.6	82.4	73.6	28.2	30.5
4	IR90688-19-1-1-1-1-1	146.7	94.1	19.9	14.5	56.8	56.1	19.0	11.9
5	IR90688-20-1-1-1-1-1	158.7	97.3	23.0	15.1	54.1	47.1	18.5	14.0
6	IR90688-27-1-1-1-1-1	146.7	97.2	24.7	17.1	91.5	41.3	19.2	24.4
7	IR90688-30-1-1-1-1-1	163.7	91.5	20.7	17.3	49.3	63.2	17.0	18.1
8	IR90688-42-1-1-1-1-P1	154.7	88.2	19.6	15.8	49.9	49.7	20.1	19.3
9	IR90688-42-1-1-1-1-P2	157.7	90.9	21.8	16.9	48.9	71.4	19.3	12.5
10	IR90688-43-1-1-1-1-P1	147.2	74.6	22.1	23.2	45.1	39.0	17.4	16.9
11	IR90688-52-1-1-1-1-1	161.2	84.8	22.2	13.5	65.2	88.3	18.7	23.2
12	IR90688-54-1-1-1-1-1	151.2	102.4	24.6	13.2	54.0	33.4	15.4	17.6
13	IR90688-56-1-1-1-1-1	154.2	99.9	24.0	29.8	46.9	37.5	19.6	15.7
14	IR90688-62-1-1-1-1-1	147.2	92.5	23.2	17.1	71.7	65.1	20.6	27.6
15	IR90688-64-1-1-1-1-1	149.2	81.7	23.2	15.5	118.7	37.8	16.5	22.2
16	IR90688-73-1-1-1-1-P1	153.2	86.6	23.5	15.7	44.0	49.2	18.6	11.8
17	IR90688-74-1-1-1-1-1	151.2	84.7	23.6	19.0	96.0	73.2	20.2	27.6
18	IR90688-77-1-1-1-1-1	155.2	82.9	23.3	22.3	44.3	42.1	25.0	20.7
19	IR90688-81-1-1-1-1-1	155.4	107.0	23.6	15.9	65.6	65.2	23.9	12.8
20	IR90688-82-1-1-1-1-1	144.4	98.2	25.4	19.3	87.9	50.9	15.6	13.0
21	IR90688-91-1-1-1-1-1	143.4	105.7	22.3	15.3	107.1	53.5	17.4	17.3
22	IR90688-92-1-1-1-1-1	138.4	93.5	14.2	15.7	160.2	60.2	16.1	23.8
23	IR90688-94-1-1-1-1-1	149.4	99.3	25.8	19.4	80.4	30.3	25.8	18.1
24	IR90688-95-1-1-1-1-1	152.4	79.7	21.0	18.8	80.8	33.5	18.3	12.4
25	IR90688-96-1-1-1-1-1	146.4	104.0	24.4	21.7	68.3	49.7	20.7	20.5
26	IR90688-103-1-1-1-1-1	149.4	90.2	20.8	21.9	73.1	16.0	22.1	14.1
27	IR90688-104-1-1-1-1-1	145.4	100.1	22.3	18.1	82.0	68.9	19.1	23.0
28	IR90688-105-1-1-1-1-1	151.7	83.3	19.0	18.2	51.0	29.5	16.4	9.8
29	IR90688-106-1-1-1-1-1	148.7	84.0	20.9	19.6	57.2	22.0	17.3	19.6
30	IR90688-108-1-1-1-1-1	144.7	84.7	19.7	17.4	65.0	33.5	21.4	17.4
31	IR90688-109-1-1-1-1-1	139.7	82.3	16.9	19.6	64.0	14.8	17.9	7.7
32	IR90688-114-1-1-1-1-1	148.7	88.9	21.9	19.7	35.7	23.1	16.4	8.1
33	IR90688-118-1-1-1-1-1	146.7	90.3	18.1	17.3	68.9	28.5	23.4	13.4
34	IR90688-120-1-1-1-1-1	153.7	72.3	18.9	18.7	50.8	53.3	18.1	12.0
35	IR90688-124-1-1-1-1-P2	146.7	91.3	22.5	17.6	95.6	81.2	17.0	30.6
36	IR90688-125-1-1-1-1-1	143.7	89.2	22.4	20.3	86.7	63.8	17.9	22.6
37	BR1 (Ck)	148.8	77.9	23.1	24.1	57.3	59.8	18.6	18.1
38	BR18 (Ck)	160.8	89.0	24.9	23.1	73.3	67.8	19.5	20.3
39	BRRI dhan28 (Ck)	139.8	99.2	24.8	12.2	143.7	77.0	18.4	18.2
40	BRRI dhan29 (Ck)	157.5	93.8	22.8	19.1	92.9	65.0	17.7	32.6
LSD (Ck Mean)		1.19	2.69	0.36	0.74	8.24	4.10	1.27	1.80
LSD (Ck Mean vs. NIL)		2.11	4.75	0.64	1.30	14.57	7.25	2.24	3.19
Range of NILs		138.4–163.7	72.3–107	14.2–25.8	13.2–29.8	45.1–160.2	14.8–97.4	15.4–28.2	7.7–30.6
CV (%)		1.11	4.22	2.13	5.31	12.7	8.61	9.67	11.43

GD = growth duration, PH = plant height, PL = panicle length, ET = effective tiller, FG = fertile grain, SF = spikelet fertility, TGW = thousand-grain weight, Ck = check variety, CV = coefficient of variation.

The NILs in this study were at an acceptable range in plant height except for one line (Table 1). The genotypes showing significantly higher yield than BR18 and BRRI dhan28 had an intermediate type plant height. Not only that, these lines had an acceptable range of ineffective tillers per hill and filled grain per panicle (82–160), but also had a good seed setting rate under natural field conditions. In addition, they had almost a similar amount of thousand-grain weight to that of BRRI dhan28 and BRRI dhan29, which are the most widely grown varieties in Bangladesh. Principal component analysis of the agronomic traits with the cold-related seedling traits showed that grain yield was highly associated with PH and FG than other traits measured in this study.

3.3. Genetic Analysis of NILs of BRRI dhan29 Using Microsatellite Markers

Out of 88 SSRs used in this study, 58 markers were found polymorphic, with an average polymorphism rate of 36.97%. Among the 58 SSRs, 11 markers were present on chromosome 1, 6 markers on each of chromosomes 2 and 3, 4 markers on each of chromosomes 4, 9, and 12, 2 markers on each of chromosomes 5 and 6, 5 markers on chromosome 7, 3 markers on each of chromosome 8 and 10, and 8 markers on chromosome 11. A total of 186 alleles of the polymorphic SSRs were detected in the 42 genotypes, including 36 NILs (Table 3).

Table 3. Allelic variation and polymorphism information content values of different rice cultivars for 58 SSR polymorphic markers.

Marker ID	Chr. No.	Position (Mb)	Allele No.	Allele Size (bp)		Highest Frequency Allele		PIC Value
				Range	Difference	Size (bp)	Frequency %	
RM7278	1	1.79	3	173–240	67	201	57.14	0.45
RM220	1	4.42	4	110–126	16	120	71.43	0.40
RM259	1	7.45	3	158–174	16	158	52.38	0.44
RM581	1	9.11	3	134–141	7	141	45.24	0.51
RM572	1	9.87	3	165–187	22	165	61.90	0.44
RM10649	1	10.28	3	449–498	49	477	61.90	0.39
RM10800	1	12.73	2	130–144	14	144	57.14	0.37
RM5638	1	20.93	4	190–235	45	207	52.38	0.44
RM11570	1	-	4	351–424	73	351	61.90	0.52
RM297	1	32.1	4	147–187	40	147	50.00	0.47
RM11874	1	-	3	181–195	14	195	66.67	0.37
RM3703	2	3.86	4	100–118	18	100	57.14	0.51
RM12769	2	7.35	5	151–164	13	151	52.38	0.47
RM424	2	11.39	3	235–278	43	235	50.00	0.46
RM13155	2	15.26	3	450–551	101	450	57.14	0.40
RM3421	2	29.91	3	142–147	5	142	54.76	0.41
RM266	2	35.43	4	112–125	13	125	50.00	0.49
RM3202	3	0.81	4	194–211	17	194	57.14	0.48
RM546	3	6.16	3	156–167	11	156	61.90	0.39
RM14795	3	10.36	4	126–139	13	139	45.24	0.51
RM1164	3	14.86	3	192–200	8	192	83.33	0.27
RM7134	3	22.01	4	177–289	112	201	66.67	0.40
RM7097	3	26.87	3	168–176	8	176	69.05	0.38
RM16686	4	14.72	2	87–90	3	87	76.19	0.30
RM273	4	23.86	3	195–204	9	195	64.29	0.41
RM317	4	29.06	3	154–160	6	154	50.00	0.49
RM127	4	34.53	3	208–217	9	217	61.90	0.39
RM413	5	2.21	4	72–100	28	75	52.38	0.44
RM3916	5	-	3	233–289	56	289	47.62	0.47
RM588	6	1.61	4	115–128	13	115	50.00	0.44
RM6811	6	29.23	3	165–187	22	165	54.76	0.41
RM295	7	0.41	3	186–193	7	186	47.62	0.47
RM6427	7	17.45	4	156–170	14	160	50.00	0.47
RM336	7	21.87	4	145–202	57	145	52.38	0.44
RM234	7	25.47	3	131–153	22	153	66.67	0.37
RM172	7	29.56	3	158–164	6	164	50.00	0.41
RM547	8	5.59	3	197–253	56	197	54.76	0.46
RM331	8	12.29	3	150–171	21	150	57.14	0.45
RM264	8	27.92	3	165–179	14	165	50.00	0.51
RM5515	9	7.15	3	115–126	11	115	59.52	0.42
RM24087	9	-	2	246–259	13	246	64.29	0.35

Table 3. *Cont.*

Marker ID	Chr. No.	Position (Mb)	Allele No.	Allele Size (bp)		Highest Frequency Allele		PIC Value
				Range	Difference	Size (bp)	Frequency %	
RM566	9	14.74	4	239–271	32	249	50.00	0.47
RM553	9	19.32	4	160–173	13	160	42.86	0.54
RM216	10	5.35	3	133–149	16	149	54.76	0.43
RM6142	10	12.8	3	82–97	15	94	66.67	0.40
RM3773	10	19.9	3	137–162	25	150	50.00	0.44
RM26063	11	2.26	2	129–137	8	137	52.38	0.37
RM26243	11	5.56	3	150–160	10	160	64.29	0.38
RM7283	11	9.12	3	149–182	33	149	66.67	0.37
RM26501	11	11.1	2	137–146	9	137	71.43	0.32
RM3428	11	13.48	3	140–157	17	140	64.29	0.38
RM229	11	18.41	3	114–129	15	114	59.52	0.40
RM224	11	21.85	4	123–160	37	160	59.52	0.43
RM206	11	22.01	4	132–171	39	165	54.76	0.44
RM20	12	0.97	3	111–135	24	111	54.76	0.43
RM512	12	5.1	2	221–226	5	226	52.38	0.37
RM28502	12	23.41	3	124–131	7	124	50.00	0.46
RM7558	12	27.02	2	134–139	5	134	57.14	0.37
Mean			3.21				57.31	0.42
Max			5.00				83.33	0.54
Min			2.00				42.86	0.27

The number of alleles generated by each marker varied from 2 to 5 alleles and with an average of 3.21 alleles per locus. The highest number of alleles (5) was detected with RM12769 and the lowest number (2) was detected in RM10800, RM16686, RM24087, RM26063, RM26501, RM512, and RM7558. The overall size of the PCR products of the tested markers ranged from 72 bp (RM413) to 551 bp (RM13155). The molecular size difference between the smallest and the largest allele for a given locus varied widely from 3 to 112 bp (Table 3). There was a considerable range in allele frequency (42.86–83.33%). On average, 57.31% of the genotypes shared a common allele.

The PIC value, which indicates the degrees of allelic diversity [46] and frequency among the varieties, was calculated for all of the markers. The PIC values for the microsatellite loci varied from 0.27 to 0.54 with a mean of 0.42. Among the markers used in the study, RM553 on chromosome 9 had the highest PIC value (0.54). RM581 and RM11570 on chromosome 1, RM3703 on chromosome 2, RM14795 on chromosome 3, and RM264 on chromosome 8 showed PIC values greater than 0.50.

The dendrogram constructed based on Shannon's similarity index following the UPGMA method (Figure 4) showed that all of the genotypes were constellated into two distinct groups at a similarity index of 0.60. The first group housed 40 genotypes, while the second contained only 2 genotypes including an NIL and a cold-tolerant check variety, Hbj.B.VI. However, at the 0.67 similarity index, the genotypes were grouped into four major clusters, although the genotypes within the clusters were further subdivided into many subclusters based on the genetic similarity between them.

The genetic similarity within the individual cluster/group ranges from 0.60 to 0.96. Cluster 1, Cluster 2, Cluster 3, and Cluster 4 housed 25, 11, 4, and 2 genotypes, respectively. Housing the maximum number of genotypes, Cluster 1 was the most diverged cluster. The genotypes within Cluster 1 were further divided into many subgroups until 95% genetic similarity was reached. At 95% genetic similarity, IR90688-54-1-1-1-1-1 (G12) and IR90688-56-1-1-1-1-1 (G13) were grouped together. Cluster 1 was subdivided into 2, 3, 4, 5, 6, 7, 8, 9, 10, 12, 14, 16, 17, 18, 20, 23, 24, and 25 subgroups at a genetic similarity of 67.5, 68, 70, 71, 71.5, 72.75, 73, 73.5, 74.5, 75, 76, 77, 78, 78.5, 79.5, 85, 91, and 95%, respectively. Contrarily, Cluster 2 contained 11 genotypes. The genetic similarity within Cluster 2 varied from 68 to 92%. IR90688-42-1-1-1-1-P1 (G8), IR90688-42-1-1-1-1-P2 (G9), IR90688-73-1-1-1-P1 (G16), and IR90688-82-1-1-1-1-1 (G20) were housed in Cluster 3. In Cluster 4, two genotypes were grouped at 67.5% genetic similarity.

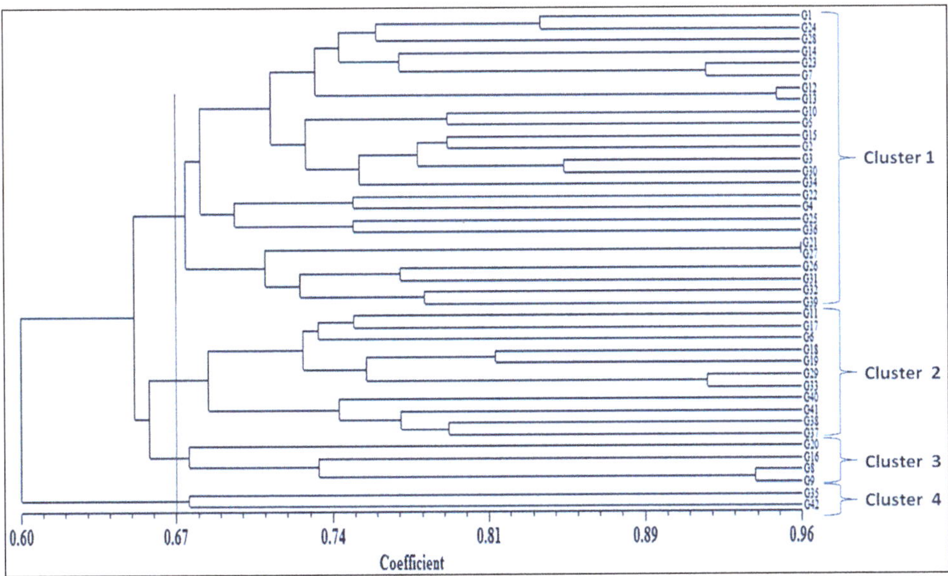

Figure 4. Dendrogram showing the genetic relationships among 42 rice cultivars using UP-GMA cluster analysis of Shanondice genetic similarity coefficients generated from 58 SSR markers. G01 = IR90688-5-1-1-1-1-1; G02 = IR90688-13-1-1-1-1-P1; G03 = IR90688-15-1-1-1-1-1; G04 = IR90688-19-1-1-1-1-1; G05 = IR90688-20-1-1-1-1-1; G06 = IR90688-27-1-1-1-1-1; G07 = IR90688-30-1-1-1-1-1; G08 = IR90688-42-1-1-1-1-P1; G09 = IR90688-42-1-1-1-1-P2; G10 = IR90688-43-1-1-1-1-P1; G11 = IR90688-52-1-1-1-1-1; G12 = IR90688-54-1-1-1-1-1; G13 = IR90688-56-1-1-1-1-1; G14 = IR90688-62-1-1-1-1-1; G15 = IR90688-64-1-1-1-1-1; G16 = IR90688-73-1-1-1-1-P1; G17 = IR90688-74-1-1-1-1-1; G18 = IR90688-77-1-1-1-1-1; G19 = IR90688-81-1-1-1-1-1; G20 = IR90688-82-1-1-1-1-1; G21 = IR90688-91-1-1-1-1-1; G22 = IR90688-92-1-1-1-1-1; G23 = IR90688-94-1-1-1-1-1; G24 = IR90688-95-1-1-1-1-1; G25 = IR90688-96-1-1-1-1-1; G26 = IR90688-103-1-1-1-1-1; G27 = IR90688-104-1-1-1-1-1; G28 = IR90688-105-1-1-1-1-1; G29 = IR90688-106-1-1-1-1-1; G30 = IR90688-108-1-1-1-1-1; G31 = IR90688-109-1-1-1-1-1; G32 = IR90688-114-1-1-1-1-1; G33 = IR90688-118-1-1-1-1-1; G34 = IR90688-120-1-1-1-1-1; G35 = IR90688-124-1-1-1-1-P2; G36 = IR90688-125-1-1-1-1-1; G37 = BR1; G38 = BR18; G39 = BRRI dhan28; G40 = BRRI dhan29; G41 = Jinbubyeo; G42 = Hbj.B.VI.

4. Discussion

Cold tolerance at the seedling and reproductive stage is very important for growing rice in the winter season or in cold stress conditions. The early vegetative stages, particularly at the germination and crop establishment stage and also the booting stages of rice, are very sensitive to LTS (Figure 1). An ambient temperature below 10–12 °C significantly increases seedling mortality in rice [47], resulting in poor crop establishment, which eventually turns into low yield. By assigning LD scores [48] upon exposure to LTS for seven days in artificially induced low temperature [32] and measuring the survivability and recovery rate, the genotypes with seedling-stage cold tolerance can easily be identified. In this study, we evaluated 36 NILs with cold tolerance and susceptible check varieties for seedling-stage cold tolerance. The genotypes showed a wide variation in LD scores, ranging from 2.2 to 8.2 with an average of 3.49 (Table 1). Suh et al. [49] also reported varying LD scores (3 to 8 with a mean value of 5) among 23 elite rice cultivars grown in a cold water-irrigated plot. In this study, the highest LD value was found with the cold-susceptible check variety BR1 (8.2). IR90688-120-1-1-1-1-1, IR90688-124-1-1-1-1-P2, and BRRI dhan28 showed no significant difference from BR1 in LD score at $p < 0.05$, indicating them to be cold-susceptible. Contrarily, IR90688-103-1-1-1-1-1, IR90688-105-1-1-1-1-1, IR90688-5-1-1-1-1-1, IR90688-105-1-1-1-1-1, IR90688-118-1-1-1-1-1, IR90688-81-1-1-1-1-1, IR90688-20-1-1-1-1-1, and IR90688-74-1-1-1-1-1 showed significantly lower LD values than the cold-tolerant check variety BR18. These results clearly revealed the capability of tolerance of the NILs to LTS. Additionally, IR90688-62-1-1-1-1-1, IR90688-42-1-1-1-1-p2, IR90688-96-1-1-1-1-1, IR90688-54-1-1-1-1-1, and BRRI dhan29 showed LD values ranging from 3.9 to 4.5, indicating a moderate level of cold tolerance at the seedling stage. Khatun et al. [32] reported a higher average LD score for BR1, moderate for BRRI dhan29, and the lowest LD score for BR18. Kundu [50] also reported similar findings from a cold-screening trial of a set of cold-tolerant *japonica* and *indica* donors, along with cold-susceptible cultivars.

The seedling survivability and recovery rate are very important for better crop establishment under LTS. In this study, the survivability and recovery rate varied widely among the genotypes, ranging from 38.70% to 100% and 31.5% to 92.6%, respectively. The susceptible variety BR1 showed only 38.7% survivability and 31.5% recovery rate (Table 1). IR90688-81-1-1-1-1-1 showed the highest level of survivability (100%), followed by IR90688-103-1-1-1-1-1 (98.3%), IR90688-74-1-1-1-1-1 (95.0%), and IR90688-94-1-1-1-1-1 (90.9%), which were, in fact, not significantly different from the cold-tolerant check variety BR18 (76.9%), but were significantly different from the susceptible check variety BR1. A total of 23 NILs showed a significantly higher recovery rate than BR1, with 4 of them (IR90688-74-1-1-1-1-1, IR90688-103-1-1-1-1-1, IR90688-118-1-1-1-1-1, and IR90688-81-1-1-1-1-1) showing significantly higher recovery than even the cold-tolerant check variety BR18. In a similar study with 10 landraces and 5 modern varieties as check genotypes, Shoroardi et al. [51] showed a higher reduction in seedling survival rate in BRRI dhan28, LR 212- NariaBochi, and LR 25-KathiGoccha and a lower reduction in BRRI dhan36, BRRI dhan29, LR 54- Tor Balam, and LR 114-Badiabona. The magnitude of the recovery rate was a little lower than the survivability rate, but they were positively correlated. The recovery rate, in fact, pronounced the real tolerance to LTS at the seedling stage.

For direct use in variety release or as a parental line in the breeding program, a breeding line should have higher yield potential in addition to its particular trait of interest. Therefore, we evaluated the NILs for yield and other agronomic performances under natural field conditions. Although the NILs were developed in the background of BRRI dhan29, they possessed distinct genetic variations, particularly in grain yield per plant, number of fertile grains per panicle, and % spikelet fertility [52]. Days to flowering, plant height, and panicle length were also variable among the NILs. The mean values of the NILs varied significantly from the check varieties for all of the characteristics measured in this study (Table 2). Eleven NILs showed significantly higher yields than the susceptible check varieties BR1 and BRRI dhan28. Out of these NILs, IR90688-92-1-1-1-1-1, IR90688-125-1-1-1-1-1, IR90688-104-1-1-1-1-1, IR90688-124-1-1-1-1-P2, IR90688-15-1-1-1-1-1, and IR90688-27-1-1-1-1-1 had a significantly shorter growth duration than BR1, while only IR90688-92-1-1-1-1-1 and IR90688-125-1-1-1-1-1 were on par with BRRI dhan28 in terms of days to maturity.

Plant height in rice is an important trait for the better adoption of a variety. Earlier study revealed that the taller plants usually produce a low yield [53]; through another study reported that short-statured plants also give a low yield some extends due to the production of low biomass [54]. The NILs in this study were at an acceptable range of plant height, except for one line (Table 1). The genotypes showing significantly higher yield than BR18 and BRRI dhan28 had intermediate type plant height. Not only did these lines have an acceptable range of effective tillers per hill and filled grain per panicle (82–160), but they also had a good seed setting rate under natural field conditions. In addition, they had an almost similar amount of thousand-grain weight to that of BRRI dhan28 and BRRI dhan29, which are the most widely grown varieties in Bangladesh. The principal component analysis of the agronomic traits with the cold-related seedling traits showed that grain yield was more highly associated with PH and FG than other traits measured in this study.

In addition to measuring yield potential, genetic diversity in the parental lines is very important for the genetic improvement of rice through breeding. Strong genetic diversity means diverse morphological traits and potentially valuable genetic information. Rice varieties/lines with high levels of genetic variation are essential resources for broadening the genetic base of the breeding germplasm, and therefore laying a good foundation for rice breeding [55]. Hence, the assessment of genetic diversity becomes important in establishing relationships among different cultivars. SSR markers are widely used in the analysis of genetic diversity among the germplasms [56].

Usually, NILs are developed through multiple backcrossing, meaning they are very close to each other as they share the majority of the recipient parental genome. In this study, we used BC_3F_5 NILs developed from a cross between a Korean *japonica* variety Jinbubyeo and a Bangladeshi high-yielding rice variety BRRI dhan29, but they showed considerable variations in different agronomic traits. We analysed the NILs along with their parents, cold-tolerant and susceptible varieties, using 58 polymorphic SSR markers randomly distributed over 12 chromosomes to estimate the genetic similarity/dissimilarity between them. A total of 186 alleles of the polymorphic SSRs were detected in the genetic analysis of the 42 genotypes, including 36 NILs (Table 3). The number of alleles generated by each marker varied from 2 to 5 alleles and with an average of 3.21 alleles per locus. The highest number of alleles (five) was detected in the locus RM12769 and the lowest number of alleles (two) was detected in seven SSR loci, namely RM10800, RM16686, RM24087, RM26063, RM26501, RM512, and RM7558. The overall size of the PCR products of the tested markers ranged from 72 bp (RM413) to 551 bp (RM13155). The molecular size difference between the smallest and the largest allele for a given locus varied widely, from 3 to 112 bp (Table 3).

The PIC value, which indicates the degrees of allelic diversity [46], varied from 0.27 to 0.54 with a mean of 0.42. The higher the PIC value, the greater the diversity within the germplasm; contrarily, a lower PIC value indicates that less divergence between the genotypes and they are closely related [40,57] in terms of the specific marker loci. Among the markers used in this study, RM553 on chromosome 9 had the highest PIC value (0.54). Two SSRs (RM581, RM11570) on chromosome 1, one SSR (RM3703) on chromosome 2, one SSR (RM14795) on chromosome 3, and one SSR (RM264) on chromosome 8 showed PIC values greater than 0.50, indicating that these are highly informative markers. An informative marker can easily and clearly discriminate the germplasm based on its band intensity and position relative to others [58]. Thus, highly informative markers are considered to be the best markers for diversity analysis. The six informative SSR markers identified from this study could be useful for future genetic studies of rice, particularly for marker-assisted breeding and QTL identification.

The UPGMA dendrogram constructed based on Shannon's similarity index (Figure 4) grouped all of the genotypes into two distinct clusters at a similarity index of 0.60. However, at the 0.67 similarity index, the genotypes were grouped into four major clusters, although the genotypes within the clusters were further subdivided into many subclusters based on the genetic similarity between them. Cluster 1, Cluster 2, Cluster 3, and Cluster 4 housed 25, 11, 4, and 2 genotypes, respectively. Housing the maximum number of genotypes, Cluster 1 was the most diverged cluster. The genotypes within Cluster 1 were further divided into many subgroups until 95% genetic similarity was reached. At 95% genetic similarity, IR90688-54-1-1-1-1-1 (G12) and IR90688-56-1-1-1-1-1 (G13) were grouped together. The genetic similarity among the genotypes in Cluster 1 ranged from 67.5% to 95%. The majority of the cold-tolerant and high-yielding NILs were grouped into Cluster 1. Contrary, Cluster 2 contained 11 genotypes, which were mostly poor yielders except for BR1 (G37), BR18 (G38), and BRRI dhan29 (G40). In this cluster, both cold-tolerant and susceptible varieties remained together, but they entered into different subgroups at a 0.77 similarity index. The most interesting thing is that Hbj. B.VI (G42), a cold-tolerant local variety for both seedling and reproductive stage [26], and IR90688-124-1-1-1-1-P2 (G35), which showed moderate tolerance to cold stress at the seedling stage (Table 1) and high yield (Table 2), were grouped into Cluster 4 with approximately 68% genetic similarity. Therefore, the crosses between IR90688-124-1-1-1-1-P2 and Hbj.B.VI could produce high yielding and cold-tolerant progenies. Not only these combinations, but also the cross combination between the lines of Cluster 4 with the high-yielding and cold-tolerant lines in Cluster 1, would produce cold-tolerant and high-yielding progeny.

5. Conclusions

Highly significant variation among the genotypes was observed for three cold-related traits at the seedling stage, namely, leaf discolouration score, survivability, and recovery rate. Nine NILs showed strong cold tolerance in at least one of three cold-related traits. Three NILs, IR90688-74-1-1-1-1-1, IR90688-81-1-1-1-1-1, and IR90688-103-1-1-1-1-1, showed tolerance to cold stress with all three traits, while IR90688-118-1-1-1-1-1 showed cold tolerance with LD score and recovery rate. In a field experiment, 11 NILs showed significantly higher yields than the susceptible check varieties BR1 and BRRI dhan28. Out of these 11 NILs, IR90688-92-1-1-1-1-1, IR90688-125-1-1-1-1-1, IR90688-104-1-1-1-1-1, IR90688-124-1-1-1-1-P2, IR90688-15-1-1-1-1-1, and IR90688-27-1-1-1-1-1 had significantly shorter growth duration than BR1, while only IR90688-92-1-1-1-1-1 and IR90688-125-1-1-1-1-1 were on par with BRRI dhan28 in terms of days to maturity. In addition, these lines had grain quality similar to BRRI dhan28 and BBRI dhan29. In the genetic analysis with 58 polymorphic SSRs, a total of 186 alleles were detected and the number of alleles per marker varied from 2 to 5, with an average of 3.21 alleles per locus. Six SSRs showed PIC values greater than 0.50, indicating that these are the best markers in discriminating the NILs derived from the cross between Jinbubyeo and BRRI dhan29. The Shannon similarity index-based dendrogram grouped the NILs at 67% genetic similarity into four major distinct clusters, in which 25 genotypes, including the high-yielding NILs and high-yielding variety BRRI dhan28, constellated together. One high-yielding NIL, IR90688-124-1-1-1-1-P2, with a moderate level of cold tolerance was grouped; however, genotype Hbj. Boro VI has been found to be a cold-tolerant local variety for both the seedling and reproductive stages, as it is a member of the two-genotype cluster. Therefore, the cold-tolerant and high-yielding NILs identified in this study could be a potential genetic resource to develop high-yielding, cold-tolerant rice varieties. The QTL mapping from the local variety Hbj. Boro VI may be considered to generate marker information, aiming to introgress into the locally adapted existing rice varieties for the sustainability of rice production during extremely cold stress conditions.

Supplementary Materials: The following supporting information can be downloaded at: https://www.mdpi.com/article/10.3390/su14094871/s1, Table S1: List of 88 SSR markers used for diversity analysis of rice cultivars.

Author Contributions: Conceptualization: N.A., N.A.I., M.A.S. and P.S.B.; methodology: N.A., N.A.I., M.A.S. and P.S.B.; formal analysis: N.A., N.A.I., M.A.S. and P.S.B.; data curation: N.A., M.A.S. and A.H.; statistical expertise: N.A., M.A.S. and A.H.; writing—original draft preparation: N.A., N.A.I., M.A.S. and P.S.B.; writing—review and editing: A.G., A.M.A. and A.H.; visualization: N.A., N.A.I., M.A.S. and P.S.B.; supervision: P.S.B.; funding acquisition: A.M.A., A.G. and A.H. All authors have read and agreed to the published version of the manuscript.

Funding: The work was financially supported by three institutes, namely, the Bangladesh Agricultural Research Institute, Bangabandhu Sheikh Mujibur Rahman Agricultural University, and the Bangladesh Rice Research Institute, Gazipur, Bangladesh. The research program also partially supported Princess Nourah bint Abdulrahman University Researchers Supporting Project number (PNURSP2022R65), Princess Nourah bint Abdulrahman University, Riyadh, Saudi Arabia.

Institutional Review Board Statement: Not applicable.

Informed Consent Statement: Not applicable.

Data Availability Statement: Data are available upon request.

Acknowledgments: The authors wish to acknowledge the financial support provided by Bangladesh Agricultural Research Institute, Bangabandhu Sheikh Mujibur Rahman Agricultural University, and the Bangladesh Rice Research Institute, Gazipur, Bangladesh. The authors appreciated Princess Nourah bint Abdulrahman University Researchers Supporting Project number (PNURSP2022R65), Princess Nourah bint Abdulrahman University, Riyadh, Saudi Arabia, for providing financial support for the publication of this research.

Conflicts of Interest: The authors declare no conflict of interest.

References

1. Saito, K.; Miura, K.; Nagano, K.; Saito, Y.H.; Araki, H.; Kato, A. Identification of two closely linked quantitative trait loci for cold tolerance on chromosome 4 of rice and their association with anther length. *Theor. Appl. Genet.* **2001**, *103*, 862–868. [CrossRef]
2. Nahar, K.; Biswas, J.K.; Shamsuzzaman, A.M.M.; Hasanuzzaman, M.; Barman, H.N. Screening of Indica Rice (*Oryza sativa* L.) genotypes against low temperature stress. *Bot. Res. Int.* **2009**, *2*, 295–303.
3. Zeng, Y.; Zhang, Y.; Xiang, J.; Uphoff, N.T.; Pan, X.; Zhu, D. Effects of low temperature stress on spikelet-related parameters during anthesis in *indica–japonica* hybrid rice. *Front. Plant Sci.* **2017**, *8*, 1350. [CrossRef] [PubMed]
4. Biswas, P.S.; Rashid, M.A.; Khatun, H.; Yasmeen, R.; Biswas, J.K. Scope and progress of rice research harnessing cold tolerance. In *Advances in Rice Research for Abiotic Stress Tolerance*; Hasanuzzaman, M., Fujita, M., Nahar, K., Biswas, J.K., Eds.; Elsevier Inc. & Woodland Publishing: Cambridge, UK, 2019; pp. 225–280.
5. Ranawake, A.L.; Manangkil, O.E.; Yoshida, S.; Ishii, T.; Mori, N.; Nakamura, C. Mapping QTLs for cold tolerance at germination and the early seedling stage in rice (*Oryza sativa* L.). *Biotechnol. Biotechnol. Equip.* **2014**, *28*, 989–998. [CrossRef] [PubMed]
6. Martínez-Eixarch, M.; Ellis, R.H. Temporal sensitivities of rice seed development from spikelet fertility to viable mature seed to extreme-temperature. *Crop Sci.* **2015**, *55*, 354–364. [CrossRef]
7. Schläppi, M.R.; Jackson, A.K.; Eizenga, G.C.; Wang, A.; Chu, C.; Shi, Y.; Shimoyama, N.; Boykin, D.L. Assessment of five chilling tolerance traits and GWAS mapping in rice using the USDA mini-core collection. *Front. Plant Sci.* **2017**, *8*, 957. [CrossRef] [PubMed]
8. Shakiba, E.; Edwards, J.D.; Jodari, F.; Duke, S.E.; Baldo, A.M.; Korniliev, P.; McCouch, S.; Eizenga, G.C. Genetic architecture of cold tolerance in rice (*Oryza sativa*) determined through high resolution genome-wide analysis. *PLoS ONE* **2017**, *12*, e0172133. [CrossRef] [PubMed]
9. Liang, Y.; Meng, L.; Lin, X.; Cui, Y.; Pang, Y.; Xu, J.; Li, Z. QTL and QTL networks for cold tolerance at the reproductive stage detected using selective introgression in rice. *PLoS ONE* **2018**, *13*, e0200846. [CrossRef]
10. Xiao, N.; Gao, Y.; Qian, H.; Gao, Q.; Wu, Y.; Zhang, D.; Li, A. Identification of genes related to cold tolerance and a functional allele that confers cold tolerance. *Plant Physiol.* **2018**, *177*, 1108–1123. [CrossRef]
11. Rashid, M.M.; Yasmeen, R. Cold Injury and Flash Flood Damage in Boro Rice Cultivation in Bangladesh, A Review. *Bangladesh Rice J.* **2017**, *21*, 13–25. [CrossRef]
12. Biswas, J.K.; Mahbub, M.A.A.; Kabir, M.S. Critical temperatures and their probabilities on important growth stages of rice. In *Annual Report of Bangladesh Rice Research Institute, 2008–2009*; Bashar, M.K., Biswas, J.C., Kashem, M.A., Eds.; Bangladesh Rice Research Institute (BRRI): Gazipur, Bangladesh, 2011; pp. 127–129.
13. Fuentes, J.L.; Escobar, F.; Alvarez, A.; Gallego, G.; Duque, M.C.; Ferrer, M.; Deus, J.E.; Tohme, J. Analyses of genetic diversity in Cuban rice varieties using isozyme, RAPD and AFLP markers. *Euphytica* **1999**, *109*, 107–115. [CrossRef]
14. Rongwen, J.; Akakya, M.S.; Bhagwat, A.A.; Lavi, U.; Cregan, P.B. The use of microsatellite DNA markers for soybean genotype identification. *Theor. Appl. Genet.* **1995**, *90*, 43–48. [CrossRef]
15. Sebastian, L.S.; Hipolito, L.R.; Tabanao, D.A.; Maramara, G.V.; Caldo, R.A. Molecular diversity of Philippines-based rice cultivars. *SABRAO J. Breed. Genet.* **1998**, *30*, 83–90.
16. Chen, X.; Temnykh, S.; Xu, Y.; Choand, Y.G.; McCouch, S.R. Development of a microsatellite framework map providing genome wide coverage in rice (*Oryza sativa* L.). *Theor. Appl. Genet.* **1997**, *95*, 553–567. [CrossRef]
17. Keurentjes, J.J.; Bentsink, L.; Alonso-Blanco, C.; Hanhart, C.J.; Blankestijn-De Vries, H.; Effgen, S.; Vreugdenhil, D.; Koornneef, M. Development of a Near-Isogenic Line Population of Arabidopsis Thaliana and Comparison of Mapping Power with a Recombinant Inbred Line Population. *Genetics* **2007**, *175*, 891–905. [CrossRef] [PubMed]
18. Andaya, V.C.; Mackill, D.J. Mapping of QTLs associated with cold tolerance during the vegetative stage in rice. *J. Exp. Bot.* **2003**, *54*, 2579–2585. [CrossRef] [PubMed]
19. Lou, Q.; Chen, L.; Sun, Z.; Xing, Y.; Li, J.; Xu, X.; Mei, H.; Luo, L. A major QTL associated with cold tolerance at seedling stage in rice (*Oryza sativa* L.). *Euphytica* **2007**, *158*, 87–94. [CrossRef]
20. Zhi, J.I.; Zeng, Y.X.; Zeng, D.I.; Ma, L.Y.; Li, X.M.; Liu, B.X.; Yang, C.D. Identification of QTLs for rice cold tolerance at plumule and 3-leaf-seedling stages by using QTL Network software. *Rice Sci.* **2010**, *17*, 282–287.
21. Suh, J.; Lee, C.; Lee, J.; Kim, J.; Kim, S.; Cho, Y.; Park, S.; Shin, J.; Kim, Y.; Jena, K.K. Identification of quantitative trait loci for seedling cold tolerance using RILs derived from a cross between *Japonica* and tropical *Japonica* rice cultivars. *Euphytica* **2012**, *184*, 101–108. [CrossRef]

22. Su, C.F.; Wang, Y.C.; Hsieh, T.H.; Lu, C.A.; Tseng, T.H.; Yu, S.M. A novel MYBS3-dependent pathway confers cold tolerance in rice. *Plant Physiol.* **2010**, *153*, 145–158. [CrossRef]

23. Yang, Z.; Huang, D.; Tang, W.; Zheng, Y.; Liang, K. Mapping of quantitative trait loci underlying cold tolerance in rice seedlings via high-throughput sequencing of pooled extremes. *PLoS ONE* **2013**, *8*, e68433. [CrossRef]

24. Park, I.K.; Oh, C.S.; Kim, D.M.; Yeo, S.M.; Ahn, S.N. QTL mapping for cold tolerance at the seedling stage using introgression lines derived from an inter sub-specific cross in rice. *Plant Breed. Biotech.* **2013**, *1*, 18. [CrossRef]

25. Zhang, S.; Zheng, J.; Liu, B.; Peng, S.; Leung, H.; Zhao, J.; Huang, Z. Identification of QTLs for cold tolerance at seedling stage in rice (*Oryza sativa* L.) using two distinct methods of cold treatment. *Euphytica* **2014**, *195*, 95–104. [CrossRef]

26. Biswas, P.S.; Khatun, H.; Das, N.; Sarker, M.M.; Anisuzzaman, M. Mapping and validation of QTLs for cold tolerance at seedling stage in rice from an indica cultivar Habiganj Boro VI (Hbj. B. VI). *3 Biotech.* **2017**, *7*, 359. [CrossRef]

27. Andaya, V.; Tai, T. Fine mapping of the *qCTS12* locus, a major QTL for seedling cold tolerance in rice. *Theor. Appl. Genet.* **2003**, *113*, 467–475. [CrossRef] [PubMed]

28. Glaszmann, J.C.; Kaw, R.N.; Khush, G.S. Genetic divergence among cold tolerant rices (*Oryza sativa* L.). *Euphytica* **1990**, *5*, 95–104. [CrossRef]

29. Pan, Y.; Zhang, H.; Zhang, D.; Li, J.; Xiong, H.; Yu, J.; Li, J.; Rashid, M.A.R.; Li, G.; Ma, X.; et al. Genetic analysis of cold tolerance at the germination and booting stages in rice by association mapping. *PLoS ONE* **2015**, *10*, e0120590. [CrossRef] [PubMed]

30. Lv, Y.; Guo, Z.; Li, X.; Ye, H.; Xiong, L. New insights into the genetic basis of natural chilling and cold shock tolerance in rice by genome-wide association analysis. *Plant Cell Environ.* **2016**, *39*, 556–570. [CrossRef]

31. Kabir, M.S.; Salam, M.U.; Chowdhury, A.; Rahman, N.M.F.; Iftekharuddaula, K.M.; Rahman, M.S.; Rashid, M.H.; Dipti, S.S.; Islam, A.; Latif, M.A.; et al. Rice Vision for Bangladesh, 2050 and Beyond. *Bangladesh Rice J.* **2015**, *19*, 1–18. [CrossRef]

32. Khatun, H.; Biswas, P.S.; Hwang, H.G.; Kim, K.M. A Quick and Simple In-house Screening Protocol forCold-Tolerance at Seedling Stage in Rice. *Plant Breed. Biotech.* **2016**, *4*, 373–378. [CrossRef]

33. Federer, W.T. Augmented Design. *Hawaiin Plant. Rec.* **1956**, *40*, 191–207. [CrossRef]

34. BRRI (Bangladesh Rice Research Institute). *Annual Research Review*; Bangladesh Rice Research Institute: Gazipur, Bangladesh, 2013.

35. IRRI (International Rice Research Institute). *Minimum List of Descriptors and Descriptor-States for Rice* (Oryza sativa L.); IRRI: Los Baños, Philippines, 1980; p. 20.

36. Federer, W.T.; Raghavarao, D. An augmented designs. *Biometrics* **1975**, *31*, 29–35. [CrossRef]

37. Panse, V.G.; Sukhatme, P.V. *Statistical Methods for Agricultural Workers*, 2nd ed.; ICAR: New Delhi, India, 1964.

38. Sharma, J.R. *Augmented Design Model-II in Statistical and Biometrical Techniques in Plant Breeding*; New Age International Publications: New Delhi, India, 1998.

39. Syed, M.A.; Iftekharuddaula, K.M.; Mian, M.A.K.; Rasul, M.G.; Rahmam, G.K.M.M.; Panaullah, G.M.; Biswas, P.S. Main effect QTLs associated with arsenic phyto-toxicity tolerance at seedling stage in rice (*Oryza sativa* L.). *Euphytica* **2016**, *209*, 805–814. [CrossRef]

40. Syed, M.A.; Iftekharuddaula, K.M.; Biswas, P.S.; Akter, N.; Hossain, M. Assessment of genetic diversity in arsenic contaminated rice using SSR markers. *Trends Appl. Sci. Res.* **2019**, *14*, 178–185.

41. Liu, K.; Muse, S.V. Power Marker, an integrated analysis environment for genetic marker analysis. *Bioin* **2005**, *21*, 2128–2129. [CrossRef]

42. Anderson, J.A.; Churchill, G.A.; Autique, J.E.; Tanksley, S.D.; Sorreils, M.E. Optimizing parental selection for genetic-linkage maps. *Genome* **1993**, *36*, 181–186. [CrossRef]

43. Nei, M. The theory and estimation of genetic distance. In *Genetic Structure of Populations*; Morton, N.E., Ed.; University of Hawaii Press: Honolulu, HI, USA, 1973; pp. 45–54.

44. Shannon, C.E. A mathematical theory of communication. *Bell Syst. Tech. J.* **1948**, *27*, 379–423. [CrossRef]

45. Rohlf, F.J. NTSYS-pc. In *Numerical Taxonomy and Multivariate Analysis System*; Exeter Software: Setauket, NY, USA, 2002.

46. Botstein, D.; White, R.L.; Skalnick, M.H.; Davies, R.W. Construction of a genetic linkage map in man using restriction fragment length polymorphism. *Am. J. Hum. Genet.* **1980**, *32*, 314–331.

47. Ye, C.; Fukai, S.; Godwin, D.I.; Reinke, R.; Snell, P.; Schiller, J.; Basnayake, J. Cold tolerance in rice varieties at different growth stages. *Crop Pasture Sci.* **2009**, *60*, 328–338. [CrossRef]

48. International Rice Research Institute (IRRI). *Standard Evaluation System (SES) for Rice*, 5th ed.; IRRI: Manila, Philippines, 2013; p. 35.

49. Suh, J.-P.; Cho, Y.-C.; Lee, J.-H.; Lee, S.-B.; Jung, J.-Y.; Choi, I.-S.; Kim, M.-K.; Kim, C.-K.; Jena, K.-K. SSR Analysis of Genetic Diversity and Cold Tolerance in Temperate Rice Germplasm. *Plant Breed. Biotech.* **2013**, *106*, 103–110. [CrossRef]

50. Kundu, A. Haplotype Diversity Analysis in Cold Tolerant Rice (*Oryza sativa* L.). Master's Thesis, Department of Biotechnology, Bangabandhu Sheikh MujiburRahman Agricultural University, Gazipur, Bangladesh, 2015.

51. Shoroardi, M.; Mortuza, M.G.; Islam, M.M.; Samih, T. Phenotypic Screening and Molecular Characterization of 10 Rice (*Oryza sativa*) Landraces for Cold Tolerance. *J. Environ. Sci. Nat. Resour.* **2017**, *10*, 85–91. [CrossRef]

52. Kim, S.M.; Suh, J.P.; Lee, C.K.; Lee, J.H.; Kim, Y.G.; Jena, K.K. QTL mapping and development of candidate gene-derived DNA markers associated with seedling cold tolerance in rice (*Oryza sativa* L.). *Mol. Genet. Genom.* **2014**, *289*, 333–343. [CrossRef] [PubMed]

53. Setter, T.L.; Laureles, E.V.; Mazaredo, A.M. Lodging reduces yield of rice by self-shading and reductions in canopy photosynthesis. *Field Crops Res.* **1997**, *49*, 95–106. [CrossRef]
54. Zhang, Y.; Yu, C.; Lin, J.; Liu, J.; Liu, B.; Wang, J.; Zhao, T. *OsMPH1* regulates plant height and improves grain yield in rice. *PLoS ONE* **2017**, *12*, e0180825. [CrossRef]
55. Upadhyay, P.; Neeraja, C.N.; Kole, C.; Singh, V.K. Population structure and genetic diversity in popular rice varieties of India as evidenced from SSR analysis. *Biochem. Genet.* **2012**, *50*, 770–783. [CrossRef]
56. Kostova, A.; Todorovska, E.; Christov, N.; Hristov, K.; Atanassov, A. Assessment of genetic variability induced by chemical mutagenesis in elite maize germplasm via SSR markers. *J. Crop Improv.* **2006**, *16*, 37–48. [CrossRef]
57. Pachauri, V.; Taneja, N.; Vikram, P.; Singh, N.K.; Singh, S. Molecular and morphological characterization of Indian farmers rice varieties (*Oryza sativa* L.). *Aust. J. Crop Sci.* **2013**, *7*, 923–932.
58. Syed, M.A.; Iftekharuddaula, K.M.; Akter, N.; Biswas, P.S. Molecular diversity analysis of some selected BBRI released rice varieties using SSR markers. *Int. Res. J. Biol. Sci.* **2019**, *1*, 51–58.

sustainability

Article

The Effect of Temperature and Water Stresses on Seed Germination and Seedling Growth of Wheat (*Triticum aestivum* L.)

Hussein Khaeim, Zoltán Kende *, István Balla, Csaba Gyuricza, Adnan Eser and Ákos Tarnawa

Agronomy Institute, Hungarian University of Agriculture and Life Sciences, 2100 Godollo, Hungary; hussein.khaeim@qu.edu.iq (H.K.); balla.istvan@uni-mate.hu (I.B.); gyuricza.csaba@uni-mate.hu (C.G.); adnaneser@hotmail.com (A.E.); tarnawa.akos@uni-mate.hu (Á.T.)
* Correspondence: kende.zoltan@uni-mate.hu

Abstract: Temperature and moisture are essential factors in germination and seedling growth. The purpose of this research was to assess the germination and growth of wheat (*Triticum aestivum* L.) seeds under various abiotic stressors. It was conducted in the Agronomy Institute of the Hungarian University of Agriculture and Life Sciences, Gödöllő, Hungary. Six distinct temperature levels were used: 5, 10, 15, 20, 25, and 30 °C. Stresses of drought and waterlogging were quantified using 25 water levels based on single-milliliter intervals and as a percentage based on thousand kernel weight (TKW). Seedling density was also tested. Temperature significantly influenced germination duration and seedling development. 20 °C was ideal with optimal range of 15 °C to less than 25 °C. Germination occurred at water amount of 75% of the TKW, and its ideal range was lower and narrower than the range for seedling development. Seed size provided an objective basis for defining germination water requirements. The current study established an optimal water supply range for wheat seedling growth of 525–825 percent of the TKW. Fifteen seeds within a 9 cm Petri dish may be preferred to denser populations.

Keywords: *Triticum aestivum* L.; wheat; seed germination; seedling development; germination time

Citation: Khaeim, H.; Kende, Z.; Balla, I.; Gyuricza, C.; Eser, A.; Tarnawa, Á. The Effect of Temperature and Water Stresses on Seed Germination and Seedling Growth of Wheat (*Triticum aestivum* L.). *Sustainability* **2022**, *14*, 3887. https://doi.org/10.3390/su14073887

Academic Editor: Balázs Varga

Received: 17 February 2022
Accepted: 23 March 2022
Published: 25 March 2022

Publisher's Note: MDPI stays neutral with regard to jurisdictional claims in published maps and institutional affiliations.

1. Introduction

Wheat is the predominant grain crop and a mainstay source of food for most of the world's population [1]. This field crop is nutritionally essential globally, ranking second only after maize production [2]. It is involved in many bakery and pastry industries [3]. It is commonly utilized to produce flour, malt, and various food products, including bread, pasta, and morning cereals [2]. Wheat is a primary source of protein and starch, providing 20–30% of daily caloric intake in most societies [4]. Because it is widespread many different ecosystems, plants face various abiotic challenges, such as drought and rising temperature, due to global warming, which results in huge yield loss [5,6].

Germination is a physiological mechanism in which sequential biological and biochemical activities occur to initiate and develop a seedling [7,8]. It is a critical stage in the life of wheat plants, as it determines seedling establishment and the effective use of available nutrients and water resources [9,10]. It is regulated by the interaction of the surrounding environmental conditions, the seeds' physiological state, and the germ [11]. Temperature, light, pH, water availability, and soil moisture most affect seed germination among abiotic factors [12]. Seeds have a specific range of environmental requirements necessary for optimal germination [13]. The success of propagation depends on the seeds' physiological response towards overlapping various external abiotic factors. Therefore, seed germination reflects inhabitants' size, abundance, and distribution [14–16].

Seed germination begins with water adsorption by the dormant dry seed and is accomplished by the commencement of radicles by lengthening the embryo's axis [17].

A sequence of ordered morphogenetic and physiological mechanisms, including energy transfer, nutrient intake, and physiological and biochemical changes, are part of this process [14,16,18]. A seed's water adsorption is a three-stage procedure, with phase I being a fast initial uptake, phase II being a plateau, and phase III being a rise in water adsorption, but only with the initiation of germination [19]. Germination phase I, called imbibition, the rapid water take-up, results in softening and swelling the seed coat close to or at an ideal temperature [20]. Seed coat rupturing permits the emergence of radicle and shoot. The seed's physiology is then activated, and it begins to respire [18,19]. Primary germination indicators include re-establishing critical processes, including translation and DNA repairing, followed by cell expansion and division [16,20]. Physically germination has a two-sequent-stages mechanism: endosperm rupture, and the evolving radicle follows the breakdown of the micropylar endosperm [14,21,22]. Enzymatic activity to degrade carbohydrate and lipid for reserved materials mobilization is essential in germination and early seedling development [23–25]. The biological, bio-chemical enzymic activity in these stages is highly affected by temperature and water availability [26].

Seed germination necessitates the presence of water. It hydrates the crucial processes of protoplasm, supplies dissolved oxygen, softens the seed's outer layer, and improves permeability [18,26]. Water stress harms germination percentage [5,27,28]. The water level has more complicated germination effects for wheat since it is the primary determinant for seed imbibition and germination. Water contributes to the subsequent germination metabolic phases [25,29]. It is vital for seed enzyme activation, breakdown, translocation, and endospermic stored materials [29,30]. Water stress declines enzymic activities that cause adverse effects on the metabolic process of carbohydrates, decrease the water potential and the amount of soluble potassium and calcium, and cause changes in seeds hormones [31,32].

The temperature has a major impact on specifying the germination time duration [32]. The temperature effect on germination can be described as a substantial temperature scale: minimal, optimal, and maximum temperatures to initiate [33]. The optimum temperature results in the most significant germination rate (%) in the shortest time. Each germination stage has its substantial temperature; due to the intricacy of the germination process, the temperature response may vary throughout the germination phases. The seeds' response to temperature relies on variety, seed quality, time from harvest, and some other factors [34]. Many studies regarding the regulation of germination by temperature have been conducted. Riley (1981) demonstrated that the energy state of cells and the behavior of certain enzymes change as the temperature changes [35]. The ATP content and protein synthesis rate increase as the temperature reaches the optimum and decreases as it goes either way [19,35]. The intra-equilibrium of reactive oxygen species indispensable for seed germination is hindered by abnormal respiration [12,35,36]. The antioxidant system highly affected wheat seed germination [37,38].

Drought, thermal, and drought + thermal stresses significantly impact the yield and quality more than each separately. For example, wheat yield losses in literature were 57% after drought stress, 76% after drought + thermal stresses, and only 31% after thermal stresses [39].

This study was conducted to investigate the performance of wheat seed germination ability under different moisture conditions, temperatures, and seed and seedling densities per Petri dish. A fundamental understanding of germination is vital to the field of crop production. The objectives of this study were: the first was (1) to determine the optimal range of water amount for germination based on water volume of single-milliliter intervals and water quantity as percentages based on thousand kernels weight (TKW). This method was used to investigate the null hypothesis "The difference in seed weight and size has no role in germination when the amount of water is constant". Therefore, part of this pre-experiment is to prove the alternative hypothesis' validity, namely that "the water requirements of the seeds differ depending on their size and weight". Theoretically, larger seeds need more water to germinate compared to smaller seeds. Therefore, if the alternative hypothesis is correct, it will be a more accurate method to find the optimal needed amount of water for germination, and the second was (2) to determine the effects of germination temperatures and germination time of wheat seeds. The third objective was (3) to investigate the effect number of seeds and seedling density in a Petri dish on germination percentage and open the Petri dish's top cover. Opened Petri dish has various side effects of moisture losses and exposure to contamination. This part was designed to answer the question of "Do the number of seeds and the density of seedlings affect the germination percentage and viability of seedlings under the same water quantity?". This research was conducted following the standard of the NO. 48 of 2004 (IV.21.) of the Ministry of Agriculture and Rural Development concerning the production and marketing of seed of agriculture crop species, Hungary, which following the International Seed Test Association (ISTA), the circumstances was changed according to our research questions.

This study yields critical information on germination demands that can evaluate tolerance to ranges of drought and temperature stresses.

2. Materials and Methods

The present study was conducted to determine the temperature and drought effect on germination vigor and seedling growth. The seeds of a regionally commonly grown variety of Hungarian wheat named Alfold 90 were obtained from a grower and used. It is a high winter tolerance wheat variety with 75–95 cm plant height, with awed spikes. Alfold 90 is sensitive to powdery mildew disease and total tolerance to stem rust disease. It is an early Hungarian variety and cultivates in dry regions. It was conducted in the Agronomy Institute/the Hungarian University of Agriculture and Life Sciences, 2100 Gödöllő, Hungary. Highly sensitive and highly accurate incubators made by Memmert, Schwabach, Germany, with natural convection or forced air circulation and double doors (interior glass, exterior stainless steel) for a clear view without a drop in temperature were used. The germination capacity can be 100%, and it declines for several factors, i.e., storage time and conditions, but it was over 98% for the used stock. These seed characterizations are: TKW is 39–44 g, test weight 78–82 kg, protein content 14–17%, wet gluten content 34–40%, and the farinograph value is 80–90%, which fall into the A2 category. Alfold 90 is described as a drought-tolerant wheat variety. The study was subdivided into three experiments conducted in the following subsections.

2.1. Temperature Experiment

This study compares wheat seed germination under six distinct temperatures, 5, 10, 15, 20, 25, and 30 °C. After Petri dishes were labeled, 20 seeds of wheat were placed in each and exposed to an identical amount of distilled water, 5 mL. The electrical conductivity of distilled water was measured 1.5 mhos/cm. Every day, four Petri dishes of each temperature level were subjected to physical measurements. The measured parameters were the number of not-germinated seeds and radicles and the first proper leaf lengths, termed a shoot in this research. The germination stander measurement is that 80% of the seedlings in Petri dishes reach the length of 1 cm.

2.2. Water Quantity Experiment

Wheat seeds were treated to a total of 25 distilled water levels, 12 levels based on milliliter intervals, and 13 levels based on TKW in sterile Petri dishes of 9 cm diameter. The Petri dishes were lined with sterile filter paper, Table 1. TKW was determined using a seed counter machine, which was 42.76 g.

$$TKW \times Seed\ n/100{,}000 = 1\%\ of\ the\ propose\ quantity\ of\ water \qquad (1)$$

$$42.76 \times 20 = 855.2,\ 855.2/100{,}000 = 0.008552$$

Table 1. Treatments of water levels experiment based on the two water basses, a single-milliliter interval and TKW.

Water Quantity Based on a Single-Milliliter Interval		Water Quantity Based on the TKW %			
[1] Treatment Number	[2] Water Quantity mL	[1] Treatment Number	[3] Proposed % of Water Quantity	[4] Quantity of Water mL	[5] Rounded Quantity of Water mL
1	0	14	75%	0.641	0.65
2	1	15	150%	1.282	1.30
3	2	16	225%	1.924	1.90
4	3	17	300%	2.565	2.55
5	4	18	375%	3.207	3.20
6	5	19	450%	3.834	3.85
7	6	20	525%	4.489	4.45
8	7	21	600%	5.131	5.15
9	8	22	675%	5.772	5.75
10	9	23	750%	6.414	6.40
11	10	24	825%	7.055	7.00
12	11	25	900%	7.696	7.50
13	12				

[1] The treatments number that total of 25 treatments, on the two water quantity bases, [2] the applied water quantity based on single-milliliter intervals, [3] the suggested percentage for water quantity application in ml regarding the TKW in (g), [4] the quantity of water in milliliter corresponds to the suggested percentage of water quantity to TKW, and [5] the rounded quantity of water to the nearest measurable digit on the pipette.

Table 1's outcome represents 1% water quantity of TKW [40]. This number was multiplied by the proposed percentage for each water treatment, as presented in Table 1. Twenty seeds were placed in each labeled Petri dish with five replications. This experiment was implied under a constant incubation chamber temperature level of 20 °C. All seedlings' radicle and shoot lengths were physically measured after ten days. After labeling, these radicles and shoots for each Petri dish were then exposed to 65 °C for 48 h in an oven. Dry weights were acquired by weighing the radicles and the shoots of the 20 seedlings of each treatment replicate.

2.3. Seed Density Experiment

This part was designed to study the seed's number effect on the germination performance under the same quantity of water. Three sets of seeds, 15, 20, and 25 per P.D, were placed in Petri dishes. First, they were treated with 6 mL of distilled water. Ten replications were implied to reduce the experimental error of the experiment and boost the accuracy. Then, they were incubated at a constant temperature of 20 °C in a controlled germination chamber. The measured traits were categorized into four groups: number of seeds that were not germinated, number of germinated seeds with the only radicle, seedling with short shoots (less than 4 cm), and normal seedling. The normal seedling term refers to the morphological stage compared to the other in the same condition: the seedling with taller shoots than 4 cm. These categories were made to achieve an aggregated value according to Equation (2) [40]:

$$\text{Aggregated value} = (\text{NOT-G} \times 0) + (\text{RO} \times 0.33) + (\text{SP} \times 0.67) + (\text{NP} \times 1) \tag{2}$$

where: NOT-G = not-germinated seeds, RO = the number of germinated seeds with only a radicle, S.P. = the number of germinated seeds with short shoots, and NP = the number of germinated seeds with the normal shoot.

2.4. Statistical Analysis

The statistics reported were presented as a mean value for each treatment of the experiment. For the water amount experiment, analysis of variance (ANOVA) and Fisher's test of least significant differences was utilized to determine significant differences at a 5% probability level (GenStat 12th edition, PL20.1 m, the United Kingdom). In addition, To fit the obtained data and depict the best fit of the temperature levels, a sigmoid curve model was made (JMP Proc 13.2.1 of SAS, Canberra, the United States, and Microsoft Excel 365). Normality check was conducted using Kolmogorov–Smirnov and Shapiro–Wilk by SPSS version 27, IBM, New York, NY, USA.

3. Results

Based on the statistically computed data (Table 2), as the Kolmogorov–Smirnov and Shapiro–Wilk are greater than 0.05, it can be stated that the normality curve is symmetric [41,42], as presented in the histogram (Figure 1).

Table 2. Data normality is determined by Kolmogorov–Smirnov and Shapiro–Wilk tests.

	[a] Kolmogorov—Smirnov			Shapiro—Wilk		
	Statistic	df	Sig.	Statistic	df	Sig.
Radicle	0.082	55	0.200 *	0.950	55	0.240
Shoot	0.090	55	0.200 *	0.973	55	0.262
Seedling	0.111	55	0.091	0.923	55	0.092

* The lower limit of the true significance; [a] Lilliefors Significance Correction; df is the degree of freedom; and Sig. is the significance.

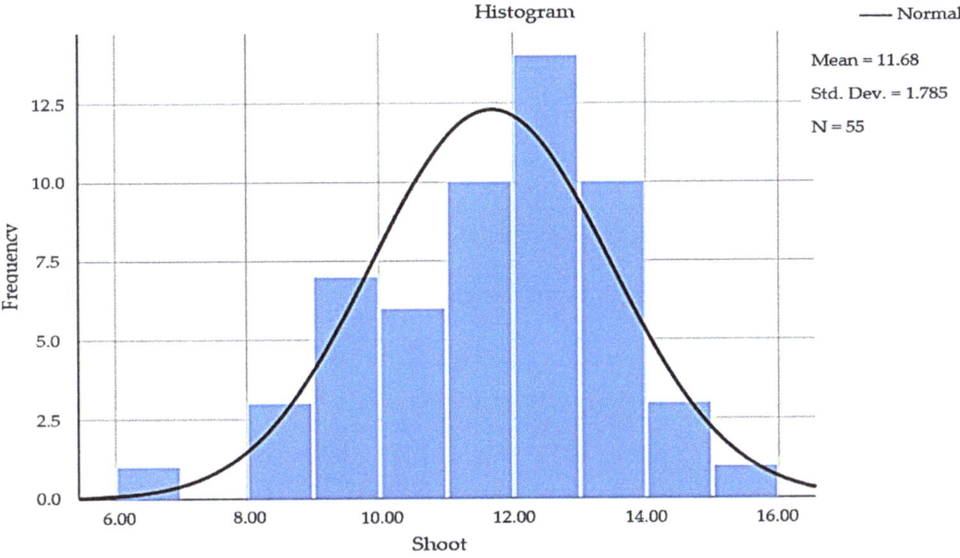

Figure 1. The distribution of the data.

3.1. Temperature Experiment

3.1.1. Germination Duration

As determined in the temperature gradient experiment, 30 °C and 25 °C, which rapidly initiated germination, were faster to reach the standard measurement point in this study of 1 cm shoot length in 48 h, Figure 2. Under 20 °C, that took longer for its germinated shoots to reach 1 cm in the temperature gradient experiment, three days after subjecting the seed to the treatment. Figure 2 illustrates that wheat seeds germinated under a constant temperature of 15 °C, which needed five days to reach the shoot measurement stander, followed by a temperature of 10 °C that needed a more prolonged duration up to 8 days in the temperature gradient experiment. Seed germination under 5 °C took the most prolonged duration to initiate germination. Germinated shoot reached the length of 1 cm after 15 days of the planting date (Figure 2). The percentage differences within the germination range between the two limits were 152.941%. Regardless of 5 °C and 10 °C, the temperatures of 20 °C, 25 °C, and 30 °C have minor margin differences in germination duration, followed by 15 °C, which took slightly longer.

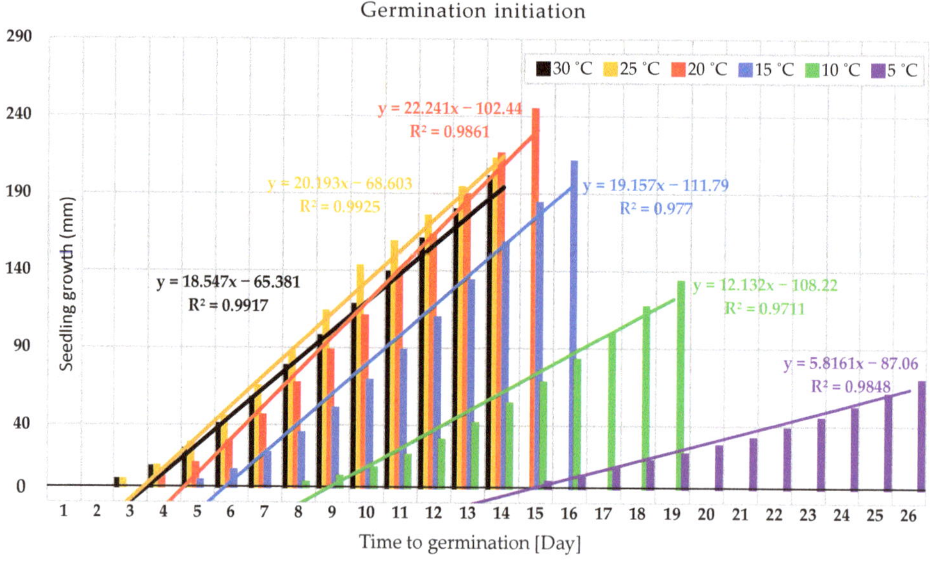

Figure 2. Germination duration from planting (day 1) to germination initiation and seedling development under different temperatures.

Temperature is critical in determining the duration of germinating [43–45]. The findings indicate that seeds accumulate the heat units from the thermal energy they absorbed, and once they reach the necessary level to initiate metabolic activity, germination begins at a pace dependent on the ambient temperature condition.

3.1.2. Seedling Development

The experiment of the temperature gradient of 6 distinct constant temperatures, 5 °C, 10 °C, 15 °C, 20 °C, 25 °C, and 30 °C, presented that seedling growth under 20 °C gave the most outstanding performance and the most salient rate of growth (Figure 3). It was the most favorable growth temperature for seedling growth with a y-value: $22.241x - 102.44$ (Figure 2). Similarly, with slightly lower development performance under 15 °C, y-value: $19.157x - 111.79$. The temperature of 10 °C shows the same pattern, but with much slower development, y-value: $12.132x - 108.22$. The experiment under 5 °C showed the most negligible seedling growth and required a longer time for development, y-value: $5.8161x - 87.06$ (Figure 2). Slightly identical performance was shown under 25 °C and a higher temperature level when they grow quicker in the very early stage of seedling development, followed by slowing the rate of the development (Figure 3). It indicates that each stage of wheat seedling development needs a different temperature. This figure illustrates that the upper germination range of 25 °C and 30 °C gives quicker germination and early seedling growth within less than five days, but not for the subsequent seedling growth phase. Thus, the optimum range for seedling growth is around 20 °C with a greater growth rate followed by 15 °C (Figure 3).

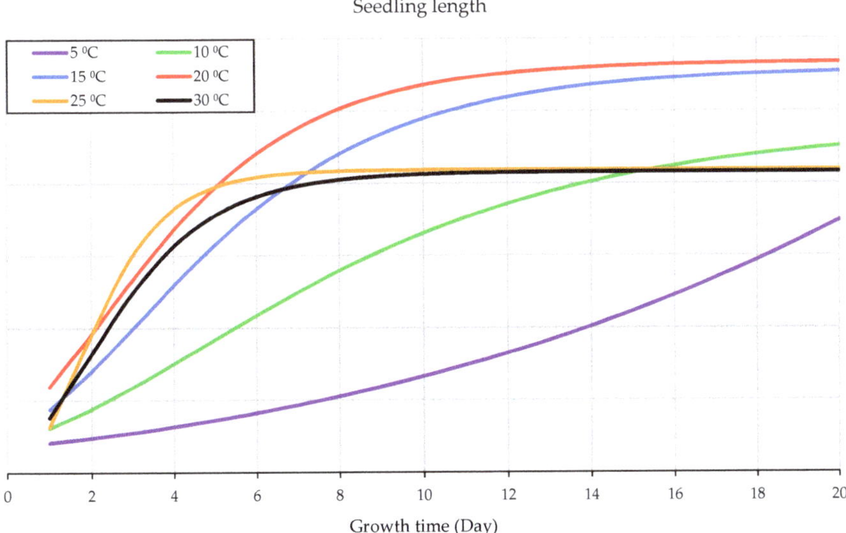

Figure 3. Patterns of seedlings development in response to temperature-induced.

The seedling's shoots and radicles showed the same development performance at and close to the optimal range of temperature; however, their growth pattern differs when the indicator arrow of the temperature crosses the optimum limits of growth range tails (Figures 4 and 5). The shoot formed and developed better, especially in the very early growth stage, than the radicle under a temperature scope higher than the optimum range (Figure 4). However, the radicle has a distinct growth pattern and need for temperature compared to the shoot because it grows better at temperatures lower than the optimum range, particularly in the later stage than the shoot (Figure 5).

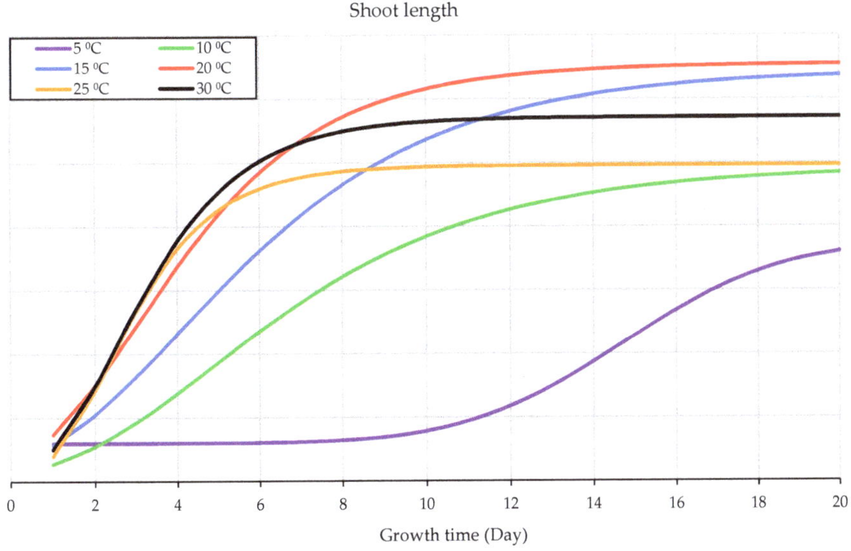

Figure 4. Patterns of shoot development in response to temperature-induced.

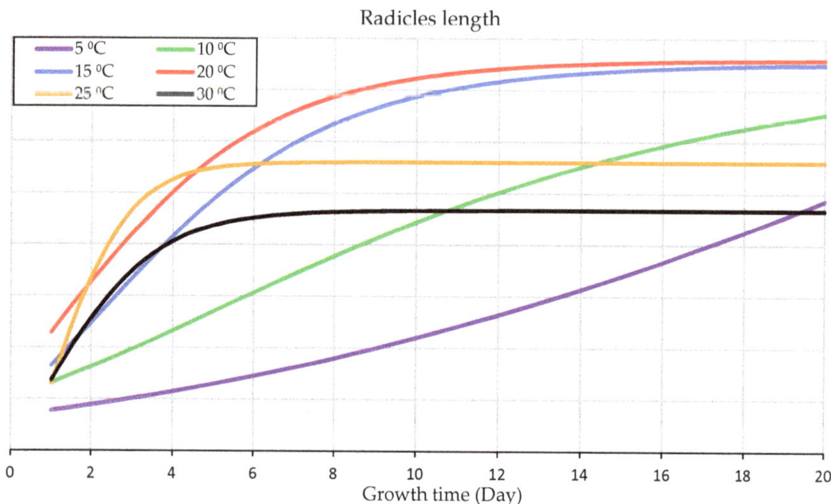

Figure 5. Patterns of radicles development in response to temperature-induced.

The highest growth and development of the shoots occurred at a temperature of 20 °C, and the least occurred at 5 °C. Therefore, the optimal temperature for the shoot and radicle development is at 20 °C, followed by 15 °C. When the temperature changes, either the shoots or the radicles will be more impacted than the other. The radicle is more resistant to cold than the shoot, and vice versa; the shoot is more resistant to temperatures over the ideal threshold than the radicle.

3.2. Water Quantity Experiment

Wheat seed quality parameters such as stress tolerance, uniformity, and germination rate can be determined using water absorption and temperature analysis. Availability of moisture reflects severe limitations on seed germination of *Triticum astivum* L. [39,40]. Therefore, seeds of the crop with stringent requirements for germination can be more effectively felicitously established than crop seeds with fewer constraints [46].

The water quantity experiment was carried out on two bases: single-milliliter intervals 0–12 mL and as percentages of TKW to develop a technique for comparing various wheat varieties with varied TKW in a future experiment. Furthermore, the two bases of water quantity application were compared and investigated simultaneously in parallel, which has a better reflectance representing water demands. Table 3 reveals significant variations among the water quantity potentials 0–12 mL for all the examined parameters: the number of non-germinated seeds, length of the radicle, length of shoot, seedling length in total, dry weight of the seedlings, seedlings' corrected dry weight that was accomplished by subtracting the ungerminated seeds, dry weight of the radicles, and dry weight of shoots.

Germination percentage increased significantly as water quantity increased up until the potential water quantity, followed by a slight decrease as the water quantity level raised due to waterlogging, as presented in Table 3, Figures 6 and A1. A similar pattern was seen when water was applied according to the TKW with statistically significant differences, Table 4 and Figure 7. Seeds of the wheat crop can be germinated under a minimal water quantity, 0.65 mL, Table 4, which represents 75% of TKW, Table 1. As water is essential for germination, seeds can germinate moistures near the wilting threshold point to activate and stimulate germination metabolic processes. The wheat seeds require internal 40% moisture to germinate [46]. If the interior moisture content were less than the critical moisture content limit, wheat seeds would not germinate. Under 0.65 mL, over half of the seeds germinated. Thus, water application based on the TKW enables better knowledge of the limitations and optimal water requirements since seed size is critical for attaining the 40% internal seed moisture level. These findings corroborate research conducted by [47], stating that "seed size has importance in predicting germination under stress conditions". The optimal range for germination is 4.45–7.00 mL, Figure 7, representing 525–825% of the TKW.

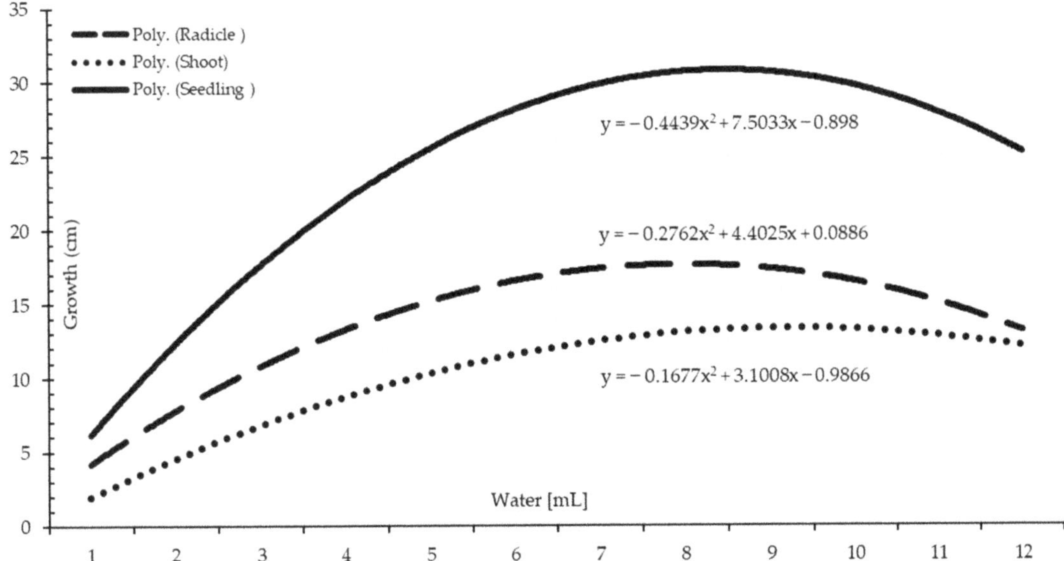

Figure 6. Seedling growth in response to water amount based on 1 mL intervals.

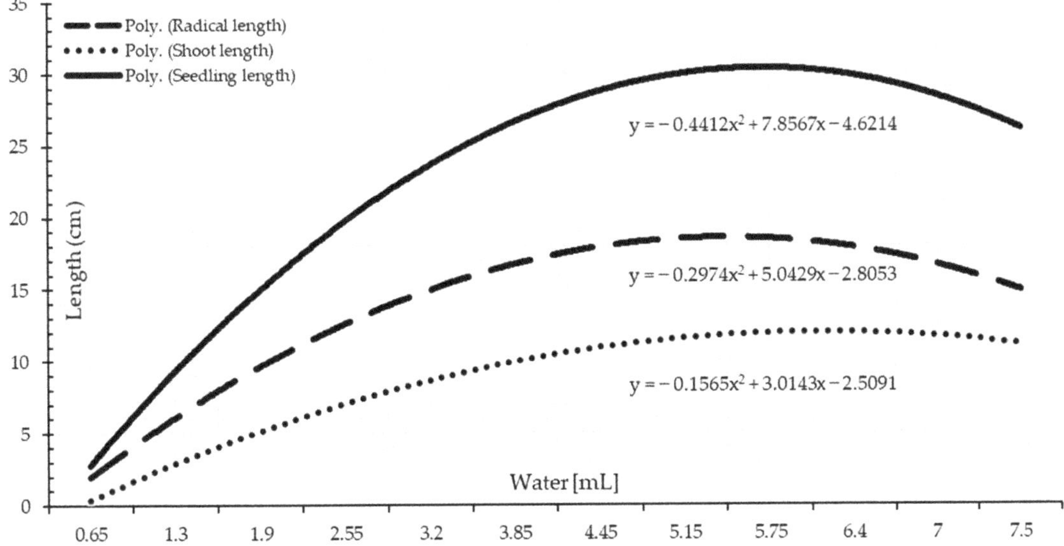

Figure 7. Seedling growth in response to the water amount based on TKW.

Table 3. Parameters relating to germination and seedling characteristics of *Triticum aestivum* L. seeds respond to the application base's water potential of single-milliliter water quantity intervals.

[1] Water mL.	[2] Not Germinated seeds	[3] Radicles (cm)	[4] Shoots (cm)	[5] Seedlings (cm)	[6] Radicles Dry W (g)	[7] Shoots Dry W (g)	[8] Seedlings Dry W (g)	[9] Corrected Dry W (g)
0	20.0 ± 0.00 a	0.000 ± 0.00 f	0.000 ± 0.00 h	0.000 ± 0.00 f	0.000 ± 0.00 e	0.000 ± 0.00 e	0.000 ± 0.00 d	0.000 ± 0.00
1	0.40 ± 0.55 bc	10.11 ± 1.16 e	5.020 ± 1.43 g	15.13± 2.56 e	0.106 ± 0.02 d	0.112 ± 0.02 d	0.218 ± 0.05 c	0.221 = 0.04 d
2	0.60 ± 0.55 bc	13.35 ± 1.44 d	8.950 ± 0.49 f	22.30 ± 1.14 d	0.123 ± 0.02 abc	0.123 ± 0.01 d	0.246 ± 0.03 bc	0.253 = 0.03 c
3	0.20 ± 0.45 c	15.72 ± 1.74 abc	9.360 ± 1.96 ef	25.08 ± 3.04 c	0.133 ± 0.01 ab	0.144 ± 0.01 c	0.277 ± 0.01 ab	0.280 ± 0.02 abc
4	0.60 ± 0.89 bc	15.18 ± 0.94 bc	11.16 ± 0.55 cd	26.34 ± 1.08 bc	0.125 ± 0.01 abc	0.155 ± 0.01 bc	0.280 ± 0.02 a	0.288 ± 0.02 ab
5	0.40 ± 0.55 bc	16.26 ± 0.98 ab	10.74 ± 0.97 de	27.00 ± 1.23 abc	0.137 ± 0.01 a	0.152 ± 0.01 bc	0.289 ± 0.01 a	0.295 ± 0.01 ab
6	0.20 ± 0.45 c	16.22 ± 1.18 ab	11.48 ± 0.92 bcd	27.70 ± 0.92 ab	0.135 ± 0.01 ab	0.157 ± 0.02 abc	0.292 ± 0.08 a	0.295 ± 0.03 ab
7	0.20 ± 0.45 c	15.85 ± 1.27 ab	12.80 ± 1.17 ab	28.65 ± 1.24 ab	0.125 ± 0.02 abc	0.167 ± 0.01 ab	0.291 ± 0.01 a	0.294 ± 0.01 ab
8	0.00 ± 0.00 c	15.41 ± 1.33 abc	12.89 ± 0.83 ab	28.30 ± 2.06 ab	0.122 ± 0.02 abc	0.170 ± 0.01 ab	0.292 ± 0.02 a	0.292 ± 0.02 ab
9	0.20 ± 0.45 c	16.80 ± 0.69 a	12.60 ± 0.46 abc	29.40 ± 0.82 a	0.123 ± 0.00 abc	0.169 ± 0.02 ab	0.292 ± 0.02 a	0.295 ± 0.01 ab
10	0.60 ± 0.89 c	15.77 ± 0.92 abc	13.31± 0.88 a	29.08 ± 1.69 a	0.120 ± 0.01 bcd	0.176 ± 0.02 a	0.295 ± 0.03 a	0.304 = 0.01 a
11	0.80 ± 0.84 bc	14.23 ± 2.09 cd	12.72 ± 2.19 ab	26.95 ± 4.01 abc	0.111 ± 0.02 cd	0.153 ± 0.01 bc	0.264 ± 0.03 ab	0.275 ± 0.03 bc
12	1.20 ± 1.30 b	15.10 ± 1.54 bc	12.47 ± 1.23 abc	27.57 ± 2.73 abc	0.105 ± 0.01 d	0.166 ± 0.02 ab	0.271 ± 0.03 ab	0.287 ± 0.02 ab
L.S.D *	0.84	1.62	1.48	2.57	0.0156	0.0204	0.2456	0.0284

* Various lowercase letters (a–h in column) reveal significant differences among the means values ($p < 0.05$), following L.S.D multiple beginnings in sequence with the latter (a) is the most significant, [1] the quantity of water application per a 9 cm Petri dish (ml), [2] the not germinated seeds number fraction of the total tested seeds, [3] the mean of each treatment's radicle lengths (cm), [4] the mean of each treatments' shoot lengths (cm), [5] the mean of each treatments' seedling lengths (cm), [6] the mean of each treatments' radicles dry weights (g), [7] the mean of each treatments' shoots dry weights (g), [8] the mean of each treatments' seedlings dry weights (g), and [9] the mean of the statistically corrected dry weight that was gained by subtracting the not germinated ones from the dry weight of the existed seedling.

Table 4. Parameters relating to germination and seedling characteristics of *Triticum aestivum* L. seeds respond to the water potential of the TKW base.

[1] Water mL	[2] Water Based on TKW	[3] Not Germinated Seeds	[4] Radicles (cm)	[5] Shoots (cm)	[6] Seedlings (cm)	[7] Radicles Dry W (g)	[8] Shoots Dry W (g)	[9] Seedlings Dry W (g)	[10] Corrected Dry W (g)
0.65	75%	11.60 ± 10.2 a	1.07 ± 1.62 e	0.19 ± 0.37 g	2.18 ± 2.30 d	0.020 ± 0.03 d	0.006 ± 0.01	0.262 ± 0.04 d	0.027 ± 0.04 d
1.30	150%	5.60 ± 8.76 b	3.16 ± 3.54 e	0.69 ± 0.98 g	3.85 ± 4.51 d	0.020 ± 0.02 d	0.009 ± 0.01	0.029 ± 0.03 d	0.029 ± 0.03 d
1.90	225%	0.60 ± 0.89 c	12.79 ± 2.25 d	7.25 ± 1.86 f	20.04 ± 4.04 c	0.113 ± 0.02 bc	0.119 ± 0.01	0.231 ± 0.03 c	0.239 ± 0.03 c
2.55	300%	0.40 ± 0.54 c	14.16 ± 1.69 bc	7.89 ± 1.01 f	22.05 ± 2.55 c	0.135 ± 0.02 abc	0.126 ± 0.02	0.261 ± 0.04 bc	0.267 ± 0.04 bc
3.20	375%	0.00 ± 0.00 c	16.14 ± 0.68 bc	9.57 ± 0.89 e	25.71 ± 1.46 b	0.141 ± 0.02 a	0.147 ± 0.01	0.289 ± 0.02 b	0.289 ± 0.02 b
3.85	450%	0.40 ± 0.55 c	18.69 ± 0.57 a	10.25 ± 0.75 cde	28.94 ± 1.08 a	0.142 ± 0.01 a	0.148 ± 0.01	0.290 ± 0.01 b	0.296 ± 0.02 b
4.45	525%	0.20 ± 0.45 c	16.50 ± 1.39 b	10.17 ± 0.94 de	26.67 ± 2.15 ab	0.136 ± 0.01 ab	0.147 ± 0.02	0.283 ± 0.03 b	0.285 ± 0.03 b
5.15	600%	0.40 ± 0.55 c	16.92 ± 1.20 ab	10.47 ± 1.05 bcde	27.40 ± 2.04 ab	0.125 ± 0.01 abc	0.147 ± 0.01	0.272 ± 0.02 b	0.277 ± 0.02 b
5.75	675%	0.40 ± 0.89 c	16.80 ± 0.68 ab	11.32 ± 0.87 abcd	28.12 ± 0.62 ab	0.133 ± 0.01 abc	0.151 ± 0.03	0.284 ± 0.03 b	0.290 ± 0.02 b
6.40	750%	0.20 ± 0.45 c	16.71 ± 0.96 ab	11.49 ± 0.20 abc	28.21 ± 1.10 ab	0.147 ± 0.04 a	0.184 ± 0.01 a	0.331 ± 0.02 a	0.334 ± 0.02 a
7.00	825%	0.40 ± 0.55 c	16.03 ± 1.22bc	12.32 ± 1.13 a	28.35 ± 0.96 ab	0.124 ± 0.01 abc	0.167 ± 0.01	0.291 ± 0.01 b	0.297 ± 0.01 b
7.50	900%	0.60 ± 0.89 c	17.42 ± 1.68 ab	11.66 ± 0.70 ab	29.04 ± 1.67 a	0.119 ± 0.01 bc	0.169 ± 0.01 ab	0.288 ± 0.03 b	0.296 ± 0.02 b
L.S.D *		4.99	2.11	1.25	2.98 0.022		0.0186	0.035	0.035

* Various lowercase letters (a–g in column) reveal significant differences among the means values ($p < 0.05$), following L.S.D multiple beginnings in sequence with the latter (a) is the most significant, [1] the quantity of water application per a 9 cm Petri dish (ml), [2] proposed percentage of water application relating to the TKW, [3] the not germinated seeds number fraction of the total tested seeds, [4] the mean of each treatment's radicle lengths (cm), [5] the mean of each treatments' shoot lengths (cm), [6] the mean of each treatments' seedling lengths (cm), [7] the mean of each treatments' radicles dry weights (g), [8] the mean of each treatments' shoots dry weights (g), [9] the mean of each treatments' seedlings dry weights (g), and [10] the mean of the statistically corrected dry weight that was gained by subtracting the not germinated ones from the dry weight of the existed seedling.

The germination test, nearly always coupled with vigor tests, such as a seedling growth test, is one of the most remarkable seed quality and seedling performance tests [48,49]. There were significant differences among the seedling performance depending on supplied moisture percentages. Under a low water quantity, the seedling length presents the least mean values, and their performance statistically significantly rose as the quantity of water increased Tables 3 and 4. The optimal range of moisture amount starts from 3.85 mL, representing 450% of TKW, Table 4. Parallel to this, significant seed length performance started under 5 mL according to water quantity applied based on single-milliliter intervals. It implies that moisture percentages calculated using TKW are more precise, consistent, and reliable. Figures 6 and 7 show that the optimal water amount range is approximately 4 to 9 mL on both water supplement bases. As a result, water stress significantly lowers seedlings' vigor; however, seedling length augmented significantly at the launching point of the optimum range.

The statistical analysis of variance revealed significant variations among the applied water quantities on each radicle length and shoot length. The radicles length progressively significantly increases as the applied water quantity increases, Tables 3 and 4. There is an optimum range for wheat radicle development, which gradually decreases as the water quantity increases than the upper limit. Shoot growth followed a different pattern than the radicle. As a result, it has the sense to measure the whole seedling, Figures 6 and 7. Water stresses of drought and waterlogging possess a greater impact on the radicles than the shoots.

The dry weight test is another laboratory test to determine the vigor of the seeds and drought tolerance of the seedlings. Significant differences among the amount of accumulated dry matter of the seedlings based on the availability of water (Tables 3 and 4). The dry matter increased gradually as the quantity of supplied water increased than the lower limit of the optimal range. Under drought stress, the creation of the dry matter is expressed on a chronological basis by accumulating the required amount to make a unite of dry matter. More water quantity had no beneficial influence on increasing the dry matter accumulation (Figures 8 and 9). Seedlings under more water supplements over the threshold range are marginally adversely impacted by surplus water compared to the required quantity. This extra water reduces the rate of dry matter accumulation in the seedlings of wheat. As the data demonstrates, a shoot-accumulated dry matter unit demands more water than a unit of radicle-accumulated dry matter. This finding corroborates the one published by [47]. Water supply had a significant effect on the proportion of seeds germinating, the rate of seedling development, and the accumulation of dry matter: The greater the degree of hydro stress, the greater the decline in these parameters' levels. Hydrological constraints and potential ranges exist at each germination and seedling development stage, which is in line with a study by [49].

3.3. Seed Number Experiment

The ANOVA revealed no significant differences among germination percentages of the seedling densities corresponding to the aggregated values of 15, 20, and 25 seeds of wheat in a Petri dish, Table 5. In addition, the subsets of the aggregated values presented no significant variations: seedlings with normal radicles, seedlings with short shoots, seedlings with only radicles, and the portion non-germinated seeds. Therefore, because increasing seedling density beyond the optimal has an opposite impact by the opening of Petri dishes lids, Figure 10, and there were no significant variations among used densities, 15 seeds per Petri dish are more appropriate than higher densities for wheat germination experiments in vitro.

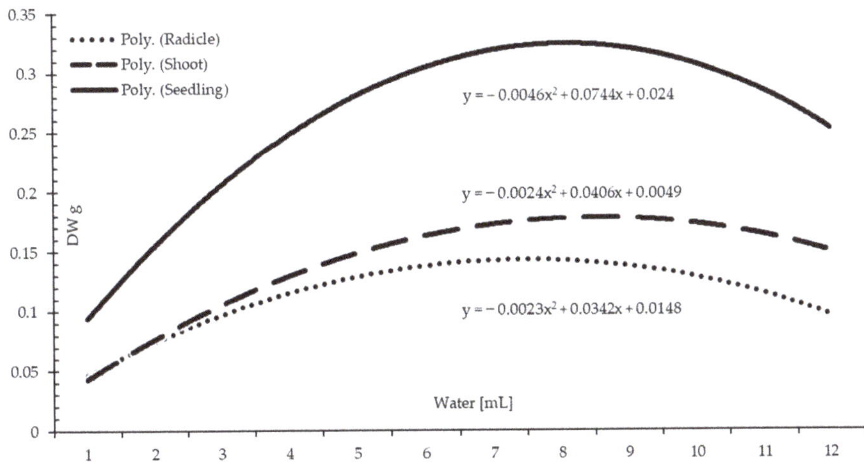

Figure 8. Seedling growth based on dry matter in response to the water amount based on 1 mL intervals.

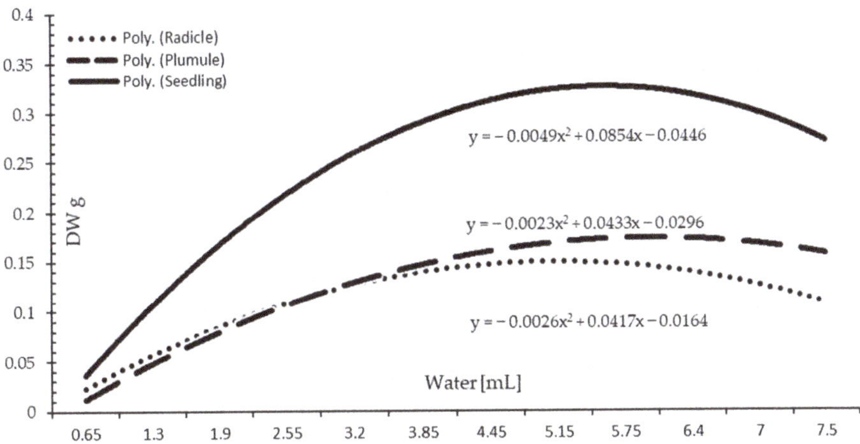

Figure 9. Seedling growth based on dry matter in response to water amount based on TKW.

Figure 10. Petri dish density of 25, 20, and 15 seedlings in a 9 cm Petri dish. The 25 seeds Petri dish shows the intensive density and growth pattern. The 20 seeds Petri dish shows a high density and growth pattern. The 15 seeds Petri dish shows the density and growth pattern.

Table 5. Parameters relating to germination and seedling characteristics of *Triticum aestivum* L. seeds respond to the number of seeds per Petri dish.

[1] Seeds n	[2] Not Germinated Seeds %	[3] Seedlings with Radicle Only %	[4] Seedlings with A Short Shoots %	[5] Seedlings with Normal Shoots %	[6] Aggregated Value %
15	0.0267 ± 0.047	0.0000 ± 0.000	0.0000 ± 0.000	0.9733 ± 0.047	0.9733 ± 0.047
20	0.0400 ± 0.046	0.0000 ± 0.000	0.0000 ± 0.000	0.9600 ± 0.046	0.9600 ± 0.046
25	0.0280 ± 0.027	0.0040 ± 0.013	0.0000 ± 0.000	0.9640 ± 0.030	0.9653 ± 0.028
L.S.D	* NS	NS	NS	NS	NS

* Indicates non-significant variations in (column) among means ($p < 0.05$), following L.S.D, [1] the number of seeds as treatment, [2] the not germinated seeds percentage of the total tested seeds, [3] percentage of seeds germinate that have only radicles, [4] percentage of seedlings that have short shoots (shorter than 4 cm). [5] Percentage of seedlings that have normal shoots length, and [6] the aggregated value of the four classified groups as determined by Equation (2): number of not germinated seeds, seedlings that have only radicle, seedlings that have a short shoot, and normal seedling length.

4. Discussion

4.1. Temperature Experiment

4.1.1. Germination Duration

Temperature is critical in regulating the length of germination duration [50,51]. The findings suggest that the crop seeds accumulate heat units from the thermal energy they absorb [52]. Once they reach the required level to commence the intracellular metabolic activity, the germination process activates at a pave depending on the ambient temperature condition [53,54]. Following the temperature gradient sub-experiment, the temperature range of 25 to 30 °C commenced germination initiation with no significant variation in the time necessary to achieve the same germination threshold and statistically followed by 20 °C, Figure 2. Seeds subjected to temperature level of 15 °C needed more time to initiate germination. However, moderate temperature of roughly 15–25 °C is demanded to commence germination of wheat seed. Changes in the state of the cell's energy supply and enzyme activity occurred, and protein synthesis was severely curtailed when the temperature rose over this threshold [35,55]. At 10 °C, germination needed a more extended time, nine days from the commencement of treatment, as validated by other investigations [56,57]. To achieve the standard measurement point of germination at 5 °C, the wheat seeds required an extended time of up to 15 days, Figure 2. According to the literature, the minimum limit of germination temperature for wheat seeds is 4 °C [56]. According to the same research, wheat seeds could not initiate germination over 37 °C. Seed can tolerate fluctuating temperature at or near the upper limit, but under constant temperature, lower temperature, suggesting 30 °C by this current research, is necessary to commence germination. Temperatures of the ideal range of germination 15° to 25 °C resulted in minor changes in germination time and internal biological activity. Although 30 °C can initiate germination rapidly, it has a reverse effect on metabolic activities in the following germination process.

4.1.2. Seedling Development

The most salient performance values and the most outstanding seedling development rate appeared at 20 °C compared to the other tested constant temperatures (Figures 2 and 3). Sigmoid curves, Figure 3, demonstrate that seedling development under 15 °C followed a similar pattern to that at 20 °C, but with a slightly lower growth rate. However, the seedling development under more than 25 °C had a reverse effect on seedling development, which is in line with other studies [57,58]. There are discrepancies with other research that reported a broader optimal range owing to day–night temperature fluctuations. However, this present research presents the accumulation of the temperature in vitro at a constant temperature, which explains the narrower. For the same reason, seedling development needed a longer time with a lower growth rate. Although nearly identical development pattern to that of the 10 °C was presented at 25 °C and 30 °C, nonetheless, it was far faster (Figure 3);

it is because of the reverse impact of the high temperature on enzymes activity, protein synthesis, and biochemical energy content increased and ROS formation, not because of the temperature accumulation [59]. The slightly analogous of seedling development under 25 °C and 30 °C presents that their growth rate was rapid in early stages but slowed in later stages (Figure 3). It indicates that each stage of wheat seedling development needs a different temperature. This figure illustrates that the upper germination range of 25 °C and 30 °C gives quicker germination and early seedling growth within less than five days, but not for the subsequent seedling growth phase. In brief, the best fitting starting point for seedling growth is 20 °C with a high development rate (Figure 3).

The shoot and radicle have comparable development patterns within and around the ideal temperature range, but their growth behavior differs when the temperature rises or lowers from this range (Figures 4 and 5). The shoot developed better than the radicle when the temperature is somewhat higher than the ideal range, particularly on the early growth seedling stages (Figure 4). Radicle development has a specific pattern in response to temperature. It grows more rapidly than the shoot at temperatures slightly below the ideal range for seedling development, especially in the later stages. Temperature deviations from the ideal range influence either the radicles or the shoots more than the other. The radicle is more tolerant of cold than the shoot, and vice versa; the shoot is more tolerant of temperatures over the ideal threshold than the radicle, which results agrees with other researchers [40,60,61].

4.2. Water Amount Experiment

Germination is one of the most eminent seed quality and performance tests linked to vigor tests, such as a seedling growth test [62–64]. Water absorption and temperature studies may offer information about wheat seeds' stress tolerance, uniformity, and germination rate. The availability of moisture may significantly impact seed germination [65,66]. Several cereal crops demand similar germination requirements as wheat [64]. To a greater extent, crop seeds with complex germination criteria will succeed in germinating than those with fewer limitations [67,68]. The water quantity experiment was carried out on two bases: single-milliliter intervals 0–12 mL and as percentages of thousand kernel weight (TKW) to develop a technique for comparing various wheat varieties with varied TKW in a future experiment. These bases of water quantity application were compared and investigated simultaneously in parallel, which has a better reflectance representing water demands. As demonstrated in the result chapter, Tables 3 and 4 reveal significant variations among water quantities potential according to both water application bases for all the examined parameters: the number of non-germinated seeds, length of the radicle, length of shoot, seedling length in total, dry weight of the seedlings, seedlings' corrected dry weight that accomplished by subtracting the ungerminated seeds, dry weight of the radicles, and dry weight of shoots. These different effects of water quantity on germination optimization agree with a study by [66].

As demonstrated by the results, germination percentage rose dramatically as water volume rose to the optimal water level, followed by a reduction somewhat as the water level increased owing to waterlogging. This is due to the demanded water limit necessary to initiate metabolic and physiological processes for germination and oxygen availability in the higher portion of water. Oxygen is critical for germination initiation [20,67]. When water levels are increased over the ideal range, oxygen availability for the seeds reduces [68]. As a result, the more water application, the less accessible oxygen availability. The wheat seeds require internal 40% moisture of the seed size to initiate germination [55]. In line with this current research, wheat seeds can initiate germination at 75% water amount of the TKW. This amount of water is sufficient to activate the metabolic processes of germination and enzymes. If the interior moisture content were less than the critical moisture content limit, wheat seeds would not germinate. Under 0.65 mL, over half of the seeds germinated. Thus, water application based on the TKW provides a better understanding of the constraints and optimal water requirements since seed size is critical for attaining the 40% internal seed moisture level. These findings corroborate research conducted by [40,57], stating that "seed size has importance in predicting germination under stress conditions". As water increases more than the upper limit of the optimal range for germination, a reverse effect of waterlogging will reduce the germination percentage because of its negative effect on the intracellular process. Therefore, this study stated that the optimal range for germination is 4.45–7.00 mL, Figure 7, representing 525–825% of the TKW, and the TKW is an objective and more accurate base for water application.

Significant differences among the seedling performance depending on supplied moisture percentages were observed. As the water was applied at the lowest quantity necessary, the seedling length had the lowest mean values, and as water quantity increased, seedling development significantly increased, as presented in Tables 3 and 4. The ideal moisture content range begins at 3.85 mL, or 450% of TKW, as shown in Table 4. Parallel to this, significant seedling length performance started under 5 mL according to water quantity applied based on single-milliliter intervals. In the same pattern, significant seedling performance started under 5 mL according to water quantity applied based on single-milliliter intervals. Water stresses of drought or waterlogging significantly reversely affect seedlings' vigor, seedling length values rises dramatically at the starting point of the ideal range of water application, and moisture percentage based on the TKW is more accurate and reliable. Water stress reduces enzymatic activity, which has a deleterious effect on glucose metabolism, decreases water potential and soluble calcium and potassium, and alters seed hormones [69–71]. Since all of these intracellular mechanisms are affected by the water stress, the seedling growth rate finally deviates from the average growth and reduces.

The radicle growth values of wheat seedlings progressively descended as the quantity of water rose over the optimal range. Shoot growth followed a distinct pattern compared to the radicle. This is in line with the finding of other research [72,73]. As a result, measuring the whole seedling is realistic; the radicle and plumule statistically remained near the ideal range as the water level rose. Either way of water stresses greatly affected the radicles more than the shoots. It can be explained in terms of the physiological aspect of radicle development; with adequate moisture availability, the seedling exhibited a slower radicle development rate and accelerated shoot development.

Significant variation among the accumulated dry matter of the seedlings is based on moisture availability, as presented in Tables 3 and 4. The dry matter values increased progressively as supplied water increased than the lower limit of the optimal range. Under drought stress, the creation of the dry matter is expressed on a chronological basis by accumulating the required amount to make a unite of dry matter. Seedlings under more water supplements over the threshold range are marginally adversely impacted by surplus water compared to the required quantity. The shoot-accumulated dry matter unit demanded a larger water quantity than a unit of radicle dry matter. This finding is in line with the result [47,74], where the researchers evaluated the effect of water potential levels 0.0 and −0.2 MPa on the germination stages of *Pinus yunnanensis* seeds. Moisture availability significantly impacted the germination rate, seedling development rate, and the accumulation of built dry matter on the radicle and the shoot: the more acute the moisture stress, the greater the reduction in dry matter accumulated values. There are water limitations and potential ranges for each germination phase and seedling development, supported by other research [47,75].

Seeds of wheat subjected to 0.65 mL of water were germinated. As a result, water application based on the TKW enables better knowledge of the limitations and optimal water requirements, as seed size is critical in achieving the required internal seed moisture. These findings corroborate research conducted by other researchers [47], stating that "seed size has importance in predicting germination under stress conditions".

4.3. Seed Number Experiment

The data demonstrate that seed numbers are a critical factor in seedling growth bioassays in vitro. Increased water volumes or denser wheat seedlings increase the amount of phytotoxin contained in each seed, hence increasing inhibition [51,52]. Greater than optimal seeding density has a detrimental effect on seedling development in Petri dishes and the opening of Petri dish lids. They are exposed to water loss, which has a detrimental effect on seedling development. Results presented no significant differences among different seeds and seedling densities of 15, 20, and 25 in a Petri dish for the measured traits, Table 5. Since high seedling density has an opposite impact by the opening of Petri dishes lids and there were no significant variations among used densities, 15 seeds per Petri dish are more appropriate than higher densities for wheat germination experiments in vitro. In plant breeding initiatives (identical breed seeds are sometimes scarce), it is critical to determine the ideal number of seeds to use in a Petri dish experiment. It is beneficial in resources optimization, particularly for plant breeders.

5. Conclusions

- Sigmoid curves provide an excellent fit for the analyzed experimental data and present increased germination temperature, generally increased germination, and seedling development rate. Overall, 20 °C was ideal for seedling development. The germination has a broader range of 20 °C to 30 °C.
- Seed size plays a role in the required amount of water for germination. Therefore, it provides a more precise impression for the application water quantity. Germination of various percentages can occur in a broad range of water quantities commencing at 0.65 mL, which represents 75% of TKW, but the optimal range for germination is 4.45–7.00 mL, representing 525–825% of the TKW.
- Dry weight can be a good indicator of seedling growth since the accumulation of dry matter is consistent and is in line with the other measurement of seedling growth, such as shoot and radicle lengths.
- No significant differences among seed densities were observed; therefore, lower density is recommended for lab experiments, 15 wheat seeds per 9 cm Petri dish. In a nutshell,

This research suggests conducting wheat seeds germination experiments under 20 °C, and water quantity as a percentage based on the TKW for optimizing the suitability of water quantity and a maximum of 15 seeds per Petri dish; more seeds have no advantage. This is common practice when seed number is limited and in breeding projects.

Author Contributions: Methodology, writing, original paper, H.K.; methodology, Z.K.; funding acquisition, I.B.; acquisition, C.G.; acquisition, A.E.; conceptualization, reviewing, Á.T. All authors have read and agreed to the published version of the manuscript.

Funding: The Hungarian University of Agriculture and Life Sciences (MATE) supported this study with financial support.

Institutional Review Board Statement: Not applicable.

Informed Consent Statement: Not applicable.

Data Availability Statement: This manuscript's data, tables, and figures are original.

Acknowledgments: The authors would like to convey their heartfelt appreciation to Zsófia Kovács, a researcher from the genetic laboratory, for her outstanding notes and work. We want to thank Katalin Dobine, the lab technician, for aiding us with the experiment.

Conflicts of Interest: The authors declare no conflict of interest.

Appendix A

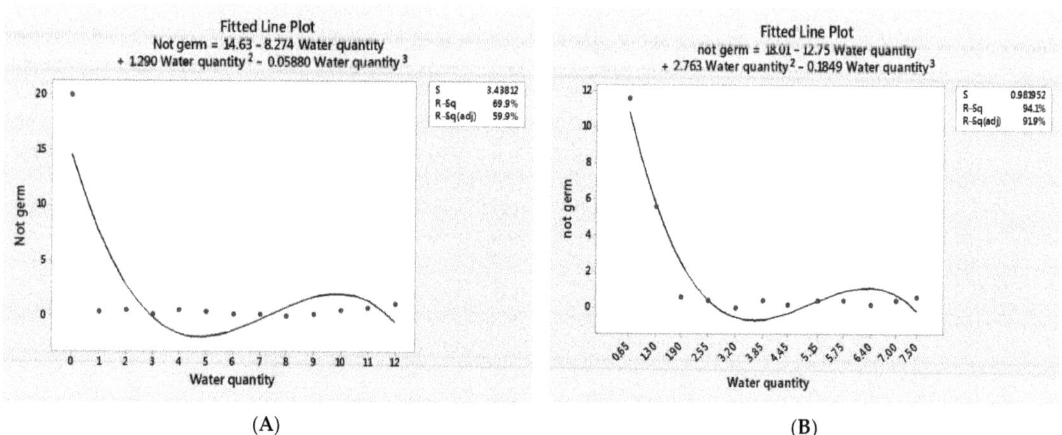

Figure A1. Seeds germination ability under different water quantities (**A**,**B**).

References

1. Lian, J.; Wu, J.; Xiong, H.; Zeb, A.; Yang, T.; Su, X.; Su, L.; Liu, W. Impact of polystyrene nanoplastics (psnps) on seed germination and seedling growth of wheat (*Triticum aestivum* L.). *J. Hazard. Mater.* **2020**, *385*, 121620. [CrossRef] [PubMed]
2. Yang, B.; Yin, Y.; Liu, C.; Zhao, Z.; Guo, M. Effect of germination time on the compositional, functional and antioxidant properties of whole wheat malt and its end-use evaluation in cookie-making. *Food Chem.* **2021**, *349*, 129125. [CrossRef] [PubMed]
3. Beta, T.; Qiu, Y.; Liu, Q.; Borgen, A.; Apea-Bah, F.B. Purple wheat (*Triticum* sp.) seeds. In *Nuts and Seeds in Health and Disease Prevention*; Elsevier: Amsterdam, The Netherlands, 2020; pp. 103–125. [CrossRef]
4. Lemmens, E.; De Brier, N.; Goos, P.; Smolders, E.; Delcour, J.A. Steeping and germination of wheat (*Triticum aestivum* L.). i. unlocking the impact of phytate and cell wall hydrolysis on bio-accessibility of iron and zinc elements. *J. Cereal Sci.* **2019**, *90*, 102847. [CrossRef]
5. Yan, M.; Xue, C.; Xiong, Y.; Meng, X.; Li, B.; Shen, R.; Lan, P. Proteomic dissection of the similar and different responses of wheat to drought, salinity and submergence during seed germination. *J. Proteom.* **2020**, *220*, 103756. [CrossRef]
6. Herman, J.J.; Sultan, S.E.; Horgan-Kobelski, T.; Riggs, C. Adaptive transgenerational plasticity in an annual plant: Grandparental and parental drought stress enhance performance of seedlings in dry soil. *Integr. Comp. Biol.* **2012**, *52*, 77–88. [CrossRef]
7. Poudel, R.; Finnie, S.; Rose, D.J. Effects of wheat kernel germination time and drying temperature on compositional and end-use properties of the resulting whole wheat flour. *J. Cereal Sci.* **2019**, *86*, 33–40. [CrossRef]
8. Khalid, N.; Tarnawa, Á.; Kende, Z.; Kassai, K.M.; Jolánkai, M. Viability of maize (*Zea mays* L.) seeds influenced by water, temperature, and salinity stress. *Acta Hydrol. Slovaca* **2021**, *22*, 113–117. [CrossRef]
9. Shah, T.; Latif, S.; Khan, H.; Munsif, F.; Nie, L. Ascorbic acid priming enhances seed germination and seedling growth of winter wheat under low temperature due to late sowing in pakistan. *Agronomy* **2019**, *9*, 757. [CrossRef]
10. Kende, Z.; Sallai, A.; Kassai, K.; Mikó, P.; Percze, A.; Birkás, M. The effects of tillage-induced soil disturbance on weed infestation of winter wheat. *Pol. J. Environ. Stud.* **2017**, *26*, 1131–1138. [CrossRef]
11. Vishal, B.; Kumar, P.P. Regulation of seed germination and abiotic stresses by gibberellins and abscisic acid. *Front. Plant Sci.* **2018**, *9*, 838. [CrossRef]
12. Rizzardi, M.A.; Luiz, A.R.; Roman, E.S.; Vargas, L. Temperatura cardeal e potencial hídrico na germinação de sementes de corda-de-viola (*Ipomoea triloba*). *Planta Daninha* **2009**, *27*, 13–21. [CrossRef]
13. Kildisheva, O.A.; Dixon, K.W.; Silveira, F.A.O.; Chapman, T.; Di Sacco, A.; Mondoni, A.; Turner, S.R.; Cross, A.T. Dormancy and germination: Making every seed count in restoration. *Restor. Ecol.* **2020**, *28*, S256–S265. [CrossRef]
14. McCormick, M.K.; Taylor, D.L.; Whigham, D.F.; Burnett, R.K. Germination patterns in three terrestrial orchids relate to abundance of mycorrhizal fungi. *J. Ecol.* **2016**, *104*, 744–754. [CrossRef]
15. Cone, J.W.; Spruit, C.J.P. Imbibition conditions and seed dormancy of *Arabidopsis thaliana*. *Physiol. Plant.* **1983**, *59*, 416–420. [CrossRef]
16. Bentsink, L.; Koornneef, M. *Seed Dormancy and Germination*; The Arabidopsis Book/American Society of Plant Biologists: Rockville, MD, USA, 2008. [CrossRef]
17. Fu, F.F.; Peng, Y.S.; Wang, G.B.; El-Kassaby, Y.A.; Cao, F.L. Integrative analysis of the metabolome and transcriptome reveals seed germination mechanism in *Punica granatum* L. *J. Integr. Agric.* **2021**, *20*, 132–146. [CrossRef]

18. Bewley, J.D.; Black, M. *Seeds: Physiology of Development and Germination*, 2nd ed.; Springer: Berlin/Heidelberg, Germany, 1994; p. 421.

19. Manz, B.; Mü, K.; Kucera, B.; Volke, F.; Leubner-Metzger, G. Water uptake and distribution in germinating tobacco seeds investigated in vivo by nuclear magnetic resonance imaging 1. *Plant Physiol.* **2005**, *138*, 1538–1551. [CrossRef]

20. Xue, X.; Du, S.; Jiao, F.; Xi, M.; Wang, A.; Xu, H.; Jiao, Q.; Zhang, X.; Jiang, H.; Chen, J.; et al. The regulatory network behind maize seed germination: Effects of temperature, water, phytohormones, and nutrients. *Crop J.* **2021**, *9*, 718–724. [CrossRef]

21. Gallardo, K.; Job, C.; Groot, S.P.C.; Puype, M.; Demol, H.; Vandekerckhove, J.; Job, D. Proteomic analysis of arabidopsis seed germination and priming. *Plant Physiol.* **2001**, *126*, 835–848. [CrossRef]

22. Masubelele, N.H.; Dewitte, W.; Menges, M.; Maughan, S.; Collins, C.; Huntley, R.; Nieuwland, J.; Scofield, S.; Murray, J.A.H. D-Type Cyclins Activate Division in the Root Apex to Promote Seed Germination in Arabidopsis. *Proc. Natl. Acad. Sci. USA* **2005**, *102*, 15694–15699. [CrossRef]

23. Krock, B.; Schmidt, S.; Hertweck, C.; Baldwin, I.T. Vegetation-derived abscisic acid and four terpenes enforce dormancy in seeds of the post-fire annual, *Nicotiana attenuata*. *Seed Sci. Res.* **2002**, *12*, 239–252. [CrossRef]

24. Petruzzelli, L.; Müller, K.; Hermann, K.; Leubner-Metzger, G. Distinct expression patterns of β-1,3-glucanases and chitinases during the germination of solanaceous seeds. *Seed Sci. Res.* **2003**, *13*, 139–153. [CrossRef]

25. Bradford, K.J.; Nonogaki, H. *Seed Development, Dormancy and Germination*; Blackwell Publishing: Hoboken, NJ, USA, 2007; pp. 1–367. [CrossRef]

26. Ozden, E.; Light, M.E.; Demir, I. Alternating temperatures increase germination and emergence in relation to endogenous hormones and enzyme activities in aubergine seeds. *S. Afr. J. Bot.* **2021**, *139*, 130–139. [CrossRef]

27. Bradford, K.J. A water relations analysis of seed germination rates. *Plant Physiol.* **1990**, *94*, 840–849. [CrossRef] [PubMed]

28. Marcos-Filho, J. Seed vigor testing: An overview of the past, present and future perspective. *Sci. Agric.* **2015**, *72*, 363–374. [CrossRef]

29. Liu, Y.; Han, C.; Deng, X.; Liu, D.; Liu, N.; Yan, Y. Integrated physiology and proteome analysis of embryo and endosperm highlights complex metabolic networks involved in seed germination in wheat (*Triticum aestivum* L.). *J. Plant Physiol.* **2018**, *229*, 63–76. [CrossRef]

30. Abido, W.A.E.; Zsombik, L. Effect of water stress on germination of some hungarian wheat landraces varieties. *Shengtai Xuebao/Acta Ecol. Sin.* **2020**, *38*, 422–428. [CrossRef]

31. Aderounmu, A.F.; Nkemnkeng, F.J.; Anjah, G.M. Effects of seed provenance and growth media on the growth performance of *Vitellaria paradoxa* c.f. gaertn. *Int. J. Biol. Chem. Sci.* **2020**, *14*, 2659–2669. [CrossRef]

32. Guan, B.; Zhou, D.; Zhang, H.; Tian, Y.; Japhet, W.; Wang, P. Germination responses of medicago ruthenica seeds to salinity, alkalinity, and temperature. *J. Arid Environ.* **2009**, *73*, 135–138. [CrossRef]

33. Ostadian Bidgoly, R.; Balouchi, H.; Soltani, E.; Moradi, A. Effect of temperature and water potential on *Carthamus tinctorius* L. seed germination: Quantification of the cardinal temperatures and modeling using hydrothermal time. *Ind. Crops Prod.* **2018**, *113*, 121–127. [CrossRef]

34. Dadach, M.; Mehdadi, Z.; Latreche, A. Effect of water stress on seed germination of *Thymus serpyllum* L. from tessala mount. *J. Plant Sci.* **2015**, *10*, 151–158. [CrossRef]

35. Riley, G.J.P. Effects of High Temperature on the Germination of Maize (*Zea Mays* L.). *Planta* **1981**, *151*, 68–74. [CrossRef] [PubMed]

36. Andronis, E.A.; Moschou, P.N.; Toumi, I.; Roubelakis-Angelakis, K.A. Peroxisomal polyamine oxidase and nadph-oxidase cross-talk for ros homeostasis which affects respiration rate in *Arabidopsis thaliana*. *Front. Plant Sci.* **2014**, *5*, 132. [CrossRef] [PubMed]

37. Bailly, C. The signalling role of ros in the regulation of seed germination and dormancy. *Biochem. J.* **2019**, *476*, 3019–3032. [CrossRef] [PubMed]

38. Kunos, V.; Cséplő, M.; Seress, D.; Eser, A.; Kende, Z.; Uhrin, A.; Bányai, J.; Bakonyi, J.; Pál, M.; Mészáros, K. The Stimulation of Superoxide Dismutase Enzyme Activity and Its Relation with the Pyrenophora teres f. teres Infection in Different Barley Genotypes. *Sustainability* **2022**, *14*, 2597. [CrossRef]

39. Balla, K.; Rakszegi, M.; Li, Z.; Békés, F.; Bencze, S.; Veisz, O. Quality of winter wheat in relation to heat and drought shock after anthesis. *Czech J. Food Sci.* **2011**, *29*, 117–128. [CrossRef]

40. Khaeim, H.; Kende, Z.; Jolánkai, M.; Kovács, G.P.; Gyuricza, C.; Tarnawa, Á. Impact of temperature and water on seed germination and seedling growth of maize (*Zea mays* L.). *Agronomy* **2022**, *12*, 397. [CrossRef]

41. Drebee, H.A.; Razak, N.A.A. Measuring the efficiency of colleges at the university of al-qadisiyah-iraq: A data envelopment analysis approach. *J. Ekon. Malays.* **2018**, *52*, 163–179. [CrossRef]

42. Bulmer, M.G.; Harrison, T.K. Principles of Statistics. *J. R. Stat. Soc. Ser. A Stat. Soc. Ser. D. Stat.* **1966**, *16*, 217. [CrossRef]

43. Seefeldt, S.S.; Kidwell, K.K.; Waller, J.E. Base growth temperatures, germination rates and growth response of contemporary spring wheat (*Triticum aestivum* L.) cultivars from the us pacific northwest. *Field Crops Res.* **2002**, *75*, 47–52. [CrossRef]

44. Sabouri, A.; Azizi, H.; Nonavar, M. Hydrotime model analysis of lemon balm (*Melissa officinalis* L.) using different distribution functions. *S. Afr. J. Bot.* **2020**, *135*, 158–163. [CrossRef]

45. Barrero, J.M.; Jacobsen, J.V.; Talbot, M.J.; White, R.G.; Swain, S.M.; Garvin, D.F.; Gubler, F. Grain dormancy and light quality effects on germination in the model grass *Brachypodium distachyon*. *New Phytol.* **2012**, *193*, 376–386. [CrossRef] [PubMed]

46. Shaban, M. Effect of water and temperature on seed germination and emergence as a seed hydrothermal time model. *Int. J. Adv. Biol. Biomed. Res.* **2013**, *1*, 1686–1691.

47. Gao, C.; Liu, F.; Zhang, C.; Feng, D.; Li, K.; Cui, K. Germination responses to water potential and temperature variation among provenances of *Pinus yunnanensis*. *Flora Morphol. Distrib. Funct. Ecol. Plants* **2021**, *276–277*, 151786. [CrossRef]

48. da Silva, L.J.; de Medeiros, A.D.; Oliveira, A.M.S. SeedCalc, a new automated r software tool for germination and seedling length data processing. *J. Seed Sci.* **2019**, *41*, 250–257. [CrossRef]

49. Eser, A.; Kassai, K.M.; Kato, H.; Kunos, V.; Tarnava, A.; Jolánkai, M. IMPACT of nitrogen topdressing on the quality parameters of winter wheat (*Triticum aestivum* L.) yield. *Acta Aliment.* **2020**, *49*, 244–253. [CrossRef]

50. Zhang, K.; Ji, Y.; Fu, G.; Yao, L.; Liu, H.; Tao, J. Dormancy-breaking and germination requirements of *Thalictrum squarrosum* stephan ex willd. seeds with underdeveloped embryos. *J. Appl. Res. Med. Aromat. Plants* **2021**, *24*, 100311. [CrossRef]

51. Contreras-Negrete, G.; Pineda-García, F.; Nicasio-Arzeta, S.; De la Barrera, E.; González-Rodríguez, A. Differences in germination response to temperature, salinity, and water potential among prosopis laevigata populations are guided by the tolerance-exploitation trade-off. *Flora* **2021**, *285*, 151963. [CrossRef]

52. Boyce, D.S. Heat and moisture transfer in ventilated grain. *J. Agric. Eng. Res.* **1966**, *11*, 255–265. [CrossRef]

53. Hu, H.; Jing, N.; Peng, Y.; Liu, C.; Ma, H.; Ma, Y. 60Coγ-ray irradiation inhibits germination of fresh walnuts by modulating respiratory metabolism and reducing energy status during storage. *Postharvest Biol. Technol.* **2021**, *182*, 111694. [CrossRef]

54. Sullivan, B.K.; Keough, M.; Laura, L.G. Copper sulphate treatment induces Heterozostera seed germination and improves seedling growth rates. *Glob. Ecol. Conserv.* **2022**, *35*, e02079. [CrossRef]

55. Gong, M.; Chen, B.O.; Li, Z.G.; Guo, L.H. Heat-shock-induced cross adaptation to heat, chilling, drought and salt stress in maize seedlings and involvement of H_2O_2. *J. Plant Physiol.* **2001**, *158*, 1125–1130. [CrossRef]

56. Schabo, D.C.; Martins, L.M.; Iamanaka, B.T.; Maciel, J.F.; Taniwaki, M.H.; Schaffner, D.W.; Magnani, M. Modeling aflatoxin b1 production by aspergillus flavus during wheat malting for craft beer as a function of grains steeping degree, temperature and time of germination. *Int. J. Food Microbiol.* **2020**, *333*, 108777. [CrossRef] [PubMed]

57. Jia, Y.; Liu, H.; Wang, H.; Zou, D.; Qu, Z.; Wang, J.; Zheng, H.; Wang, J.; Yang, L.; Mei, Y.; et al. Effects of root characteristics on panicle formation in japonica rice under low temperature water stress at the reproductive stage. *Field Crops Res.* **2022**, *277*, 108395. [CrossRef]

58. Wang, T.; Zang, Z.; Wang, S.; Liu, Y.; Wang, H.; Wang, W.; Hu, X.; Sun, J.; Tai, F.; He, R. Quaternary ammonium iminofullerenes promote root growth and osmotic-stress tolerance in maize via ros neutralization and improved energy status. *Plant Physiol. Biochem.* **2021**, *164*, 122–131. [CrossRef]

59. Cao, K.; Xie, C.; Wang, M.; Wang, P.; Gu, Z.; Yang, R. Effects of soaking and germination on deoxynivalenol content, nutrition and functional quality of fusarium naturally contaminated wheat. *LWT* **2022**, *160*, 113324. [CrossRef]

60. Bao, Y.; Pan, C.; Li, D.; Guo, A.; Dai, F. Stress response to oxytetracycline and microplastic-polyethylene in wheat (*Triticum aestivum* L.) during seed germination and seedling growth stages. *Sci. Total Environ.* **2022**, *806*, 150553. [CrossRef]

61. Cheng, X.; Tian, B.; Gao, C.; Gao, W.; Yan, S.; Yao, H.; Wang, X.; Jiang, Y.; Hu, L.; Pan, X.; et al. Identification and expression analysis of candidate genes related to seed dormancy and germination in the wheat gata family. *Plant Physiol. Biochem.* **2021**, *169*, 343–359. [CrossRef]

62. Thakur, P.S.; Chatterjee, A.; Rajput, L.S.; Rana, S.; Bhatia, V.; Prakash, S. Laser biospeckle technique for characterizing the impact of temperature and initial moisture content on seed germination. *Opt. Lasers Eng.* **2022**, *153*, 106999. [CrossRef]

63. Zhu, M.; Wang, Q.; Sun, Y.; Zhang, J. Effects of oxygenated brackish water on germination and growth characteristics of wheat. *Agric. Water Manag.* **2021**, *245*, 106520. [CrossRef]

64. Kim, Y.J.; Zhang, D.; Jung, K.H. Molecular basis of pollen germination in cereals. *Trends Plant Sci.* **2019**, *24*, 1126–1136. [CrossRef]

65. Gabriele, M.; Pucci, L. Fermentation and germination as a way to improve cereals antioxidant and antiinflammatory properties. In *Current Advances for Development of Functional Foods Modulating Inflammation and Oxidative Stress*; Academic Press: Cambridge, MA, USA, 2022; pp. 477–497. [CrossRef]

66. Campos, H.; Trejo, C.; Peña-Valdivia, C.B.; García-Nava, R.; Víctor Conde-Martínez, F.; Cruz-Ortega, R. Water availability effects on germination, membrane stability and initial root growth of *Agave lechuguilla* and *A. salmiana*. *Flora* **2020**, *268*, 151606. [CrossRef]

67. Karimmojeni, H.; Rahimian, H.; Alizadeh, H.; Yousefi, A.R.; Gonzalez-Andujar, J.L.; Mac Sweeney, E.; Mastinu, A. Competitive ability effects of *Datura stramonium* L. and *Xanthium strumarium* L. on the development of maize (*Zea mays* L.) seeds. *Plants* **2021**, *10*, 1922. [CrossRef] [PubMed]

68. Prerna, D.I.; Govindaraju, K.; Tamilselvan, S.; Kannan, M.; Vasantharaja, R.; Chaturvedi, S.; Shkolnik, D. Influence of nanoscale micro-nutrient α-fe2o3 on seed germination, seedling growth, translocation, physiological effects and yield of rice (*Oryza sativa* L.) and maize (*Zea mays* L.). *Plant Physiol. Biochem.* **2021**, *162*, 564–580. [CrossRef] [PubMed]

69. He, J.; Hu, W.; Li, Y.; Zhu, H.; Zou, J.; Wang, Y.; Meng, Y.; Chen, B.; Zhao, W.; Wang, S.; et al. Prolonged drought affects the interaction of carbon and nitrogen metabolism in root and shoot of cotton. *Environ. Exp. Bot.* **2022**, *197*, 104839. [CrossRef]

70. Thiébeau, P.; Beaudoin, N.; Justes, E.; Allirand, J.M.; Lemaire, G. Radiation use efficiency and shoot:root dry matter partitioning in seedling growths and regrowth crops of lucerne (*Medicago sativa* L.) after spring and autumn sowings. *Eur. J. Agron.* **2011**, *35*, 255–268. [CrossRef]

71. Peng, X.; Li, J.; Sun, L.; Gao, Y.; Cao, M.; Luo, J. Impacts of water deficit and post-drought irrigation on transpiration rate, root activity, and biomass yield of festuca arundinacea during phytoextraction. *Chemosphere* **2022**, *294*, 133842. [CrossRef]

72. Chen, D.; Yuan, Z.; Wei, Z.; Hu, X. Effect of maternal environment on seed germination and seed yield components of Thlaspi arvense. *Ind. Crops Prod.* **2022**, *181*, 114790. [CrossRef]
73. Weidenhamer, J.D.; Morton, T.C.; Romeo, J.T. Solution volume and seed number: Often overlooked factors in allelopathic bioassays. *J. Chem. Ecol.* **1987**, *13*, 1481–1491. [CrossRef]
74. Jolánkai, J.M.; Tarnawa, Á.; Kassai, M.K.; Eser, A.; Kende, Z. Water footprint of the protein formation of six of field crop species. *Environ. Anal. Ecol. Stud.* **2021**, *8*, 000686. [CrossRef]
75. Liu, S.; Wang, W.; Lu, H.; Shu, Q.; Zhang, Y.; Chen, Q. New perspectives on physiological, biochemical and bioactive components during germination of edible seeds: A review. *Trends Food Sci. Technol.* **2022**, *in press*. [CrossRef]

 *sustainability*

Article

The Stimulation of Superoxide Dismutase Enzyme Activity and Its Relation with the *Pyrenophora teres* f. *teres* Infection in Different Barley Genotypes

Viola Kunos [1], Mónika Cséplő [1], Diána Seress [2], Adnan Eser [3], Zoltán Kende [3], Andrea Uhrin [1], Judit Bányai [1], József Bakonyi [2], Magda Pál [1] and Klára Mészáros [1,*]

[1] Agricultural Institute, Centre for Agricultural Research, 2462 Martonvásár, Hungary; kunos.viola@atk.hu (V.K.); cseplo.monika@atk.hu (M.C.); uhrin.andrea@atk.hu (A.U.); banyai.judit@atk.hu (J.B.); pal.magda@atk.hu (M.P.)

[2] Plant Protection Institute, Centre for Agricultural Research, 1022 Budapest, Hungary; seress.diana@atk.hu (D.S.); bakonyi.jozsef@atk.hu (J.B.)

[3] Institute of Agronomy, Hungarian University of Agriculture and Life Sciences, 2100 Gödöllő, Hungary; adnaneser@hotmail.com (A.E.); kende.zoltan@uni-mate.hu (Z.K.)

* Correspondence: meszaros.klara@atk.hu

Abstract: Changes in superoxide dismutase (SOD) enzyme activity were examined in infected barley seedlings of five cultivars with the goal to study the role of SOD in the defense mechanism induced by *Pyrenophora teres* f. *teres* (PTT) infection. Our results showed that although there were differences in the responses of the cultivars, all three PTT isolates (H-618, H-774, H-949) had significantly increased SOD activity in all examined barley varieties at the early stages of the infection. The lowest SOD activity was observed in the case of the most resistant cultivar. Our results did not show a clear connection between seedling resistance of genotypes and SOD enzyme activity; however, we were able to find strong significant correlations between the PTT infection scores on the Tekauz scale and the SOD activity. The measurement of the SOD activity could offer a novel perspective to detect the early stress responses induced by PTT. Our results suggest that the resistance of varieties cannot be estimated based on SOD enzyme activity alone, because many antioxidant enzymes play a role in fine-tuning the defense response, but SOD is an important member of this system.

Keywords: barley; *Pyrenophora teres* f. *teres*; net blotch disease; biotic stress; superoxide dismutase; antioxidant enzyme

Citation: Kunos, V.; Cséplő, M.; Seress, D.; Eser, A.; Kende, Z.; Uhrin, A.; Bányai, J.; Bakonyi, J.; Pál, M.; Mészáros, K. The Stimulation of Superoxide Dismutase Enzyme Activity and Its Relation with the *Pyrenophora teres* f. *teres* Infection in Different Barley Genotypes. *Sustainability* **2022**, *14*, 2597. https://doi.org/10.3390/su14052597

Academic Editor: Balázs Varga

Received: 30 December 2021
Accepted: 21 February 2022
Published: 23 February 2022

Publisher's Note: MDPI stays neutral with regard to jurisdictional claims in published maps and institutional affiliations.

1. Introduction

Food security highly depends on successful plant breeding activity and the production of adaptive, disease-resistant crops. Cultivation of tolerant or resistant varieties is one of the most effective and eco-friendly methods of controlling plant diseases [1]; therefore, the resistance of breeding lines against pathogens is among the main selection criteria in plant breeding. Controlling foliar diseases is essential in barley (*Hordeum vulgare* L.) production, especially in the case of the damaging fungal pathogen *Pyrenophora teres* f. *teres* Drechsler (anamorph *Drechslera teres* (Sacc.) Shoem.) (PTT) [2]. This ascomycete fungus causes the typical net-like leaf symptoms of the net blotch disease and necrotic lesions with chlorotic borders on sensitive barley genotypes [3]. PTT infection could cause significant damage both in spring and winter barley genotypes with an average of 20–30% grain yield loss, especially in rainy weather [4]. In case of very sensitive barley genotypes, the damage can be up to 100% [5,6].

Pathogen infection-induced oxidative stress has been observed in several cases such as the powdery mildew on cereals [7] and the PTT and *Rhynchosporium secalis* on barley [8]; Lehmann [9] also summarizes several other relevant results. During oxidative stress, reactive oxygen species (ROS) appear in plant cells and have important roles in controlling signaling, metabolic and developmental processes [10–14]. ROS are able to act as signaling molecules at low concentrations, but they can cause oxidative stress at high concentrations [8,15–19]. The balance between ROS generation and its scavenging has a key role to combat the effects of pathogen attack in plants [20]. Production of $O_2\bullet$ is considered as the first response in a cell because it is involved in the generation of other ROS [21]. High concentrations of ROS activate ROS-dependent programmed cell death (PCD) pathways [22,23]. Avoiding cell damage and expanding the range of ROS scavenging mechanisms is necessary [24,25]. Superoxide dismutase (SOD) is one of the several defense-related antioxidant enzymes involved in the elimination of ROS [26]. The lifetime of superoxide anion ($O_2\bullet$) depends on the enzymatic activity of CuZn superoxide dismutase (SOD) [27]. SOD enzymes are responsible for the conversion of the superoxide anion (O^{2-}) to molecular oxygen and hydrogen peroxide (H_2O_2) [28]. Among the reactive oxygen species, H_2O_2 is the only one able to cross the plant membrane, therefore being essential in cell signaling [10,25,29]. Enhanced H_2O_2 production was observed as a response to pathogen attack [30]. The pathogens trigger the ROS production mainly in apoplasts [29]. Several papers have described the effects of biotic stress on the antioxidant system in cereals [31–33]. SOD is the first barrier of oxidative damage [34] and plays a significant role in the control of ROS resulting from both abiotic and biotic environmental stresses [35,36]. The role of SOD was also proven under abiotic stress conditions of cereal, such as higher or lower light intensity, drought or waterlogging [37], salinity [38], heavy metal toxicity, mineral nutrient deficiencies [26] and other environmental stresses [26,39–41]. There are two distinct peaks of ROS production during an oxidative burst caused by multiple abiotic stresses. The initial burst of ROS triggers the cell-to-cell communication and activates the specific signals. The second is the ROS burst at target [13]. Biphasic ROS accumulation with a low-amplitude, transient first phase is followed by a sustained phase of much higher magnitude that correlates with disease resistance [42,43]. Reactive oxygen species (ROS) can reduce the symptoms and block pathogen growth in plants [44]. The influence of ROS was reversed by the SOD, which indicated that O^{2-} and H_2O_2 were the most relevant reactive oxygen species during the pathogen infection [45]. Early elevations of ROS levels and increased SOD activity were observed after the appearance of net blotch symptoms in stressed barley cultivars [12,46]. In the study of Able [8], three barley cultivars were tested with virulent and avirulent PTT isolates, and an increased level of SOD activity was observed in the resistance response of barley varieties. The timely recognition of an invading microorganism coupled with the rapid and effective induction of defense responses appears to make a key difference between resistance and susceptibility [47]. The resistance of barley varieties against PTT depends on the pathotype of fungus. However, the effects of various pathotypes of PTT infection on SOD activity in the case of different barley genotypes with different resistance against PTT have not yet been investigated widely.

In this study, the aim was to investigate the relationship between SOD enzyme activity stimulated by PTT infection and the susceptibility of barley cultivars. According to these aims, we examined the change in super-oxide dismutase enzyme activity caused by the *Pyrenophora teres* f. *teres* infection in the first 72 h after infection and the subsequent two weeks on barley varieties with different PTT resistance. Five barley cultivars were inoculated in the seedling stage with three different monosporic PTT isolates from Hungary to study the defense response triggered by different PTT pathotypes in each variety. Changes in SOD enzyme activity and the severity of PTT infection were determined in inoculated barley seedlings to study the role of SOD in the defense mechanism induced by PTT pathogenesis.

2. Materials and Methods

Our experiments were performed between 2018 and 2019 in the laboratory and the greenhouse of the Agricultural Institute of the Centre for Agricultural Research, under controlled conditions.

2.1. Preparation of Inoculum

Three PTT isolates (H-618, H-774, H-949), originating from Hungary, were used in the experiment. Previously infected leaves were placed into moist chambers (water-logged filter papers in glass Petri dishes), which were incubated under white light (OSRAM model L36W/640) in a 16 h light/8 h dark cycle for 24 h at 20–22 °C to induce conidiogenesis. Monoconidial isolates were made aseptically in a laminar air flow device by transferring conidia from the leaves to V8 juice agar (V8A; 16 g agar, 3 g CaCO$_3$, 100 mL Campbell's V8 juice, 900 mL distilled water) with a sterile needle, using a Leica MZ6 stereomicroscope at 40× magnification. Plates were incubated for 10 days in the dark at 17–19 °C. To produce inoculum for the artificial inoculation experiments, isolates were grown on V8A and/or autoclaved maize leaves in 90 mm diameter plastic Petri plates under white light (OSRAM model L36W/640) in a 16 h light/8 h dark cycle at 17–19 °C for 10 days. Then, sterile distilled water containing 0.01% Tween 20 was added to the sporulating cultures (10 mL solution per plate), and conidia were removed from the conidiophores by gently agitating the mycelium mat with a sterile paint brush. Finally, the suspension was filtered through a fine sieve with 100 μm diameter pore size and the concentration of conidia was adjusted to 10.000/mL.

2.2. Plant Material

Five barley varieties were selected based on their different susceptibility to PTT, which we determined in our previous greenhouse experiments and field trials [48]. The following genotypes were selected: cv. "Canela", cv. "Harrington", cv. "Manas", cv. "Mv-Initium" and cv. "Antonella".

2.3. Setup of Treatments

Plants were grown up until the two-leaf stage at 22 °C on average in a greenhouse under a 12 h photoperiod in five replications. Plants were inoculated later with the conidial suspension of the selected PTT isolates by spraying with a hand sprayer on the surface of the leaves with a concentration of 10.000 conidia/mL until runoff. After inoculation, the plants were kept in a greenhouse chamber under a transparent plastic tent for 48 h at 22 °C and 100% humidity, which provided suitable environmental conditions and appropriate humidity for the development of the fungus.

The leaf samples were collected 0 h (as the control), 24 h, 48 h and 72 h after infection. The sampling dates were chosen based on Able's previous study [8]. In the case of treatment with H-949, infected barley seedlings were sampled on 7 and 15 DAI (days after inoculation); the longer incubation time was based on the work of Pál et al. [49] to investigate the SOD activity at a later stage in the infection. In all cases, 0.2 g of the middle part of second leaf tissue was collected in five repetitions from different pots. After that, the samples were snap-frozen in liquid nitrogen and kept at −70 °C until processing and measurement.

The evaluation of the PTT responses was based on the infection recorded on the second leaves using the 10-point scale of Tekauz [50] for single plants evaluated from each pot in three replicates until the fifteenth day.

2.4. Measurement of SOD Activity

The xanthine oxidase (EC 1.1.3.22) assay, the most frequently used method [8,28], was performed for the analysis of SOD activity according to Sigma-Aldrich's manufacturer instructions, based on the work of Bergmeyer et al. [51]. Measurements were performed with a Shimadzu UV-160A UV-VIS spectrophotometer (Shimadzu, Japan) at 550 nm with 1 cm glass cuvettes and the absorbance followed for 1 min in 10 s intervals. During the data processing, the measured changes in absorbance values were converted to enzyme activity according to the work of Zhang et al. [52].

2.5. Data Analysis

One-way ANOVA was conducted to compare the effect of infection with different PTT isolates on the different barley genotypes and the length of the contagion on the

SOD activity in the tissue samples. The Tekauz scoring points were statistically evaluated. ANOVA was performed at the $p = 0.05$ level of significance. Post-hoc comparisons using the least significant difference (LSD) test were made at $p < 0.05$. For the statistical evaluation of the results, we used the Explore and ANOVA modules of the IBM SPSS V.23 software.

3. Results

3.1. Results of the SOD Activity Measurements

3.1.1. Results of the Treatment with Isolate H-618

In the case of infection with H-618, significant differences were found between the genotypes and the SOD activity in the leaf samples at 0 h [F(4, 10) = 4189, $p = 0.03$], 24 h [F(4, 10) = 4.681, $p = 0.22$], 48 h [F(4, 10) = 12.191, $p = 0.001$] and 72 h [F(4, 10) = 22.610, $p = 0.000$] (Figure 1). We recorded slight differences between the cultivars in the first 24 h, and SOD activity increased significantly in all genotypes in the 48 h after inoculation; the largest jump in SOD activity was shown by cv. "Canela" with 678.25 ΔA/min, followed by cv. "Mv-Initium" with 340.60 ΔA/min. At 72 h, the highest SOD activity (528.02 ΔA/min) was measured in cv. "Harrington", followed by cv. "Canela" (490.84 ΔA/min). On the other hand, the SOD activity of cv. "Manas", cv. "Canela" and cv. "Mv-Initium" genotypes decreased in 72 h of the measurement with 112 ΔA/min, 188 ΔA/min and 70 ΔA/min, respectively, while cv. "Antonella" and cv. "Harrington" had further increases of 128 ΔA/min and 160 ΔA/min in the last 24 h, respectively. It is clear from the data that inoculation with H-618 influenced the SOD activity of all cultivars and that cv. "Harrington" and cv. "Canela" reacted most sensitively to the isolate. The SOD activity of cv. "Harrington" ranged between 122.98 and 528.02 ΔA/min, while cv. "Canela" had minimum activity of 101.21 ΔA/min and maximum of 782.26 ΔA/min.

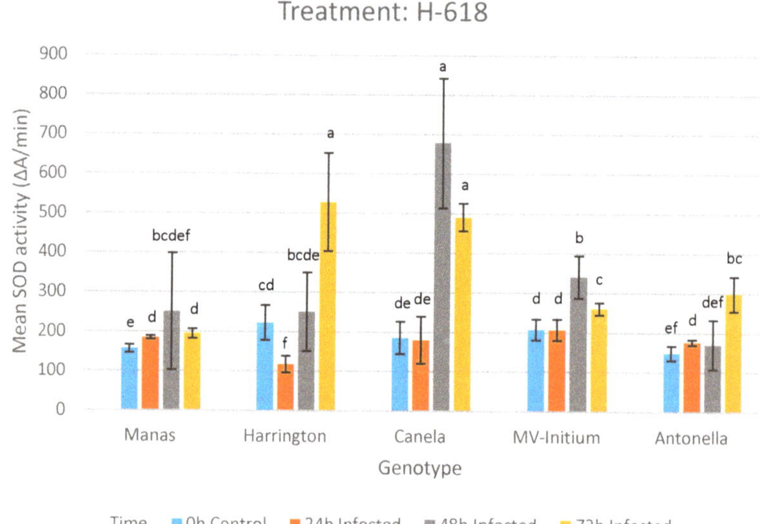

Figure 1. SOD activity of the tissue samples by the infection time in the case of H-618 treatment at 0, 1, 2 and 3 DAI. Error bars indicate the standard error (SE) of the mean instead of standard deviation (SD) at the $p = 0.05$ level. Values denoted with different letters are significantly different at $p < 0.05$.

We also performed the LSD post-hoc test on all the data to explore differences in SOD activity between barley varieties inoculated with isolate H-618 (Table 1). The post-hoc comparison showed at 0 h that the average SOD activity of cv. "Manas" (156.89 ± 8.80) was significantly lower than cv. "Harrington" (222.60 ± 38.34) and cv. "Mv-Initium" (207.71 ± 22.89). The cv. "Harrington" (222.60 ± 38.34) showed a difference only from cv.

"Antonella" (149.25 ± 15.96), which had also lower SOD activity than cv. "Mv-Initium" (207.71 ± 22.89). The genotype cv. "Canela" (185.40 ± 35.35) did not show a significant difference in SOD activity in opposition to the other examined genotypes at the beginning of the experiment at 0 h.

Table 1. Mean differences in SOD activity of genotypes after inoculation with H-618 according to LSD test for multiple comparisons.

		Manas	Harrington	Canela	MV-Initium	Antonella
0 h	Manas	-	−65.707 *	−28.511	−50.817 *	7.638
	Harrington	-	-	37.196	14.889	73.345 *
	Canela	-	-	-	−22.306	36.149
	MV-Initium	-	-	-	-	58.455 *
	Antonella	-	-	-	-	-
		Manas	Harrington	Canela	MV-Initium	Antonella
24 h	Manas	-	67.516 *	4.814	−21.997	8.355
	Harrington	-	-	−62.702 *	−89.514 *	−59.160 *
	Canela	-	-	-	−26.812	3.541
	MV-Initium	-	-	-	-	−30.353
	Antonella	-	-	-	-	-
		Manas	Harrington	Canela	MV-Initium	Antonella
48 h	Manas	-	−0.468	−427.996 *	−90.349	80.028
	Harrington	-	-	−427.527 *	−89.880	80.497
	Canela	-	-	-	337.646 *	508.025 *
	MV-Initium	-	-	-	-	170.378
	Antonella	-	-	-	-	-
		Manas	Harrington	Canela	MV-Initium	Antonella
72 h	Manas	-	−333.04 *	−295.857 *	−65.614	−103.541 *
	Harrington	-	-	37.183	267.425 *	229.498 *
	Canela	-	-	-	230.242 *	192.315 *
	MV-Initium	-	-	-	-	−37.927
	Antonella	-	-	-	-	-

* The mean difference is significant at the $p = 0.05$ level. Significant differences shown in the table are different than those shown in Figure 1 because in the figure, error bars indicate the standard error (SE) of the mean instead of standard deviation (SD) at the $p = 0.05$ level.

Measurements after 24 h showed that the average SOD activity of cv. "Manas" (184.95 ± 4.00) was significantly higher only than cv. "Harrington" (117.43 ± 18.49). In comparison, the cv. "Harrington" (117.43 ± 18.49) was significantly different from cv. "Canela" (180.14 ± 51.28), cv. "Mv-Initium" (206.95 ± 23.22) and cv. "Antonella" (176.59 ± 5.92) because of its lower SOD activity. No statistically significant change in SOD activity was found in other combinations of the genotypes after the first 24 h.

After 48 h of inoculation, only cv. "Canela" (678.25 ± 142.17) showed significantly higher differences from the other examined genotypes, the cv. "Manas" (250.25 ± 127.78), the cv. "Harrington" (250.72 ± 85.86), the cv. "Mv-Initium" (340.60 ± 46.06) and cv. "Antonella" (170.22 ± 54.28). No statistically significant change in SOD activity was found in other combinations of the genotypes after the first 48 h.

The cv. "Manas" (194.98 ± 10.66) after 72 h showed a significantly lower difference in opposition to cv. "Harrington" (528.02 ± 107.73), cv. "Canela" (490.84 ± 30.52) and cv. "Antonella" (298.52 ± 38.03), but there was no difference between cv. "Manas" and cv. "Mv-Initium" (260.59 ± 13.42). However, the cv. "Mv-Initium" showed a significant difference against cv. "Harrington" (528.02 ± 107.73) and cv. "Canela" (490.84 ± 30.52) because they showed higher SOD activity. The cv. "Antonella" had a different SOD activity than all of the other examined genotypes except the cv. "Mv-Initium". No statistically significant change in SOD activity was found in other combinations of the genotypes after 72 h.

3.1.2. Results of the Treatment with Isolate H-774

The H-774 treatment induced no significant differences between SOD activity of genotypes at 0 h [F(4, 10) = 3.026, p = 0.71], 24 h [F(4, 10) = 0.982, p = 0.46], 48 h [F(4, 10) = 1.104, p = 0.406] and 72 h [F(4, 10) = 0.673, p = 0.626]; however, it is important to note that after 48 h, SOD activity increased for all genotypes, just like in the case of isolate H-618, but it decreased in the last 24 h (Figure 2).

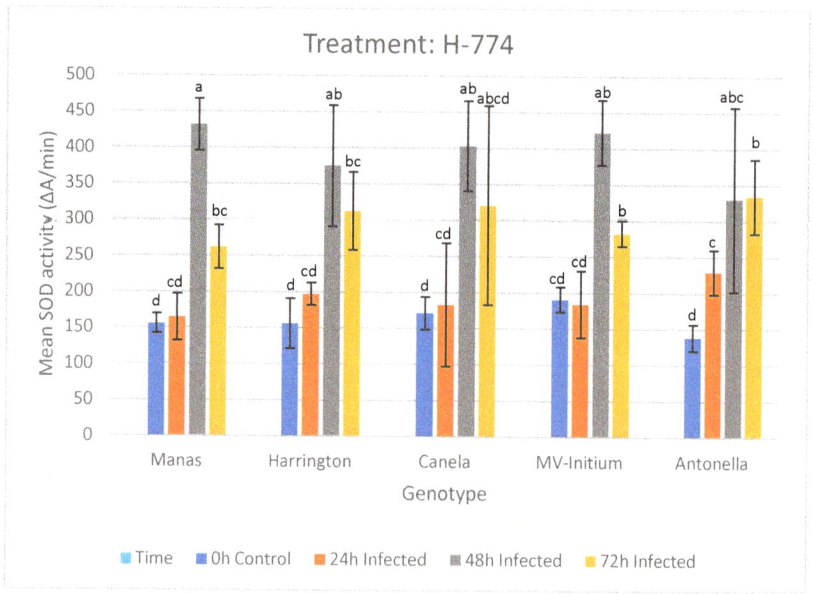

Figure 2. SOD activity of the tissue samples in the case of H-774 treatment at 0, 1, 2 and 3 DAI. Error bars indicate the standard error (SE) of the mean instead of standard deviation (SD) at p = 0.05 level. Values denoted with different letters are significantly different at p < 0.05.

For isolate H-774, the largest jump in SOD activity was measured in the 48 h after inoculation in cv. "Manas" (+267 ΔA/min), followed by cv. "Mv-Initium" (+238 ΔA/min) and cv. "Canela" (+221 ΔA/min), although the change was also significant for the varieties cv. "Harrington" (+178 ΔA/min) and cv. "Antonella" (+100 ΔA/min). The 72 h measurement showed a decrease in the SOD activity in all genotypes except cv. "Antonella", where a slight increase was observed. The largest decreases in activity of 170 ΔA/min and 140 ΔA/min occurred in the cases of cv. "Manas" and cv. "Mv-Initium", respectively.

3.1.3. Results of the Treatment with Isolate H-949

To study the changes in SOD activity in the longer term in different barley genotypes, we measured the absorbance at 0 h and on the 7th and 15th days after infection with the H-949 isolate. Although there was no significant difference between the genotypes at the beginning of the experiment at 0 h [F (4, 20) = 1.714, p = 0.186], later, a significant genotype effect was found on the SOD activity on the 7th day [F (4, 20) = 16.903, p = 0.000] and on the 15th day [F (4, 20) = 5.571, p = 0.004] (Figure 3). On the 7th day, there were significant changes in the SOD activity of the studied genotypes. The SOD activity of cv. "Manas", cv. "Harrington" and cv. "Antonella" increased greatly by 444 ΔA/min, 254 ΔA/min and 154 ΔA/min, respectively, in the first 7 days. In contrast, the cv. "Canela" and cv. "Mv-Initium" showed a significant decrease in their SOD activity by 257 ΔA/min and 322 ΔA/min, respectively.

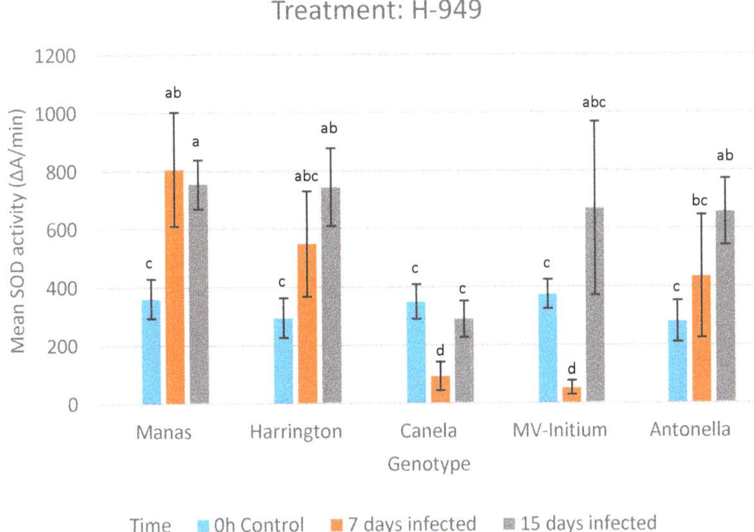

Figure 3. SOD activity of the tissue samples in the case of H-949 treatment at 0, 7 and 15 DAI. Error bars indicate the standard error (SE) of the mean instead of standard deviation (SD) at the $p = 0.05$ level. Values denoted with different letters are significantly different at $p < 0.05$.

On the 15th day, the SOD activity of cv. "Canela" almost reached its initial values, because in the last 7 days, its SOD activity increased by 198 ΔA/min, while cv. "Mv-Initium" exceeded its initial value by 285 ΔA/min.

All genotypes, except cv. "Canela", showed an increasing SOD response to isolate H-949 on the 15th day compared to the initial values. The cv. "Antonella" reached its final SOD activity at 657 ΔA/min on the 15th DAI with a total change of 376 ΔA/min. In total, cv. "MV-Initium" reached the 669 ΔA/min of SOD activity for the 15th DAI and showed a total of 296 ΔA/min change in SOD activity. The cv. "Harrington" and the cv "Manas" had 743 ΔA/min and 754 ΔA/min of SOD activity at the end of the experiment, with total changes of 448 ΔA/min and 393 ΔA/min, respectively.

The LSD post-hoc test was performed to explore differences between varieties in the case of H-949 treatment (Table 2) during the 15 days of the experiment. There were no significant differences among the SOD activities at 0 h control samples of the genotypes.

The SOD activity of cv. "Manas" (805.58 ± 219.32) at the 7th day was significantly higher from cv. "Harrington" (548.81 ± 203.48), cv. "Canela" (92.86 ± 55.15), cv. "Mv-Initium" (52.06 ± 26.74) and cv. "Antonella" (434.47 ± 235.58). Furthermore, the SOD activity of cv. "Harrington" (548.81 ± 203.48) was significantly higher than cv. "Canela" (92.86 ± 55.15) and cv. "Mv-Initium" (52.06 ± 26.74). Moreover, cv. "Antonella" (434.47 ± 235.58) had a significantly different SOD activity from cv. "Canela" (92.86 ± 55.15) and cv. "Mv-Initium" (52.06 ± 26.74) because of their lower results.

At the measurements on the 15th day, cv. "Canela" (289.68 ± 69.33) showed significantly lower SOD activity, in contrast to the genotypes of cv. "Manas" (757.25 ± 94.05), cv. "Harrington" (734.40 ± 150.40), cv. "Mv-Initium" (668.60 ± 333.44) and cv. "Antonella" (657.00 ± 128.04). There was no statistically significant difference found between the examined genotypes in any other cases.

Table 2. Post-hoc comparisons of genotypes in the case of H-949 treatment at 0, 7 and 15 DAI according to LSD test for multiple comparisons. Mean differences are shown.

0 h		Manas	Harrington	Canela	MV-Initium	Antonella
	Manas	-	66.309	11.583	−13.354	80.508
	Harrington	-	-	−54.726	−79.663	14.199
	Canela	-	-	-	24.937	93.863
	MV-Initium	-	-	-	-	−93.863
	Antonella	-	-	-	-	-

7th day		Manas	Harrington	Canela	MV-Initium	Antonella
	Manas	-	256.77 *	712.713 *	753.514 *	371.11 *
	Harrington	-	-	455.943 *	496.744 *	114.34
	Canela	-	-	-	40.8	−341.602 *
	MV-Initium	-	-	-	-	−382.403 *
	Antonella	-	-	-	-	-

15th day		Manas	Harrington	Canela	MV-Initium	Antonella
	Manas	-	10.84	464.567 *	85.650	97.25
	Harrington	-	-	453.722 *	74.805	86.405
	Canela	-	-	-	−378.916 *	−367.317 *
	MV-Initium	-	-	-	-	11.599
	Antonella	-	-	-	-	-

* The mean difference is significant at the 0.05 level. Significant differences shown in the table are different than those shown in Figure 3 because in the figure, error bars indicate the standard error (SE) of the mean instead of standard deviation (SD) at $p = 0.05$ level.

3.2. Infection of Barley Cultivars Due to Inoculation with Different PTT Isolates

We found statistically significant differences between the examined isolates while measuring on the Tekauz scoring scale the severity of the infections H-618 [$F_{(4, 10)} = 35.389$, $p = 0.000$], H-774 [$F_{(4, 10)} = 24.757$, $p = 0.000$] and H-949 [$F_{(4, 10)} = 16.346$, $p = 0.000$]. We could evaluate the degree of infection after 3 days of inoculation when the first symptoms appeared. Our results are based on the Tekauz [27] scale. The infection rates of cv. "Antonella" were significantly lower in the case of every PTT isolate, except in the case of cv. "Manas" with isolates H-618 (Figure 4).

The LSD post-hoc test was performed to explore differences between varieties in the case of isolate H-618 infection scoring on the Tekauz (1985) scale 15 days after infection (Table 3). The LSD test showed that the infection scoring was significantly different in each case. The lowest scoring points were showed by cv. "Antonella" (2.33 ± 0.577) and cv. "Manas" (4.00 ± 1.00). The only two varieties which did not differ significantly from each other were cv. "Harrington" (8.00 ± 1.00) and cv. "MV-Initium" (8.67 ± 0.578), and these two had the highest measured scoring points (9) against isolate H-618.

In the case of isolate H-774, the LSD test showed that the infection scoring was also significantly different in each case (Table 4). The only two varieties which were not significantly different from each other were cv. "Canela" (4.67 ± 1.15) and cv. "MV-Initium" (5.00 ± 1.00). The lowest scoring points were shown by cv. "Antonella" (1.00 ± 0.289) and cv. "Manas" (2.00 ± 0.575).

The results of isolate H-949 showed a slight degree of susceptibility. The cv. "Manas" (5.00 ± 1.00) did not show a significant difference from cv. "Canela" (6.00 ± 1.00) and cv. "MV-Initium" (0.56 ± 0.55), and cv. "Canela" and cv. "MV-Initium" did not differ from each other in this case. The highest scoring points were obtained by cv. "Harrington" (9). The lowest scoring points were shown by cv. "Antonella" (3) (Table 5).

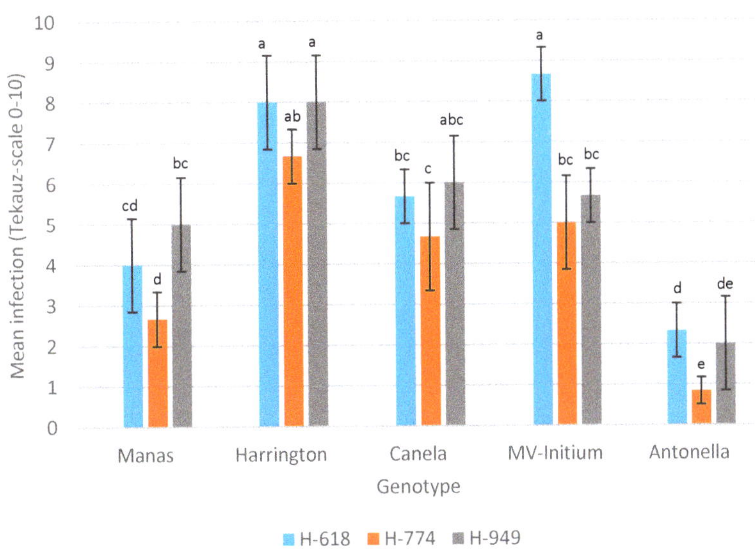

Figure 4. Infection scoring on the Tekauz scale by genotypes 15 DAI. Error bars indicate the standard error (SE) of the mean instead of standard deviation (SD) at the $p = 0.05$ level. Values denoted with different letters are significantly different at $p < 0.05$.

Table 3. Post-hoc comparisons of the isolate H-618 infection scoring on the Tekauz scale by genotypes 3 DAI using Tamhane LSD. Mean differences are shown.

	Manas	Harrington	Canela	MV-Initium	Antonella
Manas	-	−4.00 *	−1.66 *	−4.66 *	1.66 *
Harrington		-	2.33 *	−0.66	5.66 *
Canela			-	−3.00 *	3.33 *
MV-Initium				-	6.33 *
Antonella					-

* The mean difference is significant at the 0.05 level. Significant differences shown in the table are different than those shown in Figure 4 because in the figure, error bars indicate the standard error (SE) of the mean instead of standard deviation (SD) at the $p = 0.05$ level.

Table 4. Post-hoc comparisons of the isolate H-774 infection scoring on the Tekauz scale by genotypes 3 DAI using Tamhane LSD. Mean differences are shown.

	Manas	Harrington	Canela	MV-Initium	Antonella
Manas	-	−4.00 *	−2.00*	−2.33 *	1.83 *
Harrington		-	2.00 *	1.66 *	5.83 *
Canela			-	−0.33	3.83 *
MV-Initium				-	4.16 *
Antonella					-

* The mean difference is significant at the 0.05 level. Significant differences shown in the table are different than those shown in Figure 4 because in the figure, error bars indicate the standard error (SE) of the mean instead of standard deviation (SD) at the $p = 0.05$ level.

Table 5. Post-hoc comparisons of the isolate H-949 infection scoring on the Tekauz scale by genotypes 15 DAI using Tamhane LSD. Mean differences are shown.

	Manas	Harrington	Canela	MV-Initium	Antonella
Manas	-	−3.00 *	−1.00	−0.66	3.00 *
Harrington		-	2.00 *	2.33 *	6.00 *
Canela			-	0.33	4.00 *
MV-Initium				-	3.66 *
Antonella					-

* The mean difference is significant at the 0.05 level. Significant differences shown in the table are different than those shown in Figure 4 because in the figure, error bars indicate the standard error (SE) of the mean instead of standard deviation (SD) at the p = 0.05 level.

Overall, our scoring point results show that the isolates had different degrees of severity on each genotype. The most severe symptoms were caused by the H-618 (5.73 ± 2.54) followed by the H-949 (5.33 ± 2.16) and the H-774 (3.97 ± 2.19) on average. The least susceptible genotype overall was cv. "Antonella", followed by cv. "Manas", and the most susceptible genotypes were cv. "Harrington" and cv. "MV-Initium" (Figure 4).

We examined the correlation between the SOD activity and the registered Tekauz infection scoring points by genotypes and isolates (Table 6). In the case of cv. "Manas", we found a significant, positive, strong correlation between the SOD activity and the Tekauz infection scores at the treatment with the H-774 isolate (r = 0.778) and we found a strong but negative correlation (r = −0.914) at the treatment with the H-949 isolate. The cv. "Harrington" showed a strong, significant, positive correlation between the SOD activity and the Tekauz infection scoring points in the case of the treatments with H-774 and H-949 isolates. On the other hand, cv. "Canela" and cv. "MV-Initium" showed a significantly strong but inverse correlation in the case of treatments with H-618 (r_{Canela} = −0.787, $r_{Initium}$ = −0.917) and H-949 (r_{Canela} = −0.956, $r_{Initium}$ = −0.853) isolates. In relation to cv. "Antonella", we registered a strong positive correlation between the SOD activity and the Tekauz infection scorings at the treatments with isolates H-618 and H-774, which were r = 0.99 and r = 0.99, respectively.

Table 6. Pearson's correlation matrix between SOD activity (ΔA/min) and infection scoring on the Tekauz scale by genotypes after 3 DAI in the case of H-618 and H-774 and after 14 DAI in the case of H-949.

Genotype	Isolate	H-618 Tekauz	H-774 Tekauz	H-949 Tekauz
	H-618 SOD	0.143	-	-
Manas	H-774 SOD	-	0.778 *	-
	H-949 SOD	-	-	−0.914 *
	H-618 SOD	0.662	-	-
Harrington	H-774 SOD	-	0.938 *	-
	H-949 SOD	-	-	0.997 *
	H-618 SOD	−0.787 *	-	-
Canela	H-774 SOD	-	0.144	-
	H-949 SOD	-	-	−0.956 *
	H-618 SOD	−0.917 *	-	-
MV-Initium	H-774 SOD	-	−0.584	-
	H-949 SOD	-	-	−0.853 *
	H-618 SOD	0.999 *	-	-
Antonella	H-774 SOD	-	0.999 *	-
	H-949 SOD	-	-	0.588

* The correlations are significant at the 0.05 level.

We did not find significant correlations between the SOD activity and the Tekauz infection scorings in cv. "Manas" and cv. "Harrington" in the case of the isolate H-618, in cv. "Canela" and cv. "MV-Initium" in the case of the isolate H-774 or in cv. "Antonella" in the case of the isolate H-949.

4. Discussion

Changes in environmental factors, which include abiotic and biotic stress, cause modified ROS homeostasis within the plant cell. To protect plant cells from oxidative stress, ROS removal mechanisms have been developed that include antioxidant enzymes. The reactive oxygen species also play a key role as signal molecules in initiating plant defense mechanisms [19,24]. Fine-tuning this system is essential for effective defense. Although SOD, along with several other antioxidants, regulates ROS, its role is still dominant because it is the first barrier to oxidative damage. In the present study, we investigated the change in SOD activity after PTT infections in barley seedlings. Our results showed that all three PTT isolates caused significant changes in the SOD activity of all examined barley varieties in the early stages of the infection.

The treatment with H-618 resulted in significant increases in SOD activity in the examined genotypes after 48 h. The measured SOD activity highly increased in the case of every genotype except cv. "Antonella", which showed a slight decrease, but this result was not significant. Cv. "Antonella" was the most resistant genotype to PTT. Able et al. [8] observed a lower ROS production in resistant reaction to necrotrophic pathogens, and in larger amounts in susceptible plants [12]. The ROS induces the activation of different antioxidant enzymes and defense mechanism against pathogens [12,53]. On the other hand, the accumulation of ROS blocked the early growing of fungus in [23,54]. Apart from cv. "Antonella" and cv. "Harrington", the other examined genotypes showed decreases in SOD activity at the 72 h measurements. In the last 24 h, the cv. "Antonella" and cv. "Harrington" increased the measured activity significantly. Despite our results, cv. "Harrington" did not show a significant correlation between the H-618-induced PTT infection and the SOD activity; however, cv. "Antonella" had a strong significant correlation with its PTT infection with H-618, as did cv. "Canela" and cv. "MV-Initium".

The characteristic of SOD activity change was quite similar in the case of H-774 isolate to the case of H-618. For the 48 h measurements, the SOD activity reached the maximum regarding every examined genotype, expect cv. "Antonella", which showed a further increase in SOD activity in the last 24 h. However, we did not find any statistically valid differences in SOD activity between genotypes in the treatment with H-774. We found a significant correlation between the severity of the H-774-induced PTT infection and the measured SOD activity in genotypes cv. "Manas", cv. "Harrington" and cv. "Antonella".

If we compare the two isolates which were examined in the same timeframe (H-618 and H-774), we can conclude that the SOD activity was higher in cv. "Canela" in the case of H-618 infection than H-774 infection. The cv. "Harrington" was sensitive to both PTT isolates, and its SOD activity was also one of the highest in both cases on average. In contrast, the most resistant genotype to both isolates was cv. "Antonella", although its SOD activity was the lowest on average, followed by cv. "Manas" and cv. "Mv-Initium". These results support the fact that the susceptibility of barley cultivars to individual PTT isolates varies [9].

The subsequent effect of PTT infection on SOD activity was examined on days 7 and 15 after infection with the H-949 isolate. The SOD activity of cv. "MV-Initium" and cv. "Canela" decreased between 0 h and the 7th day, while significant or near significant increases in SOD activity were observed in other varieties. Further increases in activity were observed in cv. "Harrington", cv. "Canela", cv. "MV-Initium" and cv. "Antonella", and no significant decrease in SOD activity was observed in cv. "Manas". Cv. "Antonella" and cv. "MV-Initium" showed the lowest SOD activity in the case of H-949 treatment right after cv. "Canela". A negative significant correlation was observed between the infection level based on the Tekauz scale and SOD activity in the case of all cultivars expect cv. "Antonella". This result contradicts the findings of previous research that resistant varieties have higher SOD activity [8]. In our experiment, cv. "Antonella", which is the most resistant to PTT, had the lowest SOD activity in all isolates. However, this coincides with the findings that reported a positive association between the presence of ROS and disease resistance [12,55,56].

In conclusion, in our experiment, we observed a significant increase in SOD activity upon PTT inoculation, although the extent of this was genotype- and isolate-dependent. An increase in SOD activity presupposes the appearance of reactive oxygen species as a result of infection, as Able's [8] hypotheses has noted. Although he stated that a higher SOD activity suggests resistance to PTT infection, as previously advocated by Urbanek [57] and Gil-ad et al. [58], the opposite was observed in our experiment—the SOD activity of susceptible cultivars increased to a greater extent. Shetty et al. [30] state that removal of H_2O_2 by catalase at both early and late stages made wheat plants more susceptible to *Septoria tritici*, whereas H_2O_2 formation made them more resistant. Furthermore, our results also support those which show the inoculation-dependent increase in SOD activity at the 24 h stage of powdery mildew infection mentioned by Vanacker [59]. Higher SOD activity was detected in the susceptible genotype than in the resistant genotype, just like in our experiment. Increased activity of all three SOD isoforms was observed in increased *Pseudomonas* susceptible double mutant *Arabidopsis thaliana* (*At-pao1-1 x Atpao2-1*) after inoculation [55]. The previous results are in contradiction with those of Lightfoot et al. [12], who detected higher SOD activity in resistant barley genotypes, and smaller symptoms were detected in transgenic HcCSD1 knock-down lines. However, when the expression pattern of *SOD2* gene was investigated, it was found that except for clear activation of expression at 8 h after infection measured in the resistant genotype, the expression pattern fluctuated more in the sensitive ones, and there was a second expression peak at 120 h in resistance one after a decline at 48 h [56].

Based on this, our findings could offer a novel perspective of the measurement of the SOD activity to detect the early stress responses induced by PTT.

However, it should be noted that our results do not show a clear connection between seedling resistance of genotypes and SOD enzyme activity, although we did find significant correlations between the PTT infection scores and the SOD activity in several cases.

Overall, our results suggest that the SOD activity induced by the PTT infection will increase with elapsed time of infection, especially after 48 h. Increased SOD activity due to pathogenic attack promotes H_2O_2 formation and protects the plant cell from superoxide accumulation. In the case of increased production of O^{2-}, enhanced SOD activity reduces the risk of hydroxyl radical formation due to Fenton-type chemistry [57]. These results suggest that although many antioxidant enzymes play a role in fine-tuning the defense response, the resistance of varieties cannot be estimated based on SOD enzyme activity alone.

Studying the other members of the antioxidant enzyme system is necessary to clarify their role during biotic stresses, such as pathogen infection. We would like to continue our work and extend our studies beyond 3 and 15 DAI to learn more about the association between PTT infection and SOD activity, as we were unable to examine the full phenomenon because we assumed that the highest SOD activity could be observed in the early phase of the PTT infection. Furthermore, we may extend this work by involving more genotypes and isolates to understand the connection between ROS accumulation and an early pathogen infection.

Author Contributions: Conceptualization, K.M. and J.B. (József Bakonyi); methodology, K.M.; software, Z.K.; validation, M.P., V.K. and M.C.; formal analysis, A.E.; investigation, M.C., V.K. and K.M.; resources, M.P., J.B. (József Bakonyi) and D.S.; data curation, M.P. and V.K.; writing—original draft preparation, V.K.; writing—review and editing, V.K., J.B. (Judit Bányai), A.U. and K.M.; visualization, Z.K.; funding acquisition, K.M. All authors have read and agreed to the published version of the manuscript.

Funding: This research was funded by the Hungarian National Research, Development and Innovation Office, Grant Numbers: NKFIH K119276 and GINOP-2.3.2-15-2016-00029.

Institutional Review Board Statement: Not applicable.

Informed Consent Statement: Not applicable.

Data Availability Statement: The data presented in this study are available on request from the corresponding author.

Acknowledgments: We thank to Klára Illés, Ágnes Bencze and Ildiko Csorba for their help in carrying out the experiment and preparation of the semples for measuring the enzyme activity.

Conflicts of Interest: The authors declare no conflict of interest. The funders had no role in the design of the study, in the collection, analyses or interpretation of data, in the writing of the manuscript, or in the decision to publish the results.

References

1. Afanasenko, O.; Mironenko, N.; Filatova, O.; Kopahnke, D.; Krämer, I.; Ordon, F. Genetics of host-pathogen interactions in the *Pyrenophora teres* f. *teres* (net form)—Barley (*Hordeum vulgare*) pathosystem. *Eur. J. Plant Pathol.* **2007**, *117*, 267–280. [CrossRef]
2. Liu, Z.; Ellwood, S.R.; Oliver, R.P.; Friesen, T.L. *Pyrenophora teres*: Profile of an increasingly damaging barley pathogen. *Mol. Plant Pathol.* **2011**, *12*, 1–19. [CrossRef] [PubMed]
3. Smedegård-Petersen, V. *Pyrenophora teres* f. *maculata* f. *nov.* and *Pyrenophora teres* f. *teres on Barley in Denmark*; Royal Veterinary and Agricultural University; Copenhagen, Denmark, 1971; pp. 124–144.
4. Tomcsányi, A.; Szeőke, K.; Tóth, Á. TOMCSÁNYI, A., SZEŐKE, K., TÓTH, Á. (2006): Az őszi árpa védelme. *Növényvédelem* **2006**, *42*, 87–106.
5. Mathre, D.E. *Compendium of Barley Diseases*; American Phytopathological Society: St. Paul, MN, USA, 1982.
6. Steffenson, B.J. Reduction in Yield Loss Using Incomplete Resistance to *Pyrenophora teres* f. *teres* in Barley. *Plant Dis.* **1991**, *75*, 96. [CrossRef]
7. Heitefuss, R. Defence reactions of plants to fungal pathogens: Principles and perspectives, using powdery mildew on cereals as an example. *Naturwissenschaften* **2001**, *88*, 273–283. [CrossRef]
8. Able, A.J. Role of reactive oxygen species in the response of barley to necrotrophic pathogens. *Protoplasma* **2003**, *221*, 137–143. [CrossRef]
9. Lehmann, S.; Serrano, M.; L'Haridon, F.; Tjamos, S.E.; Metraux, J.P. Reactive oxygen species and plant resistance to fungal pathogens. *Phytochemistry* **2015**, *112*, 54–62. [CrossRef]
10. Baxter, A.; Mittler, R.; Suzuki, N. ROS as key players in plant stress signalling. *J. Exp. Bot.* **2014**, *65*, 1229–1240. [CrossRef]
11. Groß, F.; Durner, J.; Gaupels, F. Nitric oxide, antioxidants and prooxidants in plant defence responses. *Front. Plant Sci.* **2013**, *4*. [CrossRef]
12. Lightfoot, D.J.; Mcgrann, G.R.D.; Able, A.J. The role of a cytosolic superoxide dismutase in barley–pathogen interactions. *Mol. Plant Pathol.* **2017**, *18*, 323–335. [CrossRef] [PubMed]
13. Mittler, R.; Vanderauwera, S.; Suzuki, N.; Miller, G.; Tognetti, V.B.; Vandepoele, K.; Gollery, M.; Shulaev, V.; Van Breusegem, F. ROS signaling: The new wave? *Trends Plant Sci.* **2011**, *16*, 300–309. [CrossRef] [PubMed]
14. Torres, M.A.; Dangl, J.L. Functions of the respiratory burst oxidase in biotic interactions, abiotic stress and development. *Curr. Opin. Plant Biol.* **2005**, *8*, 397–403. [CrossRef]
15. Kranner, I.; Roach, T.; Beckett, R.P.; Whitaker, C.; Minibayeva, F.V. Extracellular production of reactive oxygen species during seed germination and early seedling growth in *Pisum sativum*. *J. Plant Physiol.* **2010**, *167*, 805–811. [CrossRef]
16. Tyagi, S.; Sharma, S.; Taneja, M.; Shumayla; Kumar, R.; Sembi, J.K.; Upadhyay, S.K. Superoxide dismutases in bread wheat (*Triticum aestivum* L.): Comprehensive characterization and expression analysis during development and, biotic and abiotic stresses. *Agri Gene* **2017**, *6*, 1–13. [CrossRef]
17. Berens, M.L.; Berry, H.M.; Mine, A.; Argueso, C.T.; Tsuda, K. Evolution of Hormone Signaling Networks in Plant Defense. *Annu. Rev. Phytopathol.* **2017**, *55*, 401–425. [CrossRef]
18. Sies, H. On the history of oxidative stress: Concept and some aspects of current development. *Curr. Opin. Toxicol.* **2018**, *7*, 122–126. [CrossRef]
19. Waszczak, C.; Carmody, M.; Kangasjärvi, J. Reactive Oxygen Species in Plant Signaling. *Annu. Rev. Plant Biol.* **2018**, *69*, 209–236. [CrossRef]
20. Wang, Y.; Ji, D.; Chen, T.; Li, B.; Zhang, Z.; Qin, G.; Tian, S. Production, signaling, and scavenging mechanisms of reactive oxygen species in fruit–pathogen interactions. *Int. J. Mol. Sci.* **2019**, *20*, 2994. [CrossRef]
21. Saed-Moucheshi, A.; Shekoofa, A.; Pessarakli, M. Reactive Oxygen Species (ROS) Generation and Detoxifying in Plants. *J. Plant Nutr.* **2014**, *37*, 1573–1585. [CrossRef]
22. Gechev, T.S.; Hille, J. Hydrogen peroxide as a signal controlling plant programmed cell death. *J. Cell Biol.* **2005**, *168*, 17–20. [CrossRef] [PubMed]
23. Künstler, A.; Bacsó, R.; Gullner, G.; Hafez, Y.M.; Király, L. Staying alive—Is cell death dispensable for plant disease resistance during the hypersensitive response? *Physiol. Mol. Plant Pathol.* **2016**, *93*, 75–84. [CrossRef]
24. García-Caparrós, P.; De Filippis, L.; Gul, A.; Hasanuzzaman, M.; Ozturk, M.; Altay, V.; Lao, M.T. Oxidative Stress and Antioxidant Metabolism under Adverse Environmental Conditions: A Review. *Bot. Rev.* **2021**, *87*, 421–466. [CrossRef]
25. Pucciariello, C.; Perata, P. New insights into reactive oxygen species and nitric oxide signalling under low oxygen in plants. *Plant Cell Environ.* **2017**, *40*, 473–482. [CrossRef] [PubMed]

26. Yu, Q.; Worth, C.; Rengel, Z. Using capillary electrophoresis to measure Cu/Zn superoxide dismutase concentration in leaves of wheat genotypes differing in tolerance to zinc deficiency. *Plant Sci.* **1999**, *143*, 231–239. [CrossRef]

27. Takagi, D.; Takumi, S.; Hashiguchi, M.; Sejima, T.; Miyake, C. Superoxide and singlet oxygen produced within the thylakoid membranes both cause photosystem I photoinhibition. *Plant Physiol.* **2016**, *171*, 1626–1634. [CrossRef]

28. Acar, O.; Türkan, I.; Özdemir, F. Superoxide dismutase and peroxidase activities in drought sensitive and resistant barley (*Hordeum vulgare* L.) varieties. *Acta Physiol. Plant.* **2001**, *23*, 351–356. [CrossRef]

29. Lam, E.; Kato, N.; Lawton, M. Programmed cell death, mitochondria and the plant hypersensitive response. *Nature* **2001**, *411*, 848–853. [CrossRef]

30. Shetty, N.P.; Mehrabi, R.; Lütken, H.; Haldrup, A.; Kema, G.H.J.; Collinge, D.B.; Jørgensen, H.J.L. Role of hydrogen peroxide during the interaction between the hemibiotrophic fungal pathogen Septoria tritici and wheat. *New Phytol.* **2007**, *174*, 637–647. [CrossRef]

31. Ivanov, S.; Miteva, L.; Alexieva, V.; Karjin, H.; Karanov, E. Alterations in some oxidative parameters in susceptible and resistant wheat plants infected with Puccinia recondita f.sp. tritici. *J. Plant Physiol.* **2005**, *162*, 275–279. [CrossRef]

32. Harrach, B.D.; Fodor, J.; Pogány, M.; Preuss, J.; Barna, B. Antioxidant, ethylene and membrane leakage responses to powdery mildew infection of near-isogenic barley lines with various types of resistance. *Eur. J. Plant Pathol.* **2008**, *121*, 21–33. [CrossRef]

33. Asthir, B.; Koundal, A.; Bains, N.S.; Mann, S.K. Stimulation of antioxidative enzymes and polyamines during stripe rust disease of wheat. *Biol. Plant.* **2010**, *54*, 329–333. [CrossRef]

34. Chung, W.H. Unraveling new functions of superoxide dismutase using yeast model system: Beyond its conventional role in superoxide radical scavenging. *J. Microbiol.* **2017**, *55*, 409–416. [CrossRef]

35. Bowler, C.; Van Camp, W.; Van Montagu, M.; Inzé, D. Superoxide Dismutase in Plants. *CRC. Crit. Rev. Plant Sci.* **1994**, *13*, 199–218. [CrossRef]

36. Monk, L.S.; Fagerstedt, K.V.; Crawford, R.M.M. Oxygen toxicity and superoxide dismutase as an antioxidant in physiological stress. *Physiol. Plant.* **1989**, *76*, 456–459. [CrossRef]

37. Deng, B.; Du, W.; Liu, C.; Sun, W.; Tian, S.; Dong, H. Antioxidant response to drought, cold and nutrient stress in two ploidy levels of tobacco plants: Low resource requirement confers polytolerance in polyploids? *Plant Growth Regul.* **2012**, *66*, 37–47. [CrossRef]

38. Miller, G.; Schlauch, K.; Tam, R.; Cortes, D.; Torres, M.A.; Shulaev, V.; Dangl, J.L.; Mittler, R. The plant NADPH oxidase RBOHD mediates rapid systemic signaling in response to diverse stimuli. *Sci. Signal.* **2009**, *2*. [CrossRef]

39. Singh, R.; Rathore, D. Oxidative stress defence responses of wheat (*Triticum aestivum* L.) and chilli (*Capsicum annuum* L.) cultivars grown under textile effluent fertilization. *Plant Physiol. Biochem.* **2018**, *123*, 342–358. [CrossRef]

40. Suzuki, N.; Miller, G.; Salazar, C.; Mondal, H.A.; Shulaev, E.; Cortes, D.F.; Shuman, J.L.; Luo, X.; Shah, J.; Schlauch, K.; et al. Temporal-spatial interaction between reactive oxygen species and abscisic acid regulates rapid systemic acclimation in plants. *Plant Cell* **2013**, *25*, 3553–3569. [CrossRef] [PubMed]

41. Tounsi, S.; Feki, K.; Kamoun, Y.; Saïdi, M.N.; Jemli, S.; Ghorbel, M.; Alcon, C.; Brini, F. Highlight on the expression and the function of a novel MnSOD from diploid wheat (*T. monococcum*) in response to abiotic stress and heavy metal toxicity. *Plant Physiol. Biochem.* **2019**, *142*, 384–394. [CrossRef]

42. Lamb, C.; Dixon, R.A. The oxidative burst in plant disease resistance. *Annu. Rev. Plant Biol.* **1997**, *48*, 251–275. [CrossRef] [PubMed]

43. Torres, M.A.; Jones, J.D.G.; Dangl, J.L. Reactive oxygen species signaling in response to pathogens. *Plant Physiol.* **2006**, *141*, 373–378. [CrossRef] [PubMed]

44. Künstler, A.; Bacsó, R.; Hafez, Y.M.; Király, L. Reactive Oxygen Species and Plant Disease Resistance. In *Reactive Oxygen Species and Oxidative Damage in Plants under Stress*; Gupta, D.K., Palma, J.M., Corpas, F.J., Eds.; Springer International Publishing: Cham, Switzerland, 2015; pp. 269–303. ISBN 978-3-319-20421-5.

45. El-Zahaby, H.M.; Hafez, Y.M.; Király, Z. Effect of Reactive Oxygen Species on Plant Pathogens in planta and on Disease Symptoms. *Acta Phytopathol. Entomol. Hung.* **2004**, *39*, 325–345. [CrossRef]

46. Abdelaal, K.A.A.; El-Shawy, E.S.A.A.; Hafez, Y.M.; Abdel-Dayem, S.M.A.; Chidya, R.C.G.; Saneoka, H.; Sabagh, A. El Nano-silver and non-traditional compounds mitigate the adverse effects of net blotch disease of barley in correlation with up-regulation of antioxidant enzymes. *Pak. J. Bot.* **2020**, *52*, 1065–1072. [CrossRef]

47. Bari, R.; Jones, J.D.G. Role of plant hormones in plant defence responses. *Plant Mol. Biol.* **2009**, *69*, 473–488. [CrossRef] [PubMed]

48. Mészáros, K.; Cséplő, M.; Kunos, V.; Búza, Z.; Bányai, J.; Seress, D.; Csorba, I.; Pál, M.; Vida, G.; Bakonyi, J. Investigation of net blotch resistance of barley and preliminary data on Hungarian pathotypes of *Pyrenophora teres* f. *teres*. In *Proceedings of the Resistance Breeding: From Pathogen Epidemiology to Molecular Breeding*; Brandsetter, A., Geppner, M., Eds.; Druck und Verlag: Sankt Pölten, Austria, 2019; pp. 27–29.

49. Pál, M.; Kovács, V.; Vida, G.; Szalai, G.; Janda, T. Changes induced by powdery mildew in the salicylic acid and polyamine contents and the antioxidant enzyme activities of wheat lines. *Eur. J. Plant Pathol.* **2013**, *135*, 35–47. [CrossRef]

50. Tekauz, A. A numerical scale to classify reactions of barley to *Pyrenophora teres*. *Can. J. Plant Pathol.* **1985**, *7*, 181–183. [CrossRef]

51. Bergmeyer, H.U.; Gawehn, K.; Grassl, M. Methods of Enzymatic Analysis. *Verlag Chemie* **1974**, *1*, 481–482.

52. Zhang, C.; Bruins, M.E.; Yang, Z.Q.; Liu, S.T.; Rao, P.F. A new formula to calculate activity of superoxide dismutase in indirect assays. *Anal. Biochem.* **2016**, *503*, 65–67. [CrossRef]

53. O'Brien, J.A.; Daudi, A.; Butt, V.S.; Bolwell, G.P. Reactive oxygen species and their role in plant defence and cell wall metabolism. *Planta* **2012**, *236*, 765–779. [CrossRef]

54. Liu, X.; Williams, C.E.; Nemacheck, J.A.; Wang, H.; Subramanyam, S.; Zheng, C.; Chen, M.S. Reactive oxygen species are involved in plant defense against a gall midge. *Plant Physiol.* **2010**, *152*, 985–999. [CrossRef] [PubMed]

55. Jasso-Robles, F.I.; Gonzalez, M.E.; Pieckenstain, F.L.; Ramírez-García, J.M.; de la Luz Guerrero-González, M.; Jiménez-Bremont, J.F.; Rodríguez-Kessler, M. Decrease of Arabidopsis PAO activity entails increased RBOH activity, ROS content and altered responses to *Pseudomonas*. *Plant Sci.* **2020**, *292*, 110–372. [CrossRef]

56. Pandey, C.; Großkinsky, D.K.; Westergaard, J.C.; Jørgensen, H.J.L.; Svensgaard, J.; Christensen, S.; Schulz, A.; Roitsch, T. Identification of a bio-signature for barley resistance against *Pyrenophora teres* infection based on physiological, molecular and sensor-based phenotyping. *Plant Sci.* **2021**, *313*, 111072. [CrossRef] [PubMed]

57. Urbanek, H.; Gajewska, E.; Karwowska, R.; Wielanek, M. Generation of superoxide anion and induction of superoxide dismutase and peroxidase in bean leaves infected with pathogenic fungi. *Acta Biochim. Pol.* **1996**, *43*, 679–685. [CrossRef]

58. Gil-ad, N.L.; Bar-Nun, N.; Noy, T.; Mayer, A.M. Enzymes of *Botrytis cinerea* capable of breaking down hydrogen peroxide. *FEMS Microbiol. Lett.* **2000**, *190*, 121–126. [CrossRef] [PubMed]

59. Vanacker, H.; Harbinson, J.; Ruisch, J.; Carver, T.L.W.; Foyer, C.H. Antioxidant defences of the apoplast. *Protoplasma* **1998**, *205*, 129–140. [CrossRef]

sustainability

Article

New Methods for Testing/Determining the Environmental Exposure to Glyphosate in Sunflower (*Helianthus annuus* L.) Plants

Dóra Farkas [1,*], Katalin Horotán [2], László Orlóci [1], András Neményi [1] and Szilvia Kisvarga [1]

1 Research Group of Ornamental Horticulture and Green System, Institute of Landscape Architecture, Urban Planning and Garden Art, Hungarian University of Agriculture and Life Sciences (MATE), 1118 Budapest, Hungary; Orloci.Laszlo@uni-mate.hu (L.O.); Nemenyi.Andras.Bela@uni-mate.hu (A.N.); Kisvarga.Szilvia@uni-mate.hu (S.K.)
2 Zoological Department, Institute of Biology, Eszterházy Károly Catholic University, 3300 Eger, Hungary; horotan.katalin@uni-eszterhazy.hu
* Correspondence: Farkas.Dora@uni-mate.hu

Abstract: Glyphosate is still the subject of much debate, as several studies report its effects on the environment. Sunflower (GK Milia CL) was set up as an experimental plant and treated with glyphosate concentrations of 500 ppm and 1000 ppm in two treatments. Glyphosate was found to be absorbed from the soil into the plant organism through the roots, which was also detectable in the leaf and root. Glyphosate was also significantly detected in the plant 5 weeks after treatment and in plants that did not receive glyphosate treatment directly, so it could be taken up through the soil. Based on the morphological results, treatment with higher concentrations (1000 ppm) of glyphosate increased the dried mass and resulted in shorter, thicker roots. Histological results also showed that basal and transporter tissue distortions were observed in the glyphosate-treated plants compared to the control group. Cells were distorted with increasing concentration, vacuoles formed, and the cell wall was weakened in both the leaf-treated and inter-row-treated groups. In the future, it will be worth exploring alternative agricultural technologies that can reduce the risk of glyphosate while increasing economic outcomes. This may make the use of glyphosate more environmentally conscious.

Keywords: *Helianthus*; sunflower; morphological; sustainable; glyphosate; pesticide; residue; pollution; weed control; organic plant production

Citation: Farkas, D.; Horotán, K.; Orlóci, L.; Neményi, A.; Kisvarga, S. New Methods for Testing/ Determining the Environmental Exposure to Glyphosate in Sunflower (*Helianthus annuus* L.) Plants. *Sustainability* **2022**, *14*, 588. https:// doi.org/10.3390/su14020588

Academic Editor: Balázs Varga

Received: 10 December 2021
Accepted: 4 January 2022
Published: 6 January 2022

Publisher's Note: MDPI stays neutral with regard to jurisdictional claims in published maps and institutional affiliations.

1. Introduction

Glyphosate (NA-(phosphono-methyl) glycine) was introduced commercially in 1974 [1] and is considered to be the most widely used herbicide in the history of agriculture [2,3], a postemergent, systemic, non-selective chemical [4,5] that poses a high risk to human health and the environment [4]. In 2014, farmers in the United States sprayed enough glyphosate to reach about 1 kg per hectare of cultivated land, amounting to nearly 0.53 kg/ha for all crops in the world [6]. In 2017, glyphosate accounted for 33% of herbicide sales in Europe. One-third of the sown area of annual crop systems and half of the area of trees received glyphosate annually. Glyphosate is widely used for at least eight agricultural purposes, including weed control, plant drying, sealing of cover crops, temporary grassland removal, and permanent grassland renewal [7]. It has been widely used for the past 40 years, with the assumption that side effects are minimal [8]. Since the mid-1990s, there have been significant changes in the timing and manner of application of glyphosate herbicides, resulting in a drastic increase in total application rates [6]. There are currently no public data on the use of glyphosate in Europe [7].

1.1. Glyphosate

Glyphosate is a highly potent broad-spectrum herbicide that targets 5-enolpyruvyl shikimate-3-phosphate synthase (EPSPS) [9]. The majority of soybeans (*Glycine max* L.) planted in the United States are resistant to glyphosate due to the introduction of a gene encoding glyphosate-insensitive 5-enolipyruvylsikimate 3 phosphate synthase [10]. Gene expression studies have also shown that the gene encoding EPSPS is maximally expressed in meristems, so glyphosate must migrate to plant growth sites to be effective [11]. Glyphosate was metabolized to the same extent in *Sorghum halepense* (L.) Pers. in both glyphosate-resistant and -sensitive populations, and there was no significant difference in the inhibition of 5-enol-pyruvylsikimate-3-phosphate synthase (EPSPS), nor in the basic activity [12].

It is used primarily against deep-rooted perennial weeds in agriculture, forestry, and wetlands and parks. The focus is on its agricultural use, and the possible presence of residues that also accumulate in crops and human and animal tissues. According to a World Health Organization (WHO) report, the residue content in plant organs is negligible [4], yet several studies point to the opposite [13]. The effect of low-dose glyphosate on animals and humans has recently been documented, suggesting changes and shifts in the composition of microbial communities in plants and the animal gut [8]. For spring wheat (*Triticum aestivum* L.), a number of laboratories are currently investigating both its accumulation and its toxicity in both animals and plants [14].

Glyphosate-resistant, genetically modified crops have been used since 1996 and are becoming more widespread worldwide [9]. Commercially available glyphosate-based formulations typically contain 41% or more of the active ingredient, while household formulations contain 1% glyphosate [15]. However, glyphosate can also have extensive, undesirable effects on nutrient efficiency, compromising agricultural sustainability. It weakens the defenses of plants against pathogens and pests as it also accumulates in meristem tissues [2]. The toxicity formulations vary due to the value being influenced by multiple factors. The toxicity of the preparation depends on the quality and quantity of surfactants. Experimental results suggest that the toxicity of the polyoxyethyleneamine (POEA) surfactant is greater than that of glyphosate alone or the average toxicity of most commercially available agents [15]. Few plant species are inherently resistant to glyphosate. In addition, measurements of glyphosate did not result in any evidence of resistance of weeds to the active substance in the field. This may be explained, among other things, by the lack of glyphosate residue in the soil [16]. Recent evidence, however, calls into question the inactivation properties and safety of glyphosate. Glyphosate can be retained and transported in soil [17].

Aminomethylphosphonic acid (AMPA) is the most frequently detected metabolite of glyphosate in plants [18]. Glyphosate and the metabolite AMPA accumulate in the environment [8]. Glyphosate-based herbicides have been found to cause abnormal growth structures and phenological changes in some agriculturally relevant plants, such as one or more morphological changes in anthers, anthers, pollen, or flowers, and to reduce the number of seeds in the glyphosate-resistant plant [19]. In soybean plants, glyphosate-containing formulations had no effect on chlorophyll content, the dry weight of roots and shoots, and the number of shoots, but reduced shoot biomass by 21–28% [20]. There is a significant difference in glyphosate and AMPA levels between some plant species [18]. In some first- and second-generation glyphosate-resistant *Glycine max* L., photosynthesis becomes inhibited [21]. The content of sikhate is twice as high for older leaves and 16 times as high for young leaves in non-glyphosate-resistant plants [22], with the highest sensitivity to glyphosate at the beginning of the vegetative stage [23]. This affects nutrient uptake, leading to a decrease in biomass. Its effect increases with increasing glyphosate levels [21]. The increasing glyphosate ratio showed a significant decrease in photosynthetic activity, accumulation of leaf tissue macro- and microelements, and decreased nutrient uptake in *Glycine max* L. individuals [24]. In *Glycine max* L. Merr. 'Williams' individuals, glyphosate also reduced chlorophyll content [25] but accumulated in nodules [26]. In *Lupinus albus* L. plants, nitrogenase activity was reduced 24 h after glyphosate treatment, even at the lowest and sublethal doses (1.25 mM). In the longer term (5 days), the starch content of the shoots and the amount of sucrose synthase also decreased, while the sucrose content of the shoots increased [27]. In *Eichhornia* Kunth. plants, glyphosate was also detectable in the plants on day 14 after treatment [28]. Glyphosate also affects the mineral content. In plants of *Glycine max* L., glyphosate has been shown to interfere with the uptake and retranslocation of Ca, Mg, Fe, and Mn, most likely to bind and thus immobilize them. The decrease in iron, manganese, calcium, and magnesium concentrations in seeds by glyphosate is very specific and may affect seed quality [22]. *Gossypium hirsutum* L. has been shown to accumulate significantly more glyphosate in reproductive tissues than in vegetative tissues [29]. *Agropyron repens* (L.) Beauv. glyphosate treatment showed that offspring closer to the mother plant survived glyphosate treatment more easily than those farther away. This suggests that higher bud death near the rhizome peak is attributable to higher glyphosate accumulation [30].

Searching major scientific databases (Google Scholar, ScienceDirect) for "glyphosate," "herbicide use," and "pesticide use" leads to just a few articles reporting global sales of glyphosate. The lack of public data on the use of glyphosate has already been highlighted by several NGOs and researchers. Based on data and additional estimates, the total volume of glyphosate sold in the EU in 2017 was 49,427 tonnes [7].

1.2. Histological and Morphological Changes

The persistence of glyphosate residues in plant tissues varies depending on the species and plant tissue type [31]. In maize (*Zea mays* L.), glyphosate reduced the efficiency of photosynthesis, causing changes in leaf anatomy and stem physical properties, leading to a decrease in grain size [32]. In *Rosa acicularis* plants, 2 years after glyphosate treatment, the effect of the chemical was also detected in pollen, petals, and flower tissues [19]. It did not affect the number of flowers, but it did affect the cumulative number of flowers [33]. Glyphosate-treated tomato seedlings (*Solanum lycopersicum* L.) showed increased root growth and elongation of chloroplasts when seeds were sown in glyphosate containing soil and if sprayed with glyphosate [34]. Glyphosate has been shown to have a detrimental effect on the yield and growth of young cocoa plants [35]. In *Sida acuta* Burm.f. specimens treated with glyphosate, the stem was swollen and bent. The leaves turned yellow, while the roots were swollen and necrotic. Vegetative growth decreased, and the plants eventually withered [36]. In plants of *Salvinia cucullata*, glyphosate was ecotoxic [37]. In this plant species, concomitant use with copper may increase the ecological risk [38]. The active substance accumulation in *Lemna minor* plants was tenfold compared to the maximum acceptable residue level (MRL) after 7 days of glyphosate treatment [39]. In wheat (*Triticum aestivum* L.), the use of glyphosate did not affect the primary and secondary structure of the proteins [40]. *Chloris elata* Desv. specimens have been shown to form wax crystals around the stomas of older plants, which contributed to a decrease in glyphosate sensitivity [41]. The use of glyphosate and paraquat negatively affected the germination of *Vicia faba*, *Phaselolus vulgaris*, and *Sorghum bicolor* seeds, photosynthetic pigments, and amino acids [42].

1.3. Effects of Glyphosate on the Human Environment

Exposure to carcinogens is responsible for a number of human health problems [43]. We are increasingly facing the severe social and economic effects of environmental degradation worldwide [44], and seasonal changes also affect soil habitats [45]. The use of glyphosate causes potential health problems [46]. Decreases in plant diversity and numbers have been widely reported in the agricultural ecosystems of North America and Europe. Intensive use of herbicides within arable land and drift from neighboring habitats are partly responsible for the change [47]. Its toxicity is not limited to plant organisms but can be clearly demonstrated in human cells, so a human health risk analysis process for glyphosate should be developed in the future [48]. The rapid transport of glyphosate is well illustrated by the detection of glyphosate residues in bottled drinking water and human urine in Mexico [49]. Glyphosate [50] is the most widely used pesticide in Indonesia. Occasionally, there are reports of contamination of drinking water sources with herbicides (including glyphosate) in Taiwan. Glyphosate is not yet a chemical in the assessment of Taiwanese drinking water quality standards [51]. Glyphosate and its metabolite, aminomethylphosphonic acid (AMPA), have recently been identified as potential contributors to the development of various diseases such as autism, Parkinson's and Alzheimer's disease, and cancer [52]. However, several researchers in Australia have a number of objections to the harmful effects of residues on human health. Glyphosate-based products are being intensively tested by governments at all levels. Some jurisdictions have already banned or restricted its use [53].

Glyphosate and AMPA are largely retained in the surface soil layer as residues [54], and AMPA persists longer than the parent compound of glyphosate [55]. It was also shown that the content of pesticides in water samples from reservoirs, including glyphosate, was higher during the agricultural season (1 April–15 September) than during the off-season [56]. Residual concentrations of herbicides and their metabolites in harvested *Zea mays* L. plants increased in direct proportion to increasing the application rate of herbicides [57]. Glyphosate was also detected in soil and rice grains in rice production [58]. The persistence of herbicides, including glyphosate, in the soil and its effect on soybean yield has also been studied in soybeans [59].

1.4. Genetic Changes Due to Glyphosate in Living Organisms

The vast majority of negative results in well-conducted bacterial reversion and in vivo mammalian micronucleus and chromosome aberration tests indicate that glyphosate and typical glyphosate-based formulations (GBF) are not genotoxic in these nuclear assays. Reports of positive results for endpoints of deoxyribonucleic acid (DNA) damage indicate that glyphosate and GBFs tend to induce DNA damage at high or toxic doses, but the data suggest that this is due to cytotoxicity rather than the GBF activity of DNA that may be associated with surfactants [60].

Glyphosate affects the composition and activity of the microbial community in the rhizosphere and increases protein metabolism and decreases amino acid synthesis [61]. Glyphosate can also genetically modify living organisms. Glyphosate resistance appears at different levels in giant ragweed (*Ambrosia trifida*). Introns show a higher expression pattern with data measured in resistant individuals following putative glyphosate treatment [62]. In *Conyza bonariensis* individuals, many genes are differentially expressed upon glyphosate treatment, so treatment involves a large number of genes [63]. Glyphosate inhibits the pathway of tryptophan biosynthesis in the apical bud of soybean (*Glycine max*) [64].

Transcripts in the microbial community were affected by glyphosate. Some cyanobacteria, such as *Synechococcus*, may use glyphosate as a source of P. In many metabolic pathways, genes are overexpressed under glyphosate stress [65]. Polymerase chain reaction (PCR) analysis of mammalian somatic cells has shown that glyphosate induces gene expression changes [66].

The aim of our study was to demonstrate that glyphosate does not degrade in soil and is incorporated into crops, even several weeks after glyphosate treatment. Glyphosate can also be taken up by the plant through the soil. The main question of our study was whether or not there was a significant difference between the amount of glyphosate absorbed through the soil and the amount of glyphosate applied to the plant through the leaf. This is demonstrated by morphological residue detection and microscopic examination. Our model plant was the sunflower (*Helianthus annuus* L.).

2. Materials and Methods

2.1. Materials Used

For our experiments we used a non-glyphosate-resistant GK Milia CL sunflower hybrid variety, which is a variety bred by Gabonakutató Nonprofit Közhasznú Ltd. (Szeged, Hungary) [67], and is also owned by the Ltd. GK Milia CL is a commercially available sunflower variety that ripens early and tolerates abiotic stress well and can be treated with the 2018 certified Clearfield® weed control technology [68]. The seed was harvested in 2020, uncoated.

Glialka, manufactured by the Monsanto Company (St. Louis, MO, USA) and owned by Bayer GMBH (Monheim am Rhein, Germany), was used for the experiment. Total herbicide with active ingredient 360 g/L glyphosate.

2.2. Experimental Conditions

The experiment was carried out at the Budatétény Station of the Institute of Landscape Architecture, Urban Planning, and Garden Art of MATE in 2021 under greenhouse conditions. The seeds were sown in a plastic propagation tray measuring 59 cm × 29 cm × 7 cm. For this purpose, a medium suitable for growing seedlings in trays (Klassmann-Deilmann TS 3 Fine, Geeste, Germany) was used, the properties of which were as follows: pH (H_2O) 6, N 140 mg L^{-1}, P (P_2O_5) 100 mg L^{-1}, K (K_2O) 180 mg L^{-1}, Mg 100 mg L^{-1}, S 150 mg L^{-1}. The trays were lined with 40 seeds in sowing and 5 replicates in a randomized block arrangement. Germinating weeds were mechanically removed in half of the boxes, and no weed removal was performed in the other half. Fourteen days before sowing, the weeds were removed in the weed trays with the Glialka chemical. The following concentrations were used in each case: 500 ppm, 1000 ppm, and 2000 ppm.

The sunflower seeds were sown on 16 September 2021 in a glyphosate-free medium and the other part in a medium sprayed with Glialka. Plants sown in the herbicide-free medium were sprayed with Glialka at 3 weeks post-sowing. The amount of spray was applied in each case at a rate of 2 L/100 m^2 [68], in the 3 concentrations already mentioned. The plants were kept at 20 °C and only irrigation water was obtained during the experiment. The control group was seeded in a clean medium and received only irrigation water. Seedlings were evaluated at 5 weeks post-sowing. The following vegetative parameters were surveyed: Root and stem length, the fresh and dried weight of root and shoot, and the number of leaves. Plants that received a concentration of 2000 ppm were killed during the experiment and were therefore not evaluated further.

2.3. Histology

Histological samples were taken from three points of the measured plants: From the root collar, from the stem 1 cm above the root collar, and from the stem above the lowest leaf. The plants were cleaned with distilled water and pruned mechanically with a hand scalpel. A Euromex bScope BS.1153-PLi biological microscope with a compatible camera (Levenhuk m1400 plus) was used for the survey. Due to the pruning procedure, the oil immersion lens arrays could not be used, so due to the nature of the sections, PLi 4/0.1 lenses were used, which provided forty-fold magnification. The eyepiece was of the WF120×/20 type and size. The samples were not stained. The images were post-corrected with GIMP 2.10 (owned by Spencer Kimball, Peter Mattis).

2.4. Residue Testing

The test method of Gonclaves and Catrinck [69] was used for the measurements. The Varian GC/MS/MS 4000 instrument was used for the tests (Table 1), using the following materials:

- EPA 547 Glyposate solution 1000 μg/mL in H_2O, Sigma-Aldrich (St. Louis, MO, USA); Lot: LRAC4997.
- BFTFA + 1% TMCS; Sigma-Aldrich; Lot: BCCD0447.
- Acetonitrile; Merck (Budapest, Hungary); Lot: I09677130 829.
- Pyridine; Scharlau (Barcelona, Spain); Lot: PI0123.

Table 1. Chromatographic conditions used for residue testing [69].

Parameters	Values	
Column	Rxi-5Sil MS, 30.00 m, 0.25 mm ID, 0.25 μm	
Injector temperature:	280 °C	
Injected volume:	1 μL	
Split ratio:	splitless	
Carrier gas:	helium 5.0	
Carrier gas volume flow	constant, 2 mL/perc	
MS settings		
Ion trap temperature	150 °C	
Ion source temperature	200 °C	
Transfer line temperature	220 °C	
Ionization	EI	
Measurement mode	SIS	
SIS parameters		
Stored masses (m/z)	232	
	312	
	340	
Temperature (°C)	100	300
Scale of temperature increase (°C/min)	0.0	8.0
Time to maintain temperature (minutes)	0.00	0.00
Total time (minutes)	0.00	25.00

In the measurement procedure, approximately $1 \times g$ of sample was weighed to a centrifuge tube and 10 mL of acetonitrile was added (in the case of a smaller sample, proportions were kept). It was shaken for 1 h. It was then centrifuged for approximately $3000 \times g$ and filtered through a syringe filter. Then, 300 μL of this clear solution was taken and evaporated to dryness at 60 °C. After cooling, 60 μL of pyridine was added to the dry residue and, after waiting for 5 min, 100 μL of the N, O-bis (trimethylsilyl)trifluoroacetamide (BSTFA) + 1% trimethylsilyl chloride (TMCS) silylating agent was added. This mixture was heated at 60 °C for 30 min and then measured by Gas Chromatography—Mass Spectrometry. The procedure was the same when creating a calibration line.

2.5. Statistical Measurements

The samples were independent, and the correct experimental design and correctness of the sampling were ensured during the sampling. We used Microsoft Office 365 Excel to document our measurement data and Microsoft Office 365 Word to edit the text. The processing, comparison, and analysis of our measurable differences were performed with IBM SPSS Statistics 25 using a one-way analysis of variance (ANOVA). The measured data were analyzed with a 95% confidence level (significance) in all cases. When evaluating the Levene test, if probability value (p) > 0.05, then the Tukey test was used, and if $p < 0.05$, the Games-Howell test was used.

3. Results

3.1. Morphology

Performing morphological measurements was an important factor in the series of measurements. In this way, we can obtain an idea of the changes at the organ level that can occur in plants that receive glyphosate through leaves or roots. This also demonstrates that glyphosate is absorbed through the root.

3.1.1. Root Morphological Changes

Root length was significantly reduced in the treated groups (Figure 1). The control stock, which did not receive any form of glyphosate treatment, had an average root length of 11.56 cm at the time of final evaluation, which is significantly different from all treated groups. For plants treated with a 500 ppm solution, the lowest value (1.873 cm) was measured in the group receiving the chemical through the root. There was no significant difference in root lengths between the 500 ppm leaf-treated and both 1000 ppm treatments.

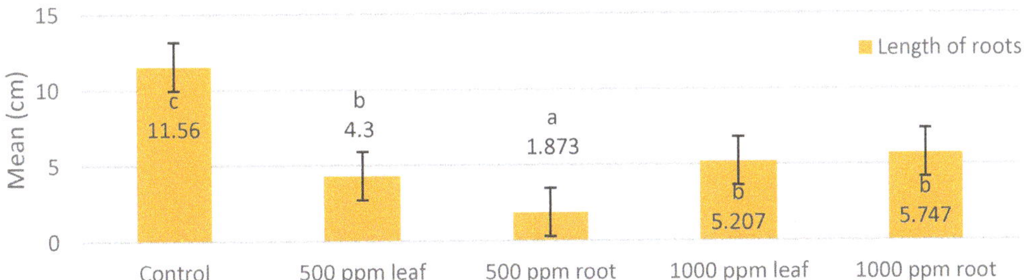

Figure 1. Length of roots on *Helianthus annuus* GK Milia CL after glyphosate treatment. Means with different letter are significantly different by Tukey's test at $p < 0.05$ (Budapest, 2021).

Comparing these data with fresh and dried root weights (Figure 2) it can be observed that the untreated group had the highest fresh root weight (0.7467 g), which was significantly different from the mean weight of the treated groups. The average fresh weight of the roots (0.6067 g) was also significantly higher in the 1000 ppm individuals treated through the root, but in this case, the root was also shorter. In the groups receiving the 500 ppm solution, the average fresh root weight of the leaf spray group was 0.41 g, from which the average fresh root weight of the glyphosate uptake (0.2253 g) was significantly different. The lowest fresh weight value was produced by the group receiving the leaf spray at a concentration of 1000 ppm (0.106 g). These data show that there are greater differences in the fresh weights of plants treated through the leaf and root than the higher concentration of glyphosate. For the groups receiving the 500 ppm solution, this difference, although showing some difference, is not significant. In summary, the effect of the 1000 ppm treatments, compared to each other, has already been shown in phenotypic properties. Compared to the control, much shorter, more fleshy roots were formed in sunflowers treated through roots at a concentration of 1000 ppm. Increased dry weight may indicate stress on the plant.

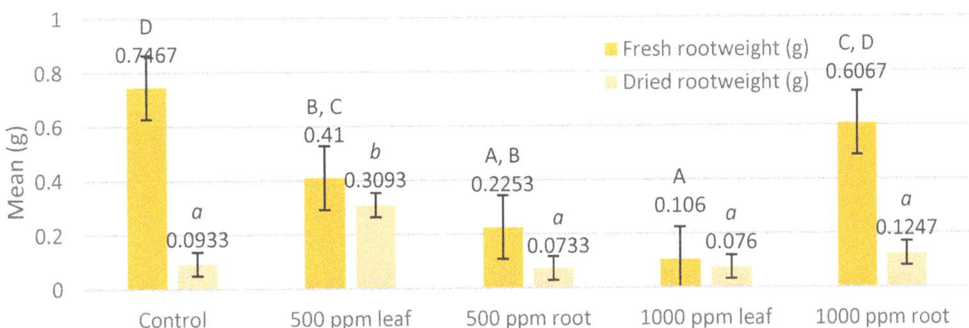

Figure 2. Parameters of roots on *Helianthus annuus* GK Milia CL after glyphosate treatment. Means with different letter are significantly different by Tukey's test at $p < 0.05$ (Budapest, 2021).

3.1.2. Stem Morphological Changes

Examining the stem weight results (Figure 3), clearly visible that the highest fresh (4.564 g) and dry (0.4253 g) stem weights were measured in the untreated individuals. These data show a significant difference with the results of all treated groups. The results of the parameters were analogous to the values measured at the roots, so the results measured in the group treated through roots with a concentration of 1000 ppm were significantly higher, both statistically, in the case of fresh (3.2447 g) and dried stems (0.336 g). This shows that absorption through the root is slower than in the case of leaf treatment with the same concentration solution, which produced almost the lowest measured average weight for fresh (0.8953 g) and dry weight (0.176 g). In the case of the treatment obtained with a solution concentration of 500 ppm, the opposite was observed, whereby the individuals receiving the leaf spray achieved a higher fresh and dried stem weight than the individuals treated through the root.

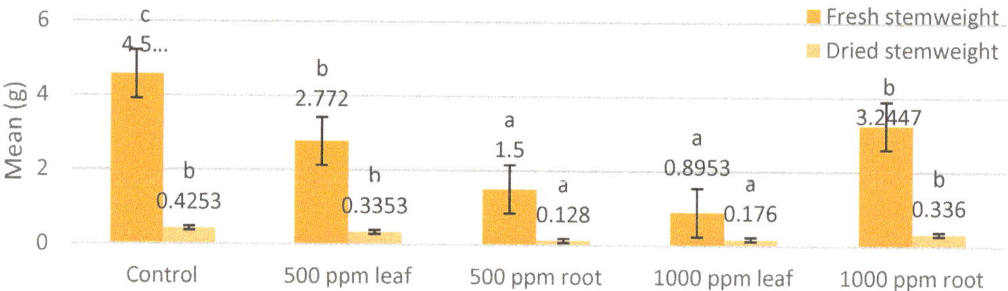

Figure 3. Rate of fresh and dried stem weight on *Helianthus annuus* GK Milia CL after glyphosate treatment. Means with different letter are significantly different by Tukey's test at $p < 0.05$ (Budapest, 2021).

3.2. Histology

Examining the stem cross-sections (Figure 4) it can be observed that the stem section shows a uniform picture in the control group (Figure 4c). The individual tissue areas are clearly visible and can be distinguished from each other. Epidermal cells close and are uniform. The location of the vascular bundles is regular and well separable from the basal tissues. The basal tissue cells are regular, and the cell wall is strong and uniform. No change or distortion is visible. As the glyphosate concentration increases, the deterioration and disintegration of the tissue structure can be clearly seen, which increases and becomes more visible with increasing concentration. Necrosis of pith cells started 4 days after treatment [70]. In the groups treated with a 500 ppm solution (Figure 4a,b), intercellular cavities (vacuoles) appear, which may indicate glyphosate-induced stress in young seedlings, as plant cell vacuoles are multifunctional organelles that play a central role in cell development strategies. They are involved in cellular responses to environmental and biotic factors that cause stress [71]. The transport vessel system is thinner, smaller in cross-section than the control group due to glyphosate, the vascular bundles are damaged, the shape of the parenchyma tissue cells changes, and the cell wall becomes thinner, which is also a phase of cell death. Glyphosate also has an effect on the disorganization of the transport tissue system [72]. This is characterized by the rupture of the plasma membrane [73]. Epidermal tissue cells adhere more loosely. The cell wall of the epidermis is thinner and not uniform in thickness.

In the groups treated with the 1000 ppm solution (Figure 4d,e), the process of cell death can be observed more and more strongly. In plants receiving this concentration, the tissues responded even more strongly. In these groups, the enlargement and disintegration of cells in the central parenchyma tissue and the weakened state of the transport tissue system became even more intense. These cells are necrotically disrupted, cell walls are not always separable, and cells are small in size. The vascular bundles were irregularly arranged, not delimited. The epithelial cells are small, do not adhere tightly, and the cell walls are weak, in several cases torn apart. Chloroplasts also disintegrated, which is also related to Lee's 1981 [74] finding that there was an inhibition of chlorophyll synthesis by glyphosate.

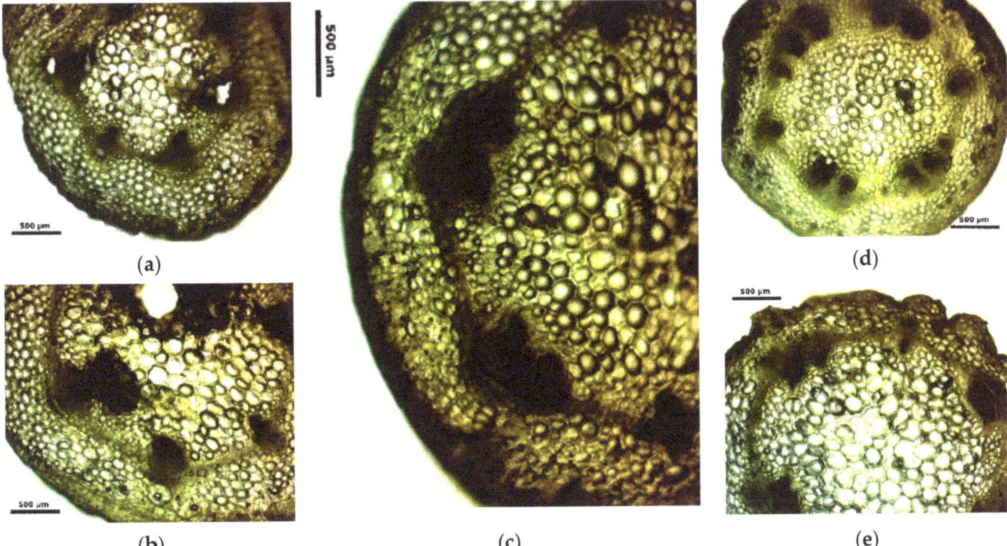

Figure 4. Root cross sections after glyphosate treatments: (**a**) 500 ppm concentration root treatment; (**b**) 500 ppm concentration leaf treatment; (**c**) control treatment; (**d**) 1000 ppm root treatment; (**e**) 1000 ppm leaf treatment (Budapest, 2021).

Comparing the histological changes caused by root and leaf treatments, it can be observed that the histological changes in the root-treated groups are weaker than in the leaf-treated groups, but the effect of glyphosate is also very pronounced in the untreated groups. A more advanced state of the glyphosate effect can be observed in the plants of the leaf-treated groups, which confirms our partial result presented earlier: The effect is stronger in the leaf-treated individuals and the absorption of the chemical is faster. This is confirmed by the worse condition of the cells, the thinner cell walls, and the greater disorder of the transport tissue system.

Overall, the effect of glyphosate can be observed in individual tissue areas, the effect of which is directly proportional to the increase in concentration.

3.3. Result of Residue Tests

In the case of residue measurements, the root and stem parts were evaluated separately (Figure 5). This was considered important because it allows the extent of glyphosate uptake by the root to be monitored even more. Glyphosate accumulated in both leaf-treated and root-treated plants when treated with 500 ppm and 1000 ppm solutions, respectively. The amount detectable for the latter is higher in both concentrations compared to the control treatment. We found that in plants treated with leaf spray, the residue value was significantly higher in the root than in the stem.

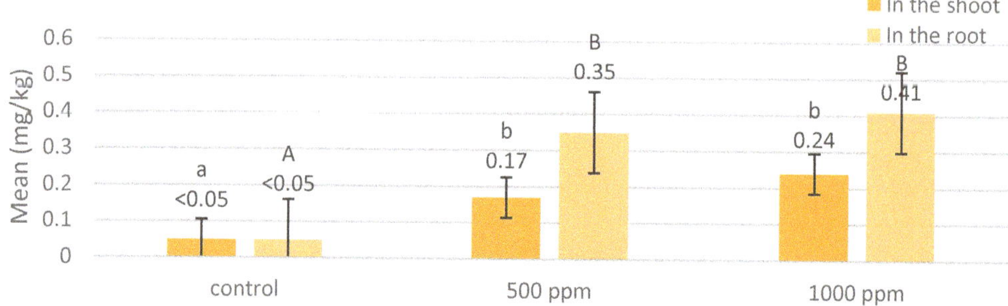

Figure 5. Glyphosate residue in leaf and root on *Helianthus annuus* GK Milia CL after glyphosate spray treatment on leaves. Means with different letter are significantly different by Tukey's test at $p < 0.05$ (Budapest, 2021).

In the groups treated through the root, it can be said that glyphosate residue was detectable in all cases (Figure 6). In the group treated with the 500 ppm solution, the residue measured in the roots (0.19 mg/kg), despite the lower concentration, is not statistically different from the results measured in the root at a concentration of 1000 ppm (0.2 mg/kg). However, when examining the shoots, there is a significant difference between the two concentrations. In the group treated with the 500 ppm solution, the residue detectable in the shoot was as low as in the control treatment (0.07 mg/kg), while in the group treated with 1000 ppm, it was more than twice as high. Overall, glyphosate was absorbed from the root into the plant organism, which was also detectable in the leaf and root.

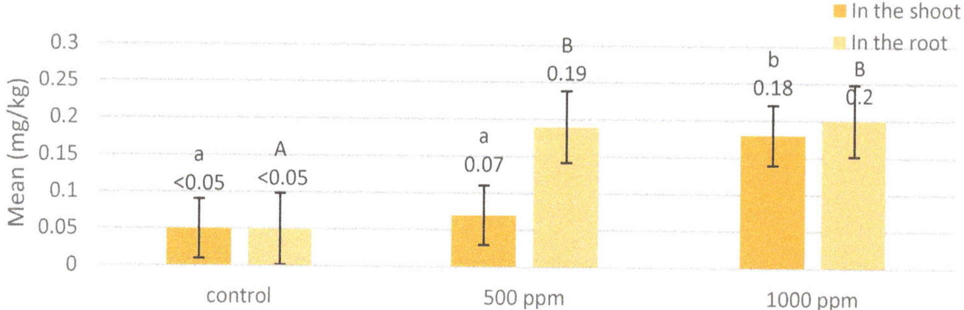

Figure 6. Glyphosate residue in leaf and root on *Helianthus annuus* GK Milia CL after glyphosate spray treatment on root. Means with different letter are significantly different by Tukey's test at $p < 0.05$ (Budapest, 2021).

4. Discussion

Glyphosate is a worldwide used herbicide [9] that, according to several studies, can both modify the gene pool of living organisms [60,63] and also have an adverse effect on the biosphere [46,47]. It may also be responsible for the development of Alzheimer's and Parkinson's disease [52].

Glyphosate is also a major herbicide in field crops. The sunflower (*Helianthus annuus* L.) cultivar GK Milia CL [67] is a very environmental stress-tolerant variety, which was used in our studies.

Although several studies have shown that glyphosate causes morphological changes in the plant [32], and glyphosate is rapidly degraded and is not transferred to other living organisms [4]. By preemergent treatment of a sunflower plant, we would have liked to investigate that glyphosate remains in the plant for several weeks and is detectable. The plants were not grown until seed ripening, so the topic of a future series of measurements will be the morphological, histological changes, and residue results that can be detected in the pre-harvest state. In the present experiment, sunflower plants were surveyed at 5 weeks of age to model the composition and properties of green manure or fodder harvested in this condition as a result of prior glyphosate treatment. The measurements were also based on experiments by [23], in which the author explains that the highest sensitivity to glyphosate can be attributed to the beginning of the vegetative stages of plants. In our series of measurements, we sought to determine whether the treatment of non-chemically treated plants with adjacent row spacing and the treatment of plants in adjacent rows with glyphosate had an effect.

Sunflower plants were evaluated by morphological and histological methods and residue measurements were performed. Morphological measurements confirmed that glyphosate, both applied to the leaf and absorbed through the root, is evident in the morphological parameters. Our results were similar to those described by [36] and contradict the results of Reddy et al. that glyphosate has no effect on root and shoot dry weight but increases shoot biomass [20].

The morphological results confirmed Zobiole et al.'s [24] measurements that biomass decreases with increasing glyphosate levels. The presence of glyphosate caused morphological and histological degeneration in the plant in all cases, as we have shown in our results. Our results confirm the findings made by Gomes et al. [32] and contradict Khan et al.'s [34] findings. The appearance of glyphosate in the tissues can be detected as early as 48 h after treatment, and these effects can be observed more and more over time [75], as was observed in our measurements. Tissue change caused by glyphosate was observed over the entire surface of the stem cross-section in all tissue areas. All tissue areas were distorted, disordered, cells were destroyed, and in several cases, vacuum cell death occurred, which is also related to van Doorn's [73] work. These tissue and cellular changes can be observed in the case of root-treated plants as well as in the case of stem-treated plants, which is an excellent example of our accumulation of glyphosate in root-treated plants.

In the residue evaluation, we found that glyphosate levels were significantly detectable at week 5 post-preemergent treatment, exceeding Wang et al. [28] finding that glyphosate was detectable for 14 days after treatment. Higher glyphosate levels were measurable in the roots, similar to the findings of Pline et al. [29] and Claus and Behrens [30]. This is also related to Sesin et al.'s [31] findings that glyphosate accumulates in different amounts in different tissues and organs. The amount of glyphosate in the shoots and roots of sunflower was significantly different in several cases. All this proved that glyphosate could be detected in the plant more than 4 weeks after treatment, and that glyphosate in the soil was absorbed through the root in statistically detectable amounts.

5. Conclusions

One of the great challenges of the present is the continued supply of food to an ever-increasing human population. This goal can only be achieved effectively if good quality, high-content, vital plants are produced. The primary task to create this is to produce and maintain a weed-free area. The use of glyphosate, although currently one of the most effective herbicides in the world, raises a number of environmental issues. As a result of our series of measurements, it can be stated that glyphosate is present in the plant even if it is not in direct contact with glyphosate.

This poses a risk, even if the fodder plant is used for human or animal use. In the case of young plants, its use as a silage plant may be of even greater concern. Another dangerous area is its use as green manure, which also means its use at a young age, and this is related to Lee's [74] measurements that the residue is more detectable in young plants. This can be particularly dangerous for the above-mentioned uses in agriculture. Glyphosate may be present in the plant in an amount that can be determined by testing.

Glyphosate therefore accumulates in the plant even if it does not come into direct contact with the plant, rather only with the soil of the adjacent row spacing or with the crop or weed population on it. This is also related to Golt and Wood's [19] results.

The use of glyphosate is not expected to decrease in the future if the plant protection product remains on the market. Due to its decades-long use, glyphosate and its excipients have been shown to accumulate in the soil, water, and living organisms. Their decomposition process is slow and carries a high risk to the environment and human health. In general, glyphosate is currently one of the most effective and widely used herbicides, the extraction of which can have significant economic effects. Therefore, it is advisable to consider alternative remediation technologies that can be used to mitigate these risks while increasing economic outcomes. Thus, it can be an effective tool for environmental protection.

Author Contributions: Conceptualization, L.O.; methodology, L.O. and S.K.; software, D.F.; validation, D.F., A.N. and K.H.; formal analysis, D.F. and A.N.; investigation, S.K. and K.H.; resources, L.O. and S.K.; data curation, D.F., S.K. and K.H.; writing—original draft preparation, S.K., D.F. and K.H.; writing—review and editing, D.F., S.K. and A.N.; visualization, D.F., K.H., L.O., A.N. and S.K.; supervision, A.N. and L.O.; project administration, D.F., K.H., S.K. and A.N.; funding acquisition, L.O. All authors have read and agreed to the published version of the manuscript.

Funding: This research received no external funding.

Institutional Review Board Statement: Not applicable.

Informed Consent Statement: Not applicable.

Data Availability Statement: The data presented in this study are available upon request from the appropriate author. The data are not public, as the series of measurements has not yet been completed, further research is underway.

Acknowledgments: Many thanks to Julianna Györgyné Gondos and Zsanett Istvánfi for their technical support during the experiment.

References

1. Duke, S.O.; Powles, S.B. Glyphosate: A once-in-a-century herbicide. *Pest Manag. Sci.* **2008**, *64*, 319–325. [CrossRef]
2. Johal, G.S.; Huber, D.M. Glyphosate effects on diseases of plants. *Eur. J. Agron.* **2009**, *31*, 144–152. [CrossRef]
3. Woodburn, A.T. Glyphosate: Production, pricing and use worldwide. *Pest Manag. Sci.* **2000**, *56*, 309–312. [CrossRef]
4. Jenkins, P.G. *Environmental Health Criteria for Glyphosate*, 1st ed.; World Health Organization (WHO): Switzerland, Geneva, 1994; ISBN 9241571594.
5. Carlisle, S.M.; Trevors, J.T. Glyphosate in the environment. *Water Air Soil Pollut.* **1988**, *39*, 409–420. [CrossRef]
6. Benbrook, C.M. Trends in glyphosate herbicide use in the United States and globally. *Environ. Sci. Eur.* **2016**, *28*, 3. [CrossRef] [PubMed]
7. Antier, C.; Kudsk, P.; Reboud, X.; Ulber, L.; Baret, P.V.; Messéan, A. Glyphosate Use in the European Agricultural Sector and a Framework for Its Further Monitoring. *Sustainability* **2020**, *12*, 5682. [CrossRef]
8. Van Bruggen, A.H.C.; He, M.M.; Shin, K.; Mai, V.; Jeong, K.C.; Finckh, M.R.; Morris, J.G. Environmental and health effects of the herbicide glyphosate. *Sci. Total Environ.* **2018**, *616*, 255–268. [CrossRef] [PubMed]
9. Duke, S.O. The history and current status of glyphosate. *Pest Manag. Sci.* **2018**, *74*, 1027–1034. [CrossRef] [PubMed]
10. Zhu, J.; Patzoldt, W.L.; Shealy, R.T.; Vodkin, L.O.; Clough, S.J.; Tranel, P.J. Transcriptome Response to Glyphosate in Sensitive and Resistant Soybean. *J. Agric. Food Chem.* **2008**, *56*, 6355–6363. [CrossRef] [PubMed]
11. Shaner, D.L. Role of Translocation as A Mechanism of Resistance to Glyphosate. *Weed Sci.* **2009**, *57*, 118–123. [CrossRef]
12. Vazquez-Garcia, J.G.; Palma-Bautista, C.; Rojano-Delgado, A.M.; De Prado, R.; Menendez, J. The First Case of Glyphosate Resistance in Johnsongrass (*Sorghum halepense* (L.) Pers.) in Europe. *Plants* **2020**, *9*, 313. [CrossRef]
13. Alarcón-Reverte, R.; García, A.; Urzúa, J.; Fischer, A.J. Resistance to Glyphosate in Junglerice (*Echinochloa colona*) from California. *Weed Sci.* **2013**, *61*, 48–54. [CrossRef]
14. Pieniazek, D.; Bukowska, B.; Duda, W. Glyphosate—A non-toxic pesticide? *Medycyna Pracy* **2003**, *54*, 579–583.
15. Bradberry, S.M.; Proudfoot, A.T.; Vale, J.A. Glyphosate Poisoning. *Toxicol. Rev.* **2004**, *23*, 159–167. [CrossRef]
16. Bradshaw, L.D.; Padgette, S.R.; Kimball, S.L.; Wells, B.H. Perspectives on Glyphosate Resistance. *Weed Technol.* **1997**, *11*, 189–198. [CrossRef]
17. Helander, M.; Saloniemi, I.; Saikkonen, K. Glyphosate in northern ecosystems. *Trends Plant Sci.* **2012**, *17*, 569–574. [CrossRef]

18. Reddy, K.N.; Rimando, A.M.; Duke, S.O.; Nandula, V.K. Aminomethylphosphonic Acid Accumulation in Plant Species Treated with Glyphosate. *J. Agric. Food Chem.* **2008**, *56*, 2125–2130. [CrossRef]

19. Golt, A.; Wood, L.J. Glyphosate-based herbicides alter the reproductive morphology of *Rosa acicularis* (Prickly Rose). *Front. Plant Sci.* **2021**, *12*, 1184. [CrossRef] [PubMed]

20. Reddy, K.N.; Zablotowicz, R.M. Glyphosate-resistant soybean response to various salts of glyphosate and glyphosate accumulation in soybean nodules. *Weed Sci.* **2003**, *51*, 496–502. [CrossRef]

21. Zobiole, L.H.S.; Kremer, R.J.; de Oliveira, R.S., Jr.; Constantin, J. Glyphosate effects on photosynthesis, nutrient accumulation, and nodulation in glyphosate-resistant soybean. *J. Plant Nutr. Soil Sci.* **2012**, *175*, 319–330. [CrossRef]

22. Cakmak, I.; Yazici, A.; Tutus, Y.; Ozturk, L. Glyphosate reduced seed and leaf concentrations of calcium, manganese, magnesium, and iron in non-glyphosate resistant soybean. *Eur. J. Agron.* **2009**, *31*, 114–119. [CrossRef]

23. Bellaloui, N.; Reddy, K.N.; Zablotowicz, R.M.; Mengistu, A. Simulated Glyphosate Drift Influences Nitrate Assimilation and Nitrogen Fixation in Non-glyphosate-Resistant Soybean. *J. Agric. Food Chem.* **2006**, *54*, 3357–3364. [CrossRef]

24. Zobiole, L.H.S.; de Oliveira Junior, R.S.; Kremer, R.J.; Muniz, A.S.; de Oliveira Junior, A. Nutrient accumulation and photosynthesis in glyphosate-resistant soybeans is reduced under glyphosate use. *J. Plant Nutr.* **2010**, *33*, 1860–1873. [CrossRef]

25. Kitchen, L.M.; Witt, W.W.; Rieck, C.E. Inhibition of Chlorophyll Accumulation by Glyphosate. *Weed Sci.* **1981**, *29*, 513–516. [CrossRef]

26. Zablotowicz, R.M.; Reddy, K.N. Impact of Glyphosate on the Bradyrhizobium japonicum Symbiosis with Glyphosate-Resistant Transgenic Soybean. *J. Environ. Qual.* **2004**, *33*, 825–831. [CrossRef] [PubMed]

27. de María, N.; Becerril, J.M.; García-Plazaola, J.I.; Hernández, A.; De Felipe, M.R.; Fernández-Pascual, M. New insights on glyphosate mode of action in nodular metabolism: Role of shikimate accumulation. *J. Agric. Food Chem.* **2006**, *54*, 2621–2628. [CrossRef] [PubMed]

28. Wang, Y.-S.; Jaw, C.-G.; Chen, Y.-L. Accumulation of 2,4-D and glyphosate in fish and water hyacinth. *Water Air Soil Pollut.* **1994**, *74*, 397–403. [CrossRef]

29. Pline, W.A.; Wilcut, J.W.; Duke, S.O.; Edmisten, K.L.; Wells, R. Tolerance and Accumulation of Shikimic Acid in Response to Glyphosate Applications in Glyphosate-Resistant and Nonglyphosate-Resistant Cotton (*Gossypium hirsutum* L.). *J. Agric. Food Chem.* **2002**, *50*, 506–512. [CrossRef] [PubMed]

30. Claus, J.S.; Behrens, R. Glyphosate Translocation and Quackgrass Rhizome Bud Kill. *Weed Sci.* **1976**, *24*, 149–152. [CrossRef]

31. Sesin, V.; Davy, C.M.; Dorken, M.E.; Gilbert, J.M.; Freeland, J.R. Variation in glyphosate effects and accumulation in emergent macrophytes. *Manag. Biol. Invasions* **2021**, *12*, 66. [CrossRef]

32. Gomes, M.P.; Rocha, D.C.; Moreira de Brito, J.C.; Tavares, D.S.; Marques, R.Z.; Soffiatti, P.; Sant'Anna-Santos, B.F. Emerging contaminants in water used for maize irrigation: Economic and food safety losses associated with ciprofloxacin and glyphosate. *Ecotoxicol. Environ. Saf.* **2020**, *196*, 110549. [CrossRef]

33. Strandberg, B.; Sørensen, P.B.; Bruus, M.; Bossi, R.; Dupont, Y.L.; Link, M.; Damgaard, C.F. Effects of glyphosate spray-drift on plant flowering. *Environ. Pollut.* **2021**, *280*, 116953. [CrossRef]

34. Khan, S.; Zhou, J.L.; Ren, L.; Mojiri, A. Effects of glyphosate on germination, photosynthesis and chloroplast morphology in tomato. *Chemosphere* **2020**, *258*, 127350. [CrossRef]

35. Konlan, S.; Quaye, A.K.; Pobee, P.; Amon-Armah, F.; Dogbatse, J.A.; Arthur, A.; Fiakpornu, R.; Dogbadzi, R. Effect of weed management with glyphosate on growth and early yield of young cocoa (*Theobroma cacao* L.) in Ghana. *AJAR* **2019**, *14*, 1229–1238. [CrossRef]

36. Maishkar, M.N. Effect of glyphosate on morphological characters of Sida acuta burm F. *Bioinfolet Q. J. Life Sci.* **2020**, *17*, 585–586.

37. Yu, H.; Peng, J.; Cao, X.; Wang, Y.; Zhang, Z.; Xu, Y.; Qi, W. Effects of microplastics and glyphosate on growth rate, morphological plasticity, photosynthesis, and oxidative stress in the aquatic species *Salvinia cucullata*. *Environ. Pollut.* **2021**, *279*, 116900. [CrossRef] [PubMed]

38. Liu, N.; Zhong, G.; Zhou, J.; Liu, Y.; Pang, Y.; Cai, H.; Wu, Z. Separate and combined effects of glyphosate and copper on growth and antioxidative enzymes in *Salvinia natans* (L.) All. *Sci. Total Environ.* **2019**, *655*, 1448–1456. [CrossRef] [PubMed]

39. Sikorski, Ł.; Baciak, M.; Bęś, A.; Adomas, B. The effects of glyphosate-based herbicide formulations on *Lemna minor*, a non-target species. *Aquat. Toxicol.* **2019**, *209*, 70–80. [CrossRef] [PubMed]

40. Malalgoda, M.; Ohm, J.-B.; Howatt, K.A.; Green, A.; Simsek, S. Effects of pre-harvest glyphosate use on protein composition and shikimic acid accumulation in spring wheat. *J. Agric. Food Chem.* **2020**, *332*, 127422. [CrossRef]

41. Placido, H.F.; Santos, R.F.; Oliveira Júnior, R.S.; Marco, L.R.; Silva, A.F.M.; Barroso, A.M.; Albrecht, A.J.P.; Victoria Filho, R. Morphological characterization of the foliar surface in glyphosate-resistant tall windmill grass. *Agron. J.* **2021**, in press. [CrossRef]

42. Maldani, M.; Aliyat, F.Z.; Cappello, S.; Morabito, M.; Giarratana, F.; Nassiri, L.; Ibijbijen, J. Effect of glyphosate and paraquat on seed germination, amino acids, photosynthetic pigments and plant morphology of *Vicia faba*, *Phaseolus vulgaris* and *Sorghum bicolor*. *Environ. Sustain.* **2021**, *4*, 723–733. [CrossRef]

43. Singh, L.; Agarwal, T. Polycyclic aromatic hydrocarbons in diet: Concern for public health. *Trends Food Sci. Technol.* **2018**, *79*, 160–170. [CrossRef]

44. Jacobs, S.; Dendoncker, N.; Martín-López, B.; Brton, D.N.; Gomez-Baggethun, E.; Boeraeve, F.; McGrath, F.L.; Vierikko, K.; Geneletti, D.; Sevecke, K.J.; et al. A new valuation school: Integrating diverse values of nature in resource and land use decisions. *Ecosyst. Serv.* **2016**, *22*, 213–220. [CrossRef]

45. Blume, E.; Bischoff, M.; Reichert, J.M.; Moorman, T.; Konopka, A.; Turco, R.F. Surface and subsurface microbial biomass, community structure and metabolic activity as a function of soil depth and sason. *Appl. Soil Ecol.* **2002**, *20*, 171–181. [CrossRef]
46. Centner, T.J.; Russell, L.; Mays, M. Viewing evidence of harm accompanying uses of glyphosate-based herbicides under US legal requirements. *Sci. Total Environ.* **2019**, *648*, 609–617. [CrossRef]
47. Boutin, C.; Strandberg, B.; Carpenter, D.; Mathiassen, S.K.; Thomas, P.J. Herbicide impact on non-target plant reproduction: What are the toxicological and ecological implications? *Environ. Pollut.* **2014**, *185*, 295–306. [CrossRef] [PubMed]
48. Agostini, L.P.; Dettogni, R.S.; dos Reis, R.S.; Stur, E.; dos Santos, E.V.W.; Ventorim, D.P.; Garcia, F.M.; Cardoso, R.C.; Graceli, J.B.; Louro, I.D. Effects of glyphosate exposure on human health: Insights from epidemiological and in vitro studies. *Sci. Total Environ.* **2020**, *705*, 135808. [CrossRef]
49. Rendon-von Osten, J.; Dzul-Caamal, R. Glyphosate Residues in Groundwater, Drinking Water and Urine of Subsistence Farmers from Intensive Agriculture Localities: A Survey in Hopelchén, Campeche, Mexico. *Int. J. Environ. Res. Public Health* **2017**, *14*, 595. [CrossRef]
50. Andriani, L.T.; Aini, L.Q.; Hadiastono, T. Glyphosate biodegradation by plant growth promoting bacteria and their effect to paddy germination in glyphosate contaminated soil. *J. Degrad. Min. Land Manag.* **2017**, *5*, 995–1000. [CrossRef]
51. Tsai, W.-T. Trends in the Use of Glyphosate Herbicide and Its Relevant Regulations in Taiwan: A Water Contaminant of Increasing Concern. *Toxics* **2019**, *7*, 4. [CrossRef]
52. Rodrigues, N.R.; de Souza, A.P.F. Occurrence of glyphosate and AMPA residues in soy-based infant formula sold in Brazil. *Food Addit. Contam. Part A* **2018**, *35*, 724–731. [CrossRef] [PubMed]
53. Beckie, H.J.; Flower, K.C.; Ashworth, M.B. Farming without Glyphosate? *Plants* **2020**, *9*, 96. [CrossRef] [PubMed]
54. Lupi, L.; Bedmar, F.; Puricelli, M.; Marino, D.; Aparicio, V.C.; Wunderlin, D.; Miglioranza, K.S.B. Glyphosate runoff and its occurrence in rainwater and subsurface soil in the nearby area of agricultural fields in Argentina. *Chemosphere* **2019**, *225*, 906–914. [CrossRef]
55. Karanasios, E.; Karasali, H.; Marousopoulou, A.; Akrivou, A.; Markellou, E. Monitoring of glyphosate and AMPA in soil samples from two olive cultivation areas in Greece: Aspects related to spray operators activities. *Environ. Monit. Assess.* **2018**, *190*, 361. [CrossRef] [PubMed]
56. Grantz, E.M.; Leslie, D.; Reba, M.; Willett, C. Residual herbicide concentrations in on-farm water storage–tailwater recovery systems: Preliminary assessment. *J. Agric. Environ.* **2020**, *5*, e20009. [CrossRef]
57. Best-Ordinioha, J.C.; Ataga, E.A.; Ordinioha, B. The effect of the application of different rates of herbicides on the residual level of the herbicides and their metabolites in harvested maize cobs. *PMJ* **2017**, *11*, 122–126. [CrossRef]
58. Chansuvarn, W.; Chansuvarn, S. Distribution of Residue Carbofuran and Glyphosate in Soil and Rice Grain. *Appl. Mech. Mater.* **2018**, *879*, 118–124. [CrossRef]
59. Nunes, A.L.; Lorenset, J.; Gubiani, J.E.; Santos, F.M. A Multy-Year Study Reveals the Importance of Residual Herbicides on Weed Control in Glyphosate-Resistant Soybean. *Planta Daninha* **2018**, *36*, e018176135. [CrossRef]
60. Kier, L.D.; Kirkland, D.J. Review of genotoxicity studies of glyphosate and glyphosate-based formulations. *Crit. Rev. Toxicol.* **2013**, *43*, 283–315. [CrossRef]
61. Newman, M.M.; Lorenz, N.; Hoilett, N.; Lee, N.R.; Dick, R.P.; Liles, M.R.; Ramsier, C.; Kloepper, J.W. Changes in rhizosphere bacterial gene expression following glyphosate treatment. *Sci. Total Environ.* **2016**, *553*, 32–41. [CrossRef]
62. Sablok, G.; Amiryousefi, A.; He, X.; Hyvönen, J.; Poczai, P. Sequencing the Plastid Genome of Giant Ragweed (*Ambrosia trifida, Asteraceae*) From a Herbarium Specimen. *Front. Plant Sci.* **2019**, *10*, 218. [CrossRef]
63. Hereward, J.P.; Werth, J.A.; Thornby, D.F.; Keenan, M.; Chauhan, B.S.; Walter, G.H. Gene expression in response to glyphosate treatment in fleabane (*Conyza bonariensis*)—Glyphosate death response and candidate resistance genes. *Pest Manag. Sci.* **2018**, *74*, 2346–2355. [CrossRef]
64. Jiang, L.-X.; Jin, L.-G.; Guo, Y.; Tao, B.; Qiu, L.-J. Glyphosate effects on the gene expression of the apical bud in soybean (Glycine max). *Biochem. Biophys. Res. Commun.* **2013**, *437*, 544–549. [CrossRef] [PubMed]
65. Lu, T.; Xu, N.; Zhang, Q.; Zhang, Z.; Debognies, A.; Zhou, Z.; Sun, L.; Qian, H. Understanding the influence of glyphosate on the structure and function of freshwater microbial community in a microcosm. *Environ. Pollut.* **2020**, *260*, 114012. [CrossRef]
66. Hokanson, R.; Fudge, R.; Chowdhary, R.; Busbee, D. Alteration of estrogen-regulated gene expression in human cells induced by the agricultural and horticultural herbicide glyphosate. *Hum. Exp. Toxicol.* **2007**, *26*, 747–752. [CrossRef]
67. GabonaKutató Zrt. (GK), GK Milia CL új Vetőmag. Available online: https://www.gabonakutato.hu/hu/vetomag/napraforgo/olajnapraforgo/gk-milia-cl (accessed on 24 November 2021).
68. Agriculture BASF (BASF), Clearfield® Production Systems. Available online: https://agriculture.basf.us/crop-protection/products/herbicides/clearfield.html (accessed on 24 November 2021).
69. Catrinck, T.C.P.G. Estudo da Derivatização de Glyphosate e AMPA Utilizando FMOC-CL e BSTFA para Análises Cromatográficas. Master's Thesis, Universidade Federal dos Vales do Jequitinhonha e Mucuri, Diamantina, MG, Brasil, 2013.
70. Lorentz, L.; Beffa, R.; Kraehmer, H. Recovery of plants and histological observations on advanced weed stages after glyphosate treatment. *Weed Res.* **2011**, *51*, 333–343. [CrossRef]
71. Francis, M. Plant vacuoles. *Plant Cell* **1999**, *11*, 587–599. [CrossRef]
72. Canal, M.J.; Albuerne, M.; Sanchez Tamés, R.; Fernandez, B. Glyphosate injury on *Cyperus esculent* us leaves and basal bulbs: Histological study. *Weed Res.* **1990**, *30*, 117–122. [CrossRef]

73. Van Doorn, W.G.; Beers, E.P.; Dangl, J.L.; Franklin-Tong, V.E.; Gallois, P.; Hara-Nishimura, I.; Jones, A.M.; Kawai-Yamada, M.; Lam, E.; Mundy, J.; et al. Morphological classification of plant cell deaths. *Cell Death Differ.* **2011**, *18*, 1241–1246. [CrossRef]
74. Lee, T.T. Effects of glyphosate on synthesis and degradation of chlorophyll in soybean and tobacco cells. *Weed Res.* **1981**, *21*, 161–164. [CrossRef]
75. Uotila, M.; Evjen, K.; Iversen, T.-H. The effects of glyphosate on the development and cell infrastructure of white mustard (*Sinapis alba* L.) seedlings. *Weed Res.* **1980**, *20*, 153–158. [CrossRef]

*sustainability*

MDPI

Article

CO$_2$ Responses of Winter Wheat, Barley and Oat Cultivars under Optimum and Limited Irrigation

Zsuzsanna Farkas [1,2,*], Angéla Anda [3], Gyula Vida [1], Ottó Veisz [1,*] and Balázs Varga [1]

[1]   Agricultural Institute, Centre for Agricultural Research, Eötvös Loránd Research Network, H-2462 Martonvásár, Hungary; gyula.vida@atk.hu (G.V.); varga.balazs@atk.hu (B.V.)
[2]   Department of Environmental Sustainability, Festetics Doctoral School, IES, Hungarian University of Agriculture and Life Sciences, H-8360 Keszthely, Hungary
[3]   Georgikon Campus, IES, Hungarian University of Agriculture and Life Sciences, H-8360 Keszthely, Hungary; anda.angela@uni-mate.hu
*   Correspondence: farkas.zsuzsanna@atk.hu (Z.F.); veisz.otto@atk.hu (O.V.); Tel.: +36-22-569-500 (Z.F.)

Abstract: Field crop production must adapt to the challenges generated by the negative consequences of climate change. Yield loss caused by abiotic stresses could be counterbalanced by increasing atmospheric CO$_2$ concentration, but C$_3$ plant species and varieties have significantly different reactions to CO$_2$. To examine the responses of wheat, barley and oat varieties to CO$_2$ enrichment in combination with simulated drought, a model experiment was conducted under controlled environmental conditions. The plants were grown in climate-controlled greenhouse chambers under ambient and enriched (700 ppm and 1000 ppm) CO$_2$ concentrations. Water shortage was induced by discontinuing the irrigation at BBCH stages 21 and 55. Positive CO$_2$ responses were determined in barley, but the CO$_2$-sink ability was low in oats. Reactions of winter wheat to enriched CO$_2$ concentration varied greatly in terms of the yield parameters (spike number and grain yield). The water uptake of all wheat cultivars decreased significantly; however at the same time, water-use efficiency improved under 1000 ppm CO$_2$. Mv Ikva was not susceptible to CO$_2$ fertilization, while no consequent CO$_2$ reactions were observed for Mv Nádor and Mv Nemere. Positive CO$_2$ responses were determined in Mv Kolompos.

Keywords: winter cereals; CO$_2$ enrichment; drought stress; WUE; climate change

Citation: Farkas, Z.; Anda, A.; Vida, G.; Veisz, O.; Varga, B. CO$_2$ Responses of Winter Wheat, Barley and Oat Cultivars under Optimum and Limited Irrigation. *Sustainability* **2021**, *13*, 9931. https://doi.org/10.3390/su13179931

Academic Editor: Roberto Mancinelli

Received: 29 July 2021
Accepted: 30 August 2021
Published: 3 September 2021

Publisher's Note: MDPI stays neutral with regard to jurisdictional claims in published maps and institutional affiliations.

1. Introduction

The Industrial Revolution had significant environmental and social impacts. Due to its enormous agricultural, hygienic and medical achievements, the human population is projected to exceed 10 billion by the end of the century [1]. Among others, the greatest challenges of the upcoming decades will be to maintain food security and to ensure drinking water supply. The inventions of the Industrial Revolution accelerated not only the rate of population growth, but also the burning of fossil fuels, increasing the concentration of atmospheric CO$_2$ from 280 ppm [2] to ~416 ppm [3]. If CO$_2$ emission remains at the current level, in 30 years its atmospheric concentration will reach 550 ppm [4]. Although CO$_2$ is part of the atmosphere and is necessary for normal plant functions, it has become one of the most significant greenhouse gases due to its level, which has almost doubled [5] since its first measurement. Increased CO$_2$ affects photosynthesis, decreases water use, improves the growth and production of the plants [6], has direct implications for plant metabolism and decreases photorespiration [7]. In relation to this, increasing carbon–nitrogen rates can be observed to have changed the chemical processes in leaves, thus also reshaping the eating habits of herbivores [8,9]. In addition, an increase in CO$_2$ level can reduce stomatal conductance, resulting in better water-use efficiency [10]. In C$_3$ crops, an elevated CO$_2$ level can stimulate net photosynthetic CO$_2$ assimilation, leading to greater biomass production and yield [11]. Although the 'CO$_2$ fertilization effect' on C$_3$ crops is a well-known phenomenon [12–14], it depends heavily on various environmental growing conditions, such as air temperature or the availability of nutrients and soil water [11,15–17].

Scenarios have forecasted that drought will be more frequent and more severe in the next decades in many crop-growing areas [18]. Drought is one of the major stress factors that limit cereal production worldwide, and may affect about 40–60% of the world's agricultural lands [19]. It can severely influence the growth and development of plants, causing various physiological and biochemical damage. For example, it can lead to stomatal closure and can reduce photosynthesis, transpiration, growth and antioxidant production, and can also change hormonal composition [20–23]. Increased atmospheric CO_2 may contribute to climate change, including changes in precipitation and evapotranspiration, and may increase the risk of drought in many areas, as seen in Central Europe [24–26]. Several studies have indicated that plants in their reproductive phases (i.e. from elongation to anthesis stages) are less tolerant to water stress [27–29]. Especially in the case of wheat, the effects of water shortage depend on onset time, duration and intensity. Aside from the developmental stages and the severity of the stress, the effects of water shortage on cereals depend on soil type, environmental conditions [30,31], the cultivated varieties or species [32,33] and also the cultivation technologies employed [34,35]. Drought can cut wheat yields by up to 92% [27], but sometimes extreme drought at the right time can lead to a total yield loss. Since wheat (*Triticum aestivum* L.) is one of the most important cereals in human and animal nutrition, and as it is one of the most extensively grown crops [36,37], drought can cause serious damage to food security. In addition to wheat, barley (*Hordeum vulgare* L.) is an important cereal, contributing nearly 157 million metric tons to cereal production worldwide [38]. The most negative correlation was observed between yield and drought stress at the heading and flowering stages in barley [39]. Although in the last few decades the demand for oats (*Avena sativa* L.) in human consumption has increased because of their dietary benefits, compared to other cereal crops, their production is more suited to marginal environments, such as cool–wet climates and soils with low fertility [40,41]. Among cereals, oats are the most sensitive regarding drought stress at germination and heading developmental stages [42]. Water-use efficiency (WUE; kg·m^{-3}) reflects the relationship between carbon and water cycles, and it is a key indicator of drought tolerance. WUE is an essential parameter for assessing the reactions of plants to climate change. It is a well-known phenomenon that there are considerable differences between the WUE values of the cereal species [43].

Although temperature, water availability and atmospheric CO_2 are important regulators of plant growth, function and development, their impacts on different species and varieties show great variability. In this study, we examined the effects of different CO_2 concentrations combined with simulated water shortage at different developmental stages on four Hungarian winter wheat varieties, one winter oat and one winter barley variety. The aims of our study were: (1) to determine how water shortage and different CO_2 concentrations influence the phenological and yield parameters of some widely cultivated cereal varieties in the Carpathian Basin, (2) to determine the water uptake and water-use efficiency of plants under different environmental conditions and (3) to quantify the specific CO_2 responses of the examined varieties.

2. Materials and Methods

2.1. Experimental Design

Four winter wheat (*Triticum aestivum* L.) varieties ('Mv Ikva', 'Mv Nádor', 'Mv Nemere', 'Mv Kolompos'), one winter barley (*Hordeum vulgare* L.) ('Mv Initium') and one winter oat (*Avena sativa* L.) ('Mv Hópehely') cultivar were examined in a model experiment at the Agricultural Institute Centre for Agricultural Research, Eötvös Loránd Research Network in Martonvásár, Hungary. All varieties were bred locally. The study was carried out in climate-controlled greenhouse chambers in 2020. The experiment was begun on 3rd February and ended at the end of June, when the plants were harvested manually. 'Mv Ikva' and 'Mv Initium' are early-ripening varieties; 'Mv Nádor' and 'Mv Nemere' are middle-ripening, while 'Mv Kolompos' and 'Mv Hópehely' are late-ripening varieties. The experimental design consisted of three water-supply treatments (control, water shortage at tillering, and water-shortage at heading developmental stage). Control ('C') plants (54 pots, 216 plants in total,) were watered until reaching 60% of soil water-holding capacity (WHC). Drought stress was simulated in one-third of the plants (54 pots, 216 plants in total) by stopping the irrigation completely at BBCH stage 21 ('T') [44], and the other one-third (54 pots, 216 plants in total) were similarly stressed at BBCH stage 55 ('H'). The WHC was determined each day at 9:00 during the stress treatments in the centre of the pot, using 5TE sensors (Decagon Devices Ltd., Pullman, WA, USA), and pots were re-watered when the soil water content dropped below 5 v/v%. In this way, the plants were continuously exposed to the same level of stress intensity. The experiment was carried out in three similar climate-controlled greenhouse chambers under three different atmospheric CO_2 levels. Aside from the ambient level chamber (~400 ppm) (control), CO_2 concentrations in the other two chambers were enriched to 700 ppm or 1000 ppm, respectively. Pure CO_2 was introduced into the chambers through a perforated pipe network placed 0.5 m above the plants. Uniform gas distribution was achieved by ventilation. Carbon dioxide concentration was controlled by the SH-VT250 device (SH-VT250 CO_2, Temperature and Humidity Transmitter, Soha Tech Co., Ltd., Soul, Korea) and the CO_2 level was measured and verified by the Wöhler CDL 210 (Wöhler CDL Serie 210 CO_2 Messgerät, Wöhler Technik GmbH, Bad Wünnenber, Germany) logger device in the chambers where the plants were grown under elevated CO_2 concentration.

Four vernalized plants of each variety were planted in plastic pots (depth: 27 cm; diameter: 24 cm) as described by Varga et al. [28]. The experimental design involved 162 pots in total; 54 pots in each of the three different greenhouse chambers with different levels of CO_2. We examined 684 plants in total. 12 plants of each variety were treated the same way at every irrigation level and every CO_2 level. At full maturity, the dry weight of the aboveground biomass (shortly biomass, BM), spike numbers and yields per pot were measured. The exact water uptake of the plants/pot was monitored by a digital balance (ICS689g-A15, Mettler Toledo Ltd., Budapest, Hungary) from the planting to the final harvest. Grain yield and biomass were measured using a digital scale (440-45N, KERN & SOHN GmbH, Balingen, Germany). The harvested aboveground biomass (BM) was oven-dried for two days at 70 °C, then the dry weight of the plant material was measured.

Water-use efficiency (*WUE*) was calculated using Equation (1)

$$WUE = \frac{GY}{WU} \tag{1}$$

where *WUE* is water-use efficiency (kg·m^{-3}), *GY* is grain yield (kg), and *WU* is water use (m^3). The harvest index (*HI*) was calculated as described in Equation (2).

$$HI = \frac{GY}{BM \times 100} \tag{2}$$

where *HI* is harvest index (%), *GY* is grain yield (kg) and *BM* is dry aboveground biomass (shortly biomass) (kg).

Relative changes of the different parameters to elevated carbon dioxide level were calculated using Equation (3)

$$\frac{Ex}{A} \text{ or } \frac{Ey}{A} \tag{3}$$

where *A* is the different parameters' values on 400 ppm CO_2 level, *Ex* is the different parameters' values at 700 ppm CO_2 level and *Ey* is the different parameters' values at 1000 ppm CO_2 level.

2.2. Plant Growth Conditions

Seeds of each variety were germinated on 14 December 2019. The seeds were kept at room temperature (22 °C) in plastic boxes in darkness for two days; after that, the plants were transferred into a vernalization chamber (temperature: 4 °C) for 48 days. Four seedlings were planted into each pot on 3 February 2020. Each pot contained 10 liters of a 3:1:1 (*v/v*) homogenous mixture of soil, sand and humus. Climatic conditions were automatically regulated using the Spring–Summer climatic program [45]. Air temperature was increased from a range of 10–12 °C to one of 24–26 °C during the growing period, and relative humidity was kept between 60% and 80%. When necessary, natural light was enhanced by artificial illumination to 500 $\mu mol \cdot m^{-2} \cdot s^{-1}$ at the beginning of the vegetation period, and gradually increased to 700 $\mu mol \cdot m^{-2} \cdot s^{-1}$. Nutrient solution was provided once a week. To each pot, 22 mL water-soluble fertilizer (14% N, 7% P_2O_5, 21% K_2O, 1% Mg, 1% B, Cu, Mn, Fe, Zn; Volldünger Classic; Kwizda Agro Ltd., Vienna, Austria) was added before irrigation. The plants were watered with tap water two times a week until the tillering stage, and three times a week afterwards. The soil was covered with non-transparent foil to prevent soil evaporation. Sulphur (Thiovit Jet) and lambda-cyhalothrin (Karate Zeon 5 CS, Syngenta Ltd. Switzerland) were applied two times.

2.3. Statistical Processing

The experimental design involved four winter wheat, one winter barley and one winter oat variety, three watering treatments and three CO_2 levels in three replicates. A multi-way ANOVA was performed to determine the effects of the tested factors (variety, water supply and CO_2) and Tukey's post hoc test was used to compare means. The SPSS 16.0 program (IBM, Armonk, NY, USA) and Microsoft Excel (Microsoft, Redmond, WA, USA) were used for the statistical analysis and visualization. The significance level was set at $p \leq 0.05$. ANOVA tables are presented in Tables A1–A6.

3. Results

At 400 ppm CO_2 level, the drought stresses (water shortage at BBCH 21 ['T'] and at BBCH 55 ['H']) caused a significant decrease in biomass compared to the control treatment in each variety, except for Mv Kolompos (Table 1). Furthermore, a significant difference was observed between the two stress treatments in Mv Hópehely: compared to the early stress, the late drought reduced biomass to a greater extent. The highest decrease in biomass compared to the control was observed for Mv Ikva (−29% and −33% in 'T' and 'H' treatments, respectively). At 700 ppm CO_2 concentration, in barley, oat, Mv Ikva and Mv Kolompos, significant differences in biomass were observed between the treatments ('C', 'T' and 'H'), and the water shortage at BBCH 55 caused significant reductions in biomass values in all examined varieties. In Mv Kolompos and Mv Nádor, both stress treatments significantly lowered the biomass of the plants compared to the control. The most pronounced decrease was observed for the oat cultivar (−19% and −48% compared with the control in 'T' and 'H' treatments, respectively) (Table 1). At 1000 ppm CO_2, water withdrawal at BBCH 21 decreased the biomass in Mv Ikva (−10%) and Mv Kolompos (−9%). The simulated drought at BBCH 55 decreased the biomass of Mv Hópehely (−35%), Mv Nemere (−13%) and Mv Kolompos (−11%) compared to the control treatment (Table 1).

Mv Initium responded positively to CO_2 enrichment under stress, but this reaction was not observed under well-watered conditions (Figure 1). The tested oat variety appeared to be susceptible to the level of atmospheric CO_2 concentration. Either in the control treatment or with water withdrawal at BBCH 21 stage, CO_2 enrichment to 700 ppm increased biomass, while 1000 ppm CO_2 concentration inhibited plant growth. When oat plants suffered from drought at BBCH stage 55, both levels of CO_2 enrichment influenced biomass production negatively. Significant and negative CO_2 reactions were observed for Mv Ikva in the control and the 'T' treatments, but no significant CO_2 responses could be observed under drought stress conditions induced at BBCH stage 55. The other three wheat varieties (Mv Nádor, Mv Kolompos and Mv Nemere) showed positive CO_2 responses in terms of biomass under 700 ppm CO_2 level in each treatment but this tendency was not detected under 1000 ppm (Figure 1).

Table 1. Biomass (g) of the tested varieties.

Variety	Treatment	~400 ppm	700 ppm	1000 ppm
Mv Initium (winter barley)	C	55.46[Ba1]	53.78[Cc1]	53.86[Aa1]
	T	47.26[Bb3]	58.59[Aa1]	54.56[Aa2]
	H	45.3[Ab2]	54.61[Ab1]	52.32[Aa3]
Mv Hópehely (winter oat)	C	61.70[Aa2]	73.67[Aa1]	55.82[Aa3]
	T	54.55[Ab2]	59.82[Ab1]	52.52[Aa2]
	H	49.05[Ac1]	38.59[Cc2]	36.08[Cb2]
Mv Ikva (winter wheat)	C	50.15[Da1]	43.41[Ea2]	36.97[Ca3]
	T	35.73[Cb1]	33.02[Ec2]	33.24[Cb2]
	H	33.82[Cb12]	36.92[Db1]	33.36[CDb2]
Mv Nádor (winter wheat)	C	40.05[Ea2]	44.82[DEa1]	38.19[Ca2]
	T	35.79[Cb12]	38.24[Cb1]	34.86[Ca2]
	H	34.62[Cb2]	37.21[Db1]	34.94[CDa2]
Mv Nemere (winter wheat)	C	40.93[Ea2]	45.86[Da1]	36.87[Ca2]
	T	37.91[Cb2]	41.32[Db1]	35.19[Cab3]
	H	36.74[Cb2]	39.34[Cb1]	31.97[Db3]
Mv Kolompos (winter wheat)	C	46.49[Ca3]	55.91[Ba1]	48.03[Ba2]
	T	45.65[Ba2]	54.83[Ba1]	43.82[Bb2]
	H	45.00[Ba2]	49.94[Bb1]	42.96[Bb2]

'C': control treatment; 'T': drought stress at BBCH 21; 'H': drought stress at BBCH 55. Capital letters indicate the statistical significance between the varieties; lowercase letters indicate the statistical significance between the treatments; the numbers show the statistical significance between the different CO_2 levels at $p \leq 0.05$ level (n = 3).

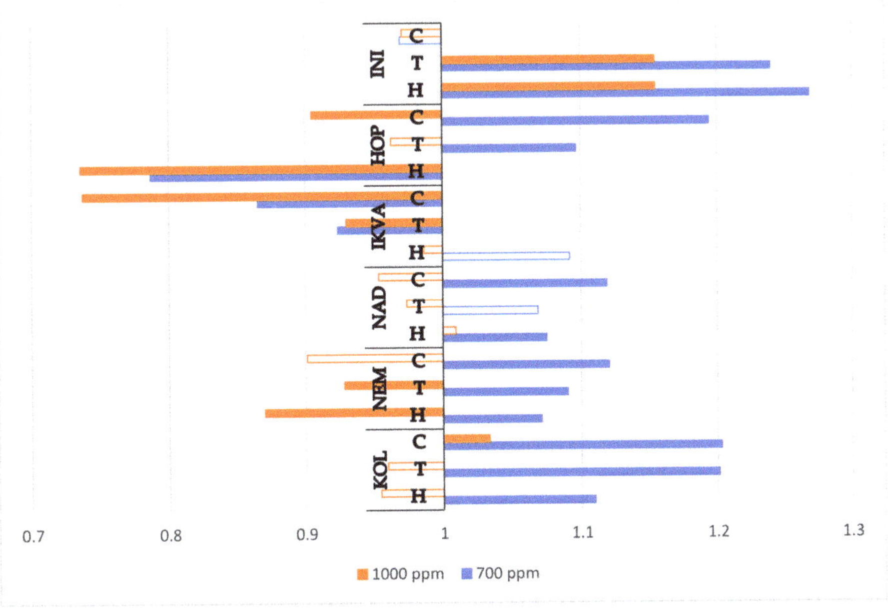

Figure 1. Relative changes in cultivars in response to elevated CO_2 in terms of biomass; 'C': control treatment; 'T': drought stress at tillering stage; 'H': drought stress at heading stage. 'INI': Mv Initium, (winter barley); 'HOP': Mv Hópehely (winter oat); 'IKVA': Mv Ikva (winter wheat); 'NAD': Mv Nádor (winter wheat); 'NEM': Mv Nemere (winter wheat); 'KOL': Mv Kolompos (winter wheat). Full bars represent significant differences compared to the control (400 ppm) $p \leq 0.05$ level (n = 3).

Under ambient CO_2 concentration, drought induced at BBCH stage 21 caused a significant increase in spike numbers in Mv Nemere (+30%) and Mv Nádor (+18%), but a decrease was observed for Mv Initium (−13%) (Table 2). Water shortage at BBCH 55 resulted in a decrease in spike numbers in Mv Ikva (−24%) and Mv Initium (−13%). At 700 ppm CO_2 level, simulated drought at the tillering stage (BBCH 21) significantly reduced spike numbers in Mv Ikva (−12%) but increased the number of productive tillers in Mv Nemere (+23%) and Mv Initium (+13%) (Table 2). Even drought stress at BBCH stage 55 influenced spike numbers by improving tillering ability. Spike numbers increased in Mv Nádor (+43%) and decreased in Mv Ikva (−12%) compared with the well-watered control. At 1000 ppm CO_2 level, water shortage at tillering stage increased spike numbers only in Mv Nemere (+30%), and at heading stage only in Mv Ikva (+31%). In Mv Ikva, Mv Nemere and Mv Kolompos, there were significant differences between the two stress treatments. Higher spike numbers were observed for Mv Nemere and Mv Kolompos when water shortage occurred at the early stage of development, while in Mv Ikva, the effect of the late drought was more pronounced (Table 2).

Table 2. Average spike number per pots of the tested varieties.

Variety	Treatment	~400 ppm	700 ppm	1000 ppm
Mv Initium (winter barley)	C	15^{ABa1}	15^{ABb1}	14^{Aa1}
	T	13^{Bb2}	17^{Aa1}	16^{Aa1}
	H	13^{Ab2}	14^{Bb12}	15^{ABa1}
Mv Hópehely (winter oat)	C	13^{BCa1}	14^{ABa1}	12^{ABa1}
	T	13^{Ba1}	14^{ABa1}	14^{Aa1}
	H	13^{Aa1}	12^{Ca1}	12^{BCa1}
Mv Ikva (winter wheat)	C	17^{Aa1}	16^{Aa1}	13^{ABb2}
	T	15^{Aa1}	14^{ABb12}	13^{Ab2}
	H	13^{Ab2}	14^{Bb2}	17^{Aa1}
Mv Nádor (winter wheat)	C	11^{Cb2}	14^{ABb1}	13^{ABa12}
	T	13^{Ba2}	16^{ABb1}	13^{Aa2}
	H	12^{Ab2}	20^{Aa1}	14^{Ba2}
Mv Nemere (winter wheat)	C	10^{Cb2}	13^{Bb1}	10^{Bb2}
	T	13^{Ba2}	16^{ABa1}	13^{Aa2}
	H	11^{Ab12}	13^{BCb1}	10^{Cb2}
Mv Kolompos (winter wheat)	C	11^{Ca1}	12^{Ba1}	11^{ABab1}
	T	11^{Ca12}	14^{Ba1}	13^{Aa12}
	H	11^{Aa1}	12^{BCa1}	10^{Cb1}

'C': control treatment; 'T': drought stress at BBCH 21; 'H': drought stress at BBCH 55. Capital letters indicate the statistical significance between the varieties; lowercase letters indicate the statistical significance between the treatments; the numbers show the statistical significance between the different CO_2 levels at $p \leq 0.05$ level (n = 3).

Generally, CO_2 enrichment had positive effects on tillering ability and spike numbers in cereals (Figure 2). Only the 1000 ppm CO_2 concentration influenced the spike numbers of Mv Ikva negatively, and this phenomenon could be observed only for the stress-treated plants. The CO_2 fertilization did not influence the spike numbers of Mv Hópehely significantly. Mv Nádor showed the most favorable CO_2 reactions among wheat varieties in terms of spike numbers: under 700 ppm CO_2 concentration, spike numbers in each treatment were significantly greater than under ambient conditions. This beneficial effect in Mv Nádor could not be detected under 1000 ppm concentration, and similar trends were observed for Mv Nemere and Mv Kolompos (Figure 2).

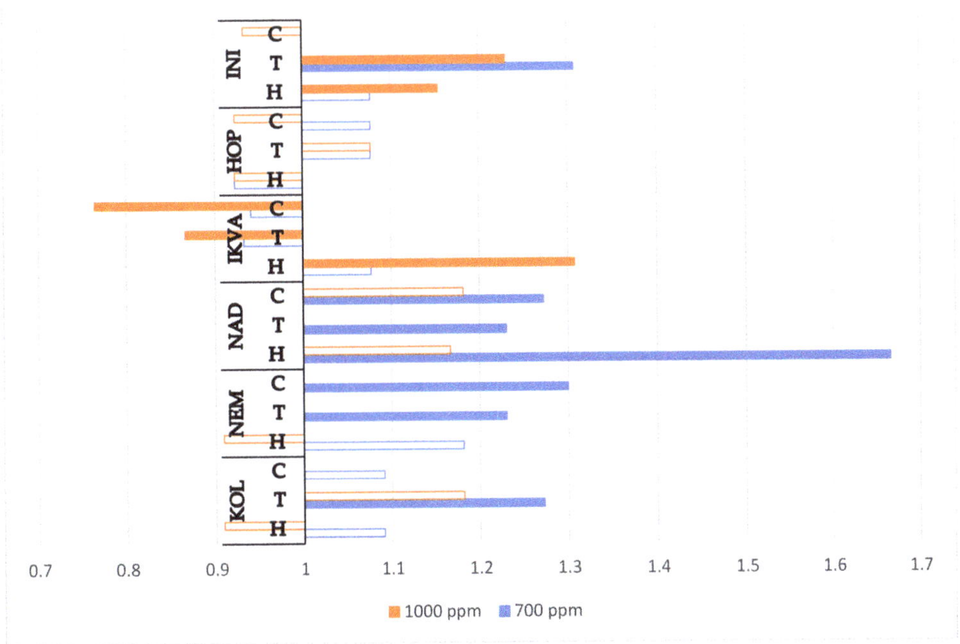

Figure 2. Relative changes in cultivars in response to elevated CO_2 in terms of spike number 'C': control treatment; 'T': drought stress at tillering stage; 'H': drought stress at heading stage. 'INI': Mv Initium (winter barley); 'HOP': Mv Hópehely (winter oat); 'IKVA': Mv Ikva (winter wheat); 'NAD': Mv Nádor (winter wheat); 'NEM': Mv Nemere (winter wheat); 'KOL': Mv Kolompos (winter wheat). Full bars represent significant differences compared to the control (400 ppm) $p \leq 0.05$ level (n = 3).

At ambient CO_2 level (~400 ppm), in the early (BBCH 21) and late (BBCH 55) developmental stages, simulated drought stress decreased the plants' grain yield in each examined variety except for Mv Kolompos (Table 3). The most considerable changes were observed for Mv Hópehely (−24% in 'T' and −54% in 'H' treatment, respectively). Furthermore, in Mv Initium, Mv Ikva and Mv Hópehely, significant differences were observed between the stress treatments; the lowest grain yield was observed for treatment 'H'. Under elevated CO_2 concentration (700 ppm), the drought stress at BBCH stage 21 decreased the grain yield of Mv Hópehely (−47%), Mv Ikva (−20%), Mv Nádor (−12%) and Mv Nemere (−10%). At the heading stage (BBCH 55), the drought stress decreased the grain yield of each cultivar except for Mv Initium. The most exposed variety was Mv Hópehely with 77% yield loss. The drought treatment significantly lowered the grain yield in Mv Nádor (−34%), Mv Kolompos (−26%), Mv Ikva (−25%) and Mv Nemere (−23%). Significant differences were observed between the two stress treatments in Mv Kolompos, Mv Nádor, Mv Nemere and Mv Hópehely: significantly lower grain yield values were observed when the drought occurred at heading (BBCH 55) (Table 3). At 1000 ppm CO_2 level, water withdrawal at BBCH stage 21 decreased the grain yield only in Mv Ikva (−15%). The late-stage stress (BBCH 55) reduced grain yield in each cultivar except for Mv Kolompos. The highest rate of yield reduction was detected in Mv Hópehely (−64%). Furthermore, significant differences were observed between the two stress treatments in the barley and oat varieties: in these cases, the consequences of late-stage drought stress were even more severe (Table 3).

Table 3. Average grain yield values per pot (g) of the tested varieties.

Variety	Treatment	~400 ppm	700 ppm	1000 ppm
Mv Initium (winter barley)	C	21.38^{Ba2}	23.01^{Ba1}	23.42^{Aa1}
	T	19.08^{ABb2}	22.68^{Aa1}	23.70^{Aa1}
	H	18.28^{ABc2}	20.86^{Aa1}	21.28^{Ab1}
Mv Hópehely (winter oat)	C	25.89^{Aa2}	31.14^{Aa1}	16.31^{Ca3}
	T	19.60^{Ab1}	16.40^{Bb2}	16.07^{Ca2}
	H	11.97^{Cc1}	7.10^{Dc2}	5.94^{Db2}
Mv Ikva (winter wheat)	C	27.72^{Aa1}	24.49^{Ba2}	22.46^{Aa2}
	T	20.41^{Ab1}	19.54^{ABb1}	19.12^{Bb1}
	H	18.63^{ABc1}	18.38^{Cb1}	18.50^{Bb1}
Mv Nádor (winter wheat)	C	21.10^{Ba1}	22.81^{Ba1}	20.61^{Aa1}
	T	18.63^{Bb2}	20.16^{Ab1}	18.51^{Bab2}
	H	18.06^{ABb1}	15.03^{Bc2}	16.02^{Cb2}
Mv Nemere (winter wheat)	C	21.16^{Ba2}	24.98^{Ba1}	19.47^{Ba2}
	T	18.39^{ABb2}	22.43^{Ab1}	18.68^{Bab2}
	H	19.22^{Ab1}	19.35^{Cc1}	17.37^{BCb2}
Mv Kolompos (winter wheat)	C	17.34^{Ca3}	23.44^{Ba1}	20.38^{Aa2}
	T	17.70^{Ba2}	22.88^{Aa1}	18.81^{Ba2}
	H	16.54^{Ba2}	17.40^{Cb12}	18.32^{Ba1}

'C': control treatment; 'T': drought stress at BBCH 21; 'H': drought stress at BBCH 55. Capital letters indicate the statistical significance between the varieties; lowercase letters indicate the statistical significance between the treatments; the numbers show the statistical significance between the different CO_2 levels at $p \leq 0.05$ level (n = 3).

Significantly positive CO_2 responses were observed for Mv Initium in terms of grain yield at each watering level, but the stimulating effects of CO_2 fertilization were more intense under drought stress conditions (Figure 3). Opposite tendencies were observed for Mv Hópehely: CO_2 fertilization (1000 ppm) reduced the yield significantly. Under 700 ppm CO_2, a positive response was observed only under well-watered conditions, but CO_2 enrichment combined with water withdrawal reduced yield more intensely in a high-CO_2 environment. Negative CO_2 responses were observed for Mv Ikva under control watering, while the CO_2 enrichment did not influence the grain yield significantly. Positive CO_2 reactions were found in Mv Nádor, Mv Nemere and Mv Kolompos under 700 ppm concentration when plants were grown under optimum irrigation or stressed at BBCH stage 21. The 1000 ppm concentration induced positive responses in Mv Kolompos, which was statistically confirmed in the control and in the stressed treatments at BBCH 55. At the heading stage, Mv Nádor was especially susceptible to CO_2 enrichment, but yield responses were negative under both concentrations (Figure 3).

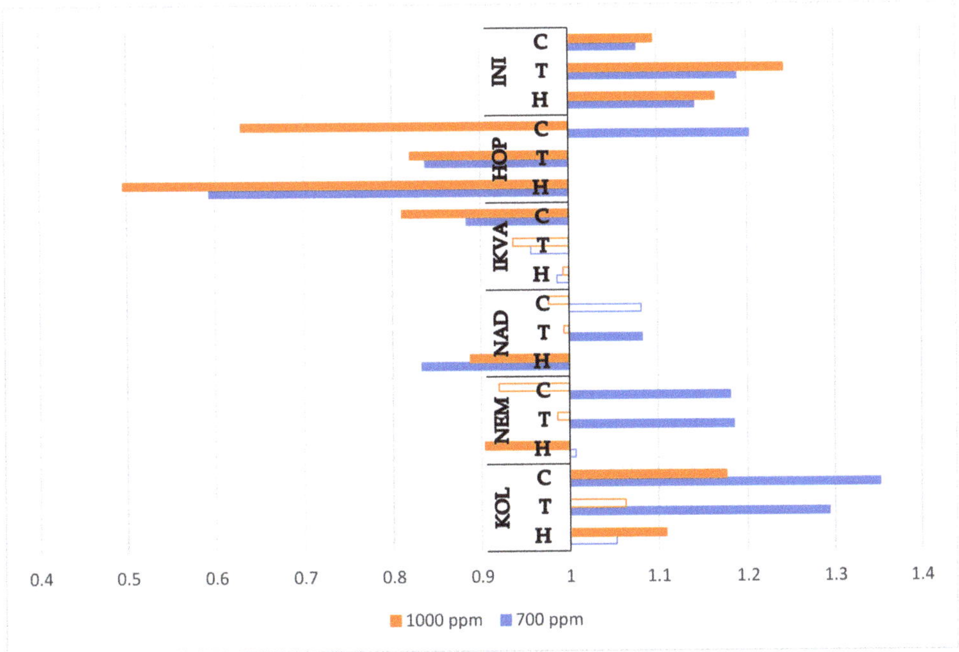

Figure 3. Relative changes in cultivars in response to elevated CO_2 in terms of grain yield 'C': control treatment; 'T': drought stress at tillering stage; 'H': drought stress at heading stage. 'INI': Mv Initium (winter barley); 'HOP': Mv Hópehely (winter oat); 'IKVA': Mv Ikva (winter wheat); 'NAD': Mv Nádor (winter wheat); 'NEM': Mv Nemere (winter wheat); 'KOL': Mv Kolompos (winter wheat). Full bars represent significant differences compared to the control (400 ppm) $p \leq 0.05$ level (n = 3).

At atmospheric CO_2 level, drought stress induced at BBCH stage 21 decreased harvest index in Mv Hópehely (−15%) and increased it in Mv Kolompos (+8%), while water shortage at BBCH stage 55 decreased the harvest index of Mv Hópehely (−45%) and increased it in Mv Initium by 7% compared to the well-watered control treatment. Furthermore, in oats, the drought stress at BBCH 55 resulted in a significant decrease in harvest index values compared to the other two treatments (Table 4). At 700 ppm CO_2 level, the drought stress at tillering (BBCH 21) increased the harvest index of Mv Ikva (+9%), Mv Nemere (+4%) and Mv Nádor (+4%) and decreased it in Mv Hópehely (−31%) and Mv Initium (−8%) significantly. At the heading stage (BBCH 55), water shortage reduced the harvest index of each variety; the most considerable reduction was observed for Mv Hópehely (−62%). Water withdrawal at BBCH stage 55 reduced the harvest index significantly in Mv Nádor, Mv Initium, Mv Kolompos, Mv Ikva and Mv Nemere by −19%, −18%, −16%, −15% and −8%, respectively, compared with the control. Furthermore, there were significant differences between the two stress treatments (drought stress at BBCH 21 and BBCH 55) in each variety: late-stage drought stress reduced the grain yield more intensively than the aboveground biomass; therefore, the reduction in harvest index was more intensive in this treatment (Table 4). Under 1000 ppm CO_2 concentration, drought stress at tillering induced no significant changes in the harvest index, but stress conditions at the heading stage significantly decreased the harvest index in Mv Hópehely (−43%), Mv Nádor (−15%) and Mv Ikva (−9%) (Table 4).

Table 4. Average harvest index values (%) of the tested varieties.

Variety	Treatment	400 ppm	700 ppm	1000 ppm
Mv Initium (winter barley)	C	38.5[Db2]	43.7[Da1]	43.5[Ca1]
	T	40.4[Cab2]	40.3[Db2]	43.4[Ca1]
	H	41.4[Ba1]	36.0[Dc2]	40.7[Ca1]
Mv Hópehely (winter oat)	C	42.0[Ca1]	42.3[Da1]	29.2[Da2]
	T	35.5[Db1]	29.3[Eb2]	30.6[Da2]
	H	22.9[Cc1]	15.9[Ec2]	16.5[Db2]
Mv Ikva (winter wheat)	C	55.3[Aa2]	56.9[Ab2]	60.8[Aa1]
	T	57.1[Aa2]	62.2[Aa1]	57.5[Aab2]
	H	55.2[Aa1]	48.4[Bc2]	55.5[Ab1]
Mv Nádor (winter wheat)	C	52.7[Ba1]	50.9[Ab1]	54.0[Aa1]
	T	52.0[Ba1]	52.7[Ca1]	53.1[Ba1]
	H	52.2[Aa1]	41.1[Cc3]	45.8[Bb2]
Mv Nemere (winter wheat)	C	51.7[Ba1]	53.1[Ab1]	53.0[Ba1]
	T	49.9[Ba2]	55.4[Ba1]	52.2[Ba12]
	H	52.3[Aa1]	49.2[Ac2]	54.3[Aa1]
Mv Kolompos (winter wheat)	C	35.8[Eb2]	41.9[Da1]	42.4[Ca1]
	T	38.8[Ca2]	42.5[Da12]	42.9[Ca1]
	H	36.7[Bab2]	35.4[Db2]	42.6[BCa1]

'C': control treatment; 'T': drought stress by tillering; 'H': drought stress by heading. Capital letters indicate the statistical significance between the varieties at $p \leq 0.05$ level; lowercase letters indicate the statistical significance between the treatments at $p \leq 0.05$ level. The numbers in the indexes indicate the statistical significance between the different CO_2 levels at $p \leq 0.05$ level (n = 3).

In terms of the harvest index, consistent CO_2 responses were observed only for Mv Hópehely (Figure 4). Under 1000 ppm CO_2, harvest index was significantly lower in each water treatment group, and even the 700 ppm level resulted in a decrease under drought stress conditions. Both enriched CO_2 concentrations had a positive influence on the harvest index of Mv Initium, but this trend could be detected only in the well-watered plants. No subsequent CO_2 responses could be observed in Mv Ikva and Mv Nemere, but CO_2 fertilization significantly decreased the harvest index of Mv Nádor when water availability was reduced at BBCH stage 55. Mv Kolompos showed positive CO_2 reactions, which were statistically significant either at 700 or 1000 ppm CO_2 under well-watered conditions, but this tendency was significant in drought stress treatments only under 1000 ppm CO_2 (Figure 4).

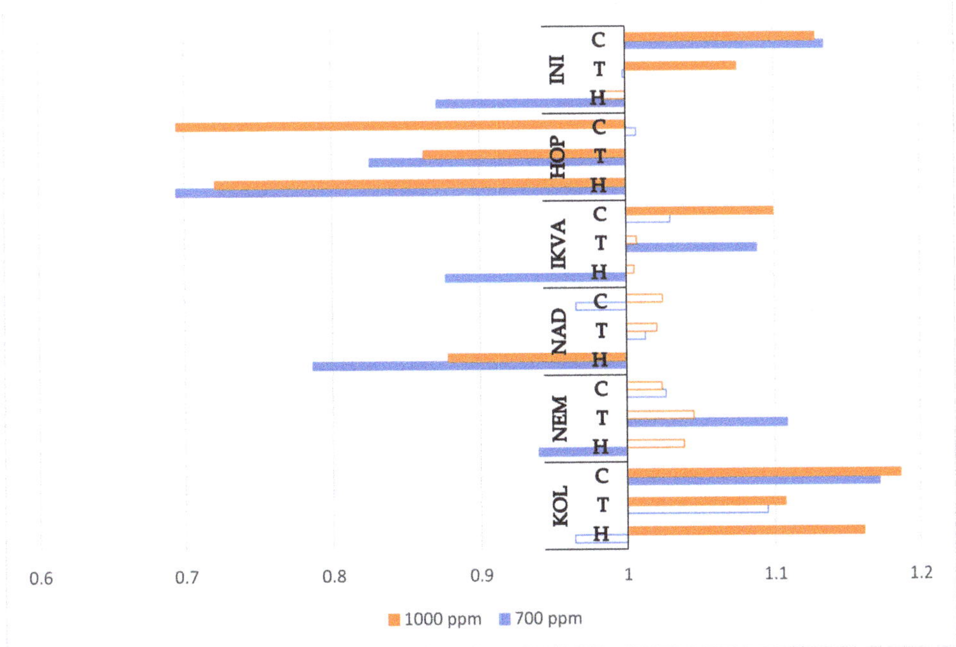

Figure 4. Relative changes in cultivars in response to elevated CO_2 in terms of harvest index 'C': control treatment; 'T': drought stress at tillering stage; 'H': drought stress at heading stage. 'INI': Mv Initium (winter barley); 'HOP': Mv Hópehely (winter oat); 'IKVA': Mv Ikva (winter wheat); 'NAD': Mv Nádor (winter wheat); 'NEM': Mv Nemere (winter wheat); 'KOL': Mv Kolompos (winter wheat). Full bars represent significant differences compared to the control (400 ppm) $p \leq 0.05$ level (n = 3).

At ambient CO_2 level, drought stress at the early developmental stage (BBCH 21) caused no significant changes in the plants' water use, but water shortage at BBCH stage 55 significantly decreased the water uptake of Mv Ikva (-24%) and Mv Nemere (-12%) compared with the control treatment. Significant differences between the two drought stress treatments were also observed for Mv Nemere (the lowest water use was observed for treatment 'H') (Figure 5). Under 700 ppm CO_2 concentration, drought stress at the early stage decreased the water use of Mv Ikva (-18%), and water withdrawal at BBCH stage 55 decreased water uptake values in each wheat variety; the highest decrease (-20%) was observed for Mv Ikva. Furthermore, significant differences were observed between the two stress treatments in Mv Kolompos (drought stress at BBCH 55 reduced water uptake by 6% compared to the water shortage simulated at BBCH 21) (Table 5). At 1000 ppm CO_2 level, both stress treatments decreased water use in Mv Ikva significantly (-15% and -12% in 'T' and 'H' treatments, respectively) (Figure 5).

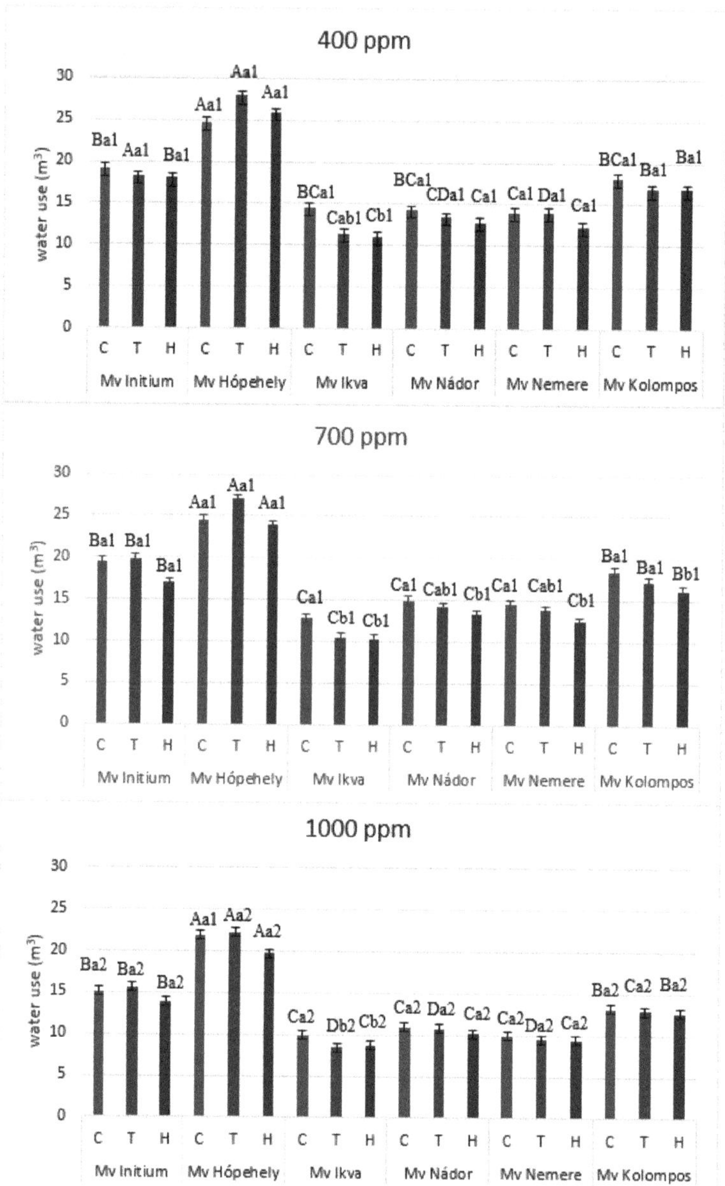

Figure 5. Average water use values (m^3) of the tested varieties. 'C': control treatment; 'T': drought stress at BBCH 21; 'H': drought stress at BBCH 55. Capital letters indicate the statistical significance between the varieties; lowercase letters indicate the statistical significance between the treatments; the numbers show the statistical significance between the different CO$_2$ levels at $p \leq 0.05$ level (n = 3) (Mv Initium is winter barley; Mv Hópehely is winter oat; Mv Ikva, Mv Nádor, Mv Nemere and Mv Kolompos are winter wheat).

Table 5. The average water-use efficiency values ($kg \cdot m^{-3}$) of the tested varieties.

Variety	Treatment	~400 ppm	700 ppm	1000 ppm
Mv Initium (winter barley)	C	1.115[Ba2]	1.227[Ca2]	1.555[Ca1]
	T	1.130[Ba2]	1.169[Da2]	1.515[CDa1]
	H	1.040[Ca2]	1.174[Ba2]	1.538[Ca1]
Mv Hópehely (winter oat)	C	0.985[Ba12]	1.217[Ca1]	0.746[Da2]
	T	0.656[Cab1]	0.628[Eb1]	0.724[Ea1]
	H	0.461[Db1]	0.291[Cc2]	0.300[Db2]
Mv Ikva (winter wheat)	C	1.762[Aa2]	1.867[Aa2]	2.271[Aa1]
	T	1.789[Aa2]	1.905[Aa2]	2.262[Aa1]
	H	1.686[Aa2]	1.745[Aa2]	2.126[Aa1]
Mv Nádor (winter wheat)	C	1.480[Aa2]	1.516[Ba2]	1.889[Ba1]
	T	1.389[Aa2]	1.422[Cb2]	1.722[BCab1]
	H	1.416[Ba2]	1.126[Bc3]	1.576[Cb1]
Mv Nemere (winter wheat)	C	1.516[Aa3]	1.720[Aa2]	1.970[Ba1]
	T	1.289[Ba2]	1.664[Ba1]	1.933[Ba1]
	H	1.567[ABa2]	1.555[Ab2]	1.853[Ba1]
Mv Kolompos (winter wheat)	C	0.957[Ba2]	1.269[Ca1]	1.556[Ca1]
	T	1.051[BCa2]	1.290[Da1]	1.453[Da1]
	H	0.927[Ca2]	1.077[Bb2]	1.445[Ca1]

'C': control treatment; 'T': drought stress at BBCH 21; 'H': drought stress at BBCH 55. Capital letters indicate the statistical significance between the varieties; lowercase letters indicate the statistical significance between the treatments; the numbers show the statistical significance between the different CO_2 levels at $p \leq 0.05$ level (n = 3).

There were differences between the plants' reactions in terms of water uptake at the two levels of CO_2 enrichment (Figure 6). No significant changes in water uptake could be observed under 700 ppm, but the 1000 ppm CO_2 level reduced the water demand of plants during vegetation in each variety and each water treatment group, except for Mv Hópehely under optimum watering. Under optimum watering, the water demand of the cultivars decreased by 24% on average, and when water shortage was simulated at BBCH stages 21 and 55, the reduction in water use was 23% less in both stress treatments than under control conditions (400 ppm) (Figure 6).

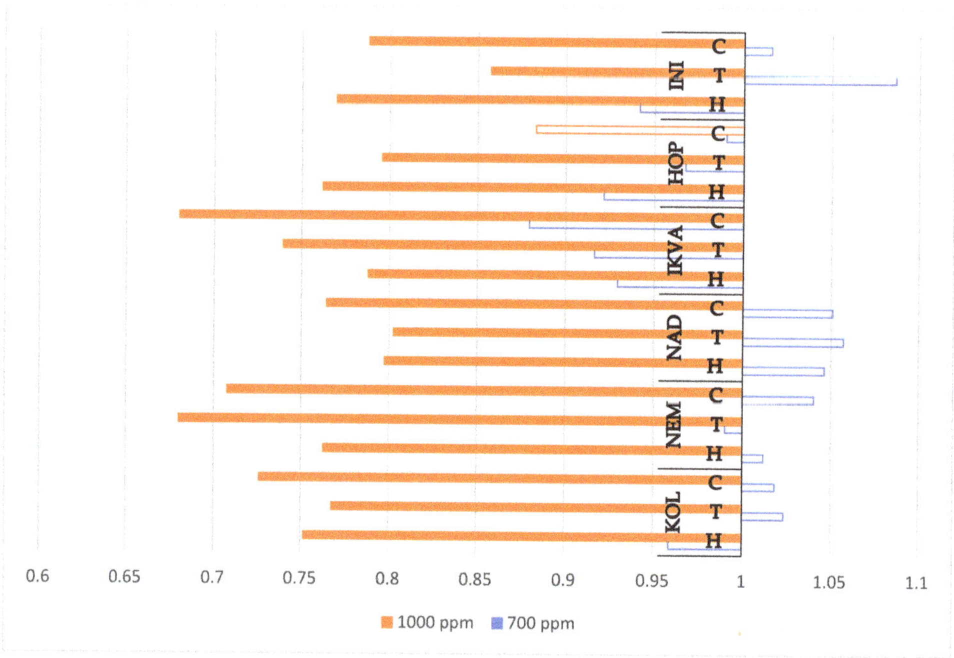

Figure 6. Relative changes in cultivars to elevated CO_2 in terms of water use 'C': control treatment; 'T': drought stress at tillering stage; 'H': drought stress at heading stage. 'INI': Mv Initium (winter barley); 'HOP': Mv Hópehely (winter oat); 'IKVA': Mv Ikva (winter wheat); 'NAD': Mv Nádor (winter wheat); 'NEM': Mv Nemere (winter wheat); 'KOL': Mv Kolompos (winter wheat). Full bars represent significant differences compared to the control (400 ppm) $p \leq 0.05$ level (n = 3).

Under 400 ppm CO_2 concentration, the water shortage induced at BBCH stage 21 caused no significant changes in the plants' water-use efficiency (WUE), but limited water availability at BBCH stage 55 decreased the WUE of Mv Hópehely from the initial 1.115 kg/m^3 (control) to 1.040 kg/m^3 (Table 5). At the 700 ppm CO_2 level, simulated drought at tillering decreased water-use efficiency in Mv Hópehely by 48% and in Mv Nádor by 6%, but no significant changes were observed for the other varieties. In Mv Hópehely, water shortage at heading (BBCH 55) had very serious consequences: its water-use efficiency (0.291 kg/m^3) decreased by 76% compared to the well-watered treatment (1.217 kg/m^3). Even the WUE of Mv Nádor, Mv Kolompos and Mv Nemere decreased by 25%, 15% and 10%, respectively, compared to the well-watered control. However, the best adaptability was observed in Mv Nemere, as its water-use efficiency was significantly higher under drought conditions at the heading stage than that of Mv Nádor and Mv Kolompos, and it did not differ from the early-ripening cultivar (Mv Ikva). Furthermore, in Mv Hópehely, Mv Nádor, Mv Nemere and Mv Kolompos, significant differences were observed between the two drought stress treatments: water shortage at the heading stage induced a more intensive decrease in water-use efficiency than the shortage simulated at tillering (Table 5). At 1000 ppm CO_2 level, water withdrawal at the tillering stage did not induce changes in the WUE, indicating that CO_2 fertilization could counterbalance the negative impacts of a limited-water environment. Water shortage at the heading stage reduced water-use efficiency significantly only in Mv Hópehely and Mv Nádor by 60% and 17%, respectively, which indicated the sensitivity of these varieties to drought stress (Table 5).

As a consequence of the positive responses in grain yield and the moderated water uptake of Mv Initium, 1000 ppm CO_2 resulted in an improved water-use efficiency in each treatment (39%, 34% and 48% in control, 'T' and 'H' treatments, respectively) (Figure 7). The negative trends determined by the grain yield of Mv Hópehely under elevated CO_2 were counterbalanced by the reduced water uptake in the control and by the water shortage at BBCH 21; therefore, CO_2 enrichment did not influence water-use efficiency significantly. The negative impacts of the late drought stress at BBCH 55 was extremely serious in the oat variety; therefore, a declined water-use efficiency was found under both elevated CO2 concentrations as well. The 1000 ppm CO_2 level improved the water-use efficiency of each wheat variety, but the best responses were determined in Mv Kolompos with 163%, 138% and 156% in the control, 'T' and 'H' treatments, respectively. In terms of CO_2 reactions, no significant differences were observed between Mv Nádor, Mv Nemere and Mv Ikva. An increased WUE was detected in Mv Kolompos under optimum irrigation, and in the drought-stressed plants at BBCH stage 21. A similar trend was observed for Mv Nemere but only in the control treatment (Figure 7).

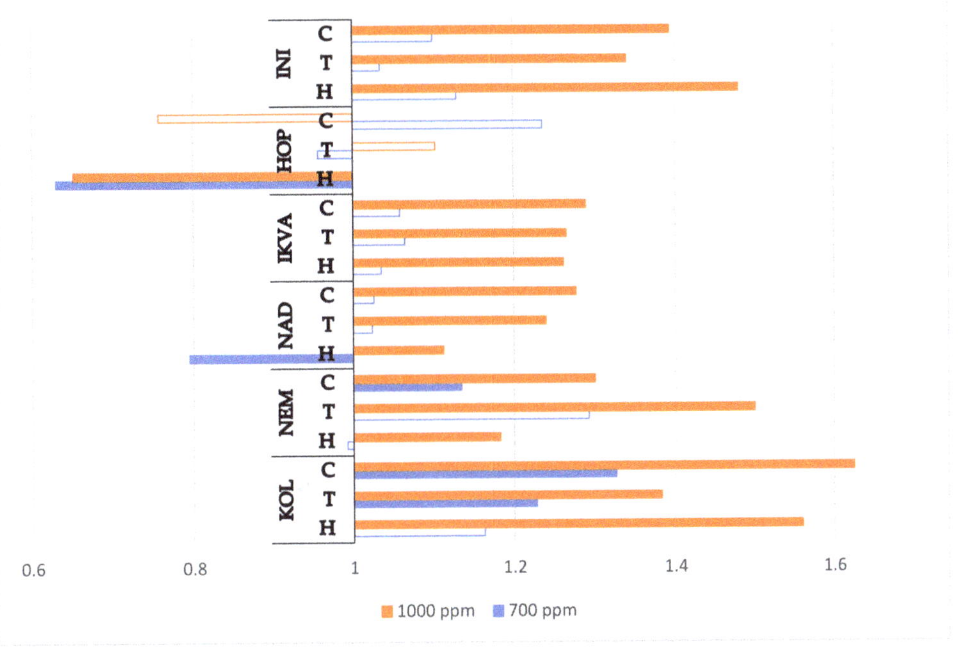

Figure 7. Relative changes in cultivars in response to elevated CO_2 in terms of water-use efficiency 'C': control treatment; 'T': drought stress at tillering stage; 'H': drought stress at heading stage. 'INI': Mv Initium (winter barley); 'HOP': Mv Hópehely (winter oat); 'IKVA': Mv Ikva (winter wheat); 'NAD': Mv Nádor (winter wheat); 'NEM': Mv Nemere (winter wheat); 'KOL': Mv Kolompos (winter wheat). Full bars represent significant differences compared to the control (400 ppm) $p \le 0.05$ level (n = 3).

4. Discussion

In our experiment, the drought stress induced in the vegetative (BBCH 21) and the generative (BBCH 55) phases of development decreased the biomass of five out of six winter cereal varieties at atmospheric CO_2 level. When the CO_2 level was enriched to 700 ppm, a similar tendency was observed in Mv Hópehely, Mv Ikva, Mv Nemere and Mv Nádor. Furthermore, the biomass of Mv Ikva and Mv Nádor decreased as an effect of drought stress at the 1000 ppm CO_2 level, regardless of the developmental stage at which the stress occurred. In the case of the examined barley cultivar, elevation of the CO_2 level (700 ppm or 1000 ppm) resulted in increased biomass values when the plants were stressed. Our result is in agreement with the findings of Dong et al. [46], Ding et al. [47] and Zhao et al. [48], who claimed that the biomass of winter wheat or oat decreases as an effect of reduced irrigation in different developmental stages at ambient CO_2 levels. Manderscheid and Weigel [49] and Li et al. [50] observed that water withdrawal initiated after steam elongation at elevated CO_2 levels (~700 ppm and 800 ppm) decreased the biomass of spring wheat. We also found that the early-stage drought stress decreased biomass at 700 ppm CO_2 level. It was found that under elevated CO_2 (700 ppm) water limitation at the terminal growing stage reduced biomass in durum wheat, according to Garmendia et al. [51], and in barley, according to Bista et al. [52]; however, in the case of the examined barley cultivar, we found increased biomass values compared to the control. In disagreement with our findings, Varga et al. [28] found no significant differences in the plants' biomass between the well-watered and drought-stressed winter wheat at different developmental stages at elevated CO_2 level (1000 ppm), which fact confirms the variety-specific CO_2 responses. In this study, we observed increased biomass values at both elevated CO_2 levels (700 ppm, 1000 ppm) in barley when the plants were stressed. At different developmental stages (tillering or heading), induced drought stress increased the biomass values in Mv Nádor and Mv Kolompos at elevated CO_2 levels (700 ppm) compared to the ambient concentration. Increased biomass production was observed in Mv Hópehely, Mv Nemere and Mv Kolompos in well-watered plants under 700 ppm CO_2 levels, and also in Mv Kolompos under 1000 ppm (compared to ~400 ppm). Ulfat et al. [53] stressed winter wheat with drought by anthesis at ambient and elevated CO_2 levels (800 ppm). They found the highest biomass values in well-watered plants at ambient CO_2 levels, and the second highest in well-watered plants under elevated CO_2. However, we observed higher biomass at elevated CO_2 levels (700 ppm) compared to the ambient level in well-watered or drought-stressed Mv Nádor, Mv Nemere and Mv Kolompos.

According to our findings, water shortage at the tillering stage under ~400 ppm increased spike numbers in Mv Nemere and decreased them in the barley variety. The drought at heading decreased spike numbers in the barley and Mv Ikva. Our findings are in agreement with other results [47,54–56]. Samarah et al. [54] also found that late-terminal drought stress decreased the spike number of barley. According to Khakwani et al. [55], water deficit at the reproductive stage reduces the number of panicles per plant in winter wheat. Ding et al. [47] claimed that the drought stress either at elongation or at the heading stage reduced the spikes per plant in winter wheat. Rollins et al. [56] also found a significantly lowered number of spikes in barley in drought-stress treatment at the generative stage compared to the control. When the CO_2 level was elevated to 700 ppm, drought at tillering increased spike numbers in the examined barley and Mv Nemere, and decreased them for Mv Ikva. The water shortage at BBCH stage 55 increased the numbers of panicles per plants in Mv Nádor and decreased them in Mv Ikva. According to the findings of Garmendia et al. [51], the number of spikes in durum wheat slightly increased as an effect of terminal drought stress and elevated CO_2 level (700 ppm). We had a similar result for one winter wheat variety (Mv Nádor). In our experiment, at 1000 ppm CO_2 level, water shortage at the tillering or heading stages increased spike numbers only in Mv Ikva. Sionit et al. [57] found the opposite effect: late-stage drought stress decreased spike numbers in spring wheat. We found that neither drought treatments nor changes in the levels of CO_2 produced significant differences in the number of panicles in the examined oat cultivar. In Mv Nádor, higher spike numbers were observed at the 700 ppm CO_2 level in every watering regime compared to the ambient CO_2 level. This was in line with the results of Thilakarathne et al. [58], who found a higher spike number in spring wheat at an elevated CO_2 level (700 ppm) compared to the ambient CO_2 level under optimum watering.

In our experiment, water shortage at BBCH 21 and BBCH 55 decreased the grain yield of Mv Ikva compared to the well-watered plants under all tested CO_2 concentrations (~400 ppm, 700 ppm, 1000 ppm). The two drought treatments also decreased the yield of Mv Hópehely, Mv Nádor and Nemere at ambient and elevated CO_2 levels (700 ppm) and for the barley at ~400 ppm. At 1000 ppm CO_2, drought stress at the heading stage decreased the grain yield of all examined varieties except for Mv Kolompos. Zhao et al. [48] also found decreased grain yield values in oats as an effect of drought-induced stress at different developmental stages. Quaseem et al. [59] found similar results in wheat as an effect of simulated water shortage at pre-anthesis. Our results are in agreement with Varga et al. [28], Manderscheid and Weigel [49] and Ulfat et al. [53]; these studies also observed a decrease in the grain yield values of wheat as an effect of early-stage or late-stage water shortage at elevated CO_2 levels (700 ppm, 800 ppm or 1000 ppm). Positive CO_2 fertilization effects in barley and negative CO_2 fertilization effects in the oat cultivar were observed for grain yield at both elevated CO_2 levels (700 ppm and 1000 ppm) in every treatment (control, drought at BBCH stage 21 and drought at BBCH stage 55). Additionally, higher grain yield values were observed for the well-watered and early-stage drought-stressed plants in Mv Nemere and Mv Kolompos varieties when the CO_2 concentration was elevated to 700 ppm (compared to the ambient CO_2 level). Thilakarathne et al. [58] also found elevated grain yield values in spring wheat at elevated an CO_2 level (700 ppm) under optimal watering treatment.

According to our results, water withdrawal at tillering increased the harvest index in Mv Kolompos at ambient concentrations and in Mv Ikva, Mv Nádor and Mv Nemere at 700 ppm CO_2 levels, and decreased it in Mv Hópehely at ambient and elevated (700 ppm) levels and in Mv Initium at 700 ppm CO_2 level. Water shortage at heading decreased the harvest index of the oat cultivar at each CO_2 level. Late-stage water shortage decreased the harvest index of every examined variety at 700 ppm CO_2 level, and of Mv Hópehely and Mv Nádor at 1000 CO_2 level. Increments in the CO_2 (700 ppm and 1000 ppm) affected the harvest index negatively in Mv Hópehely compared to the ambient CO_2. In contrast to our findings, Zhao et al. [48] observed that the harvest index of the examined oat was improved as an effect of the applied drought stress at ambient levels of CO_2. Samarah et al. [54] stated that the harvest index of barley decreased as an effect of late-terminal drought stress, however, we noted an opposite tendency. Ding et al. [47] found that drought stress at the elongation stage under ~400 ppm CO_2 improved the harvest index of winter wheat; our results were in line with this but only with one wheat cultivar (Mv Kolompos). Wu et al. [60] found slightly higher harvest index values under drought conditions in spring wheat compared with the control (80% of field water capacity) under elevated CO_2 (~700 ppm). Compared to the control treatment, Ulfat et al. [53] found decreased harvest index values for winter wheat as a result of drought stress at anthesis under 800 ppm CO_2 concentration. We also found decreased harvest index values at 700 ppm CO_2 level as a result of drought at heading. Varga et al. [28] also found lower harvest index values due to the effect of water withdrawal compared to the control under 700 ppm and 1000 ppm.

At ambient CO_2 level, drought stress at BBCH stage 21 caused no significant changes in the plants' water use, and water shortage at BBCH stage 55 significantly decreased water use of Mv Ikva and Mv Nemere compared with the control treatment. Under elevated CO_2 concentrations (700 and 1000 ppm), both drought stresses decreased the water use of Mv Ikva, and water withdrawal at BBCH stage 55 decreased water use values in each wheat cultivar at 700 ppm. Although CO_2 enrichment to 700 ppm had no significant effect on the water uptake of the examined varieties, the elevation of CO_2 to 1000 ppm caused a positive CO_2 effect. Namely, the water use of each examined variety decreased significantly in all treatments (control, drought stress at tillering and drought stress at heading stage). According to Varga et al. [28], an elevated CO_2 level (1000 ppm) can decrease the water uptake of plants in both optimum watering and drought stress at heading.

In our experiment, the drought stress simulated at BBCH 21 decreased the water-use efficiency of Mv Hópehely and Mv Nádor, but only at an elevated CO_2 (700 ppm) level. Water withdrawal at BBCH 55 decreased the WUE of the examined oat in every CO_2 treatment, of Mv Nádor under the elevated CO_2 concentrations (700 ppm and 1000 ppm) and of Mv Nemere and Mv Kolompos under 700 ppm CO_2. A positive CO_2 fertilization effect was observed also in terms of water-use efficiency in barley and all wheat varieties under all water treatments when the CO_2 level was elevated to 1000 ppm. However, this positive response was only observed for Mv Kolompos and Mv Nemere under 700 ppm CO_2, when the plants were well-watered or the drought stress was initiated at BBCH stage

21. Liu et al. [61] found increased water-use efficency values in oats when the CO_2 level was elevated (700 ppm) and the plants were grown under optimum conditions, but in our experiment, there was no significant difference between the well-watered plants' WUE at the different CO_2 levels. Li et al. [50] found that the water-use efficiency of winter wheat was slightly increased (compared to the control) by drought treatment in the elongation phase at elevated CO_2 levels (800 ppm), and WUE was significantly affected by CO_2 treatment in that experiment. In disagreement with the results of Li et al. [50], we found decreased water-use efficiency values for the examined wheat cultivars under 700 ppm when the plants were stressed at tillering or heading stage. Robredo et al. [62] observed that the highest WUE was observed when the plants (winter and spring wheat or barley) were treated with drought at an elevated CO_2 level (700 ppm). Medeiros and Ward [63] found the highest water-use efficiency under moderate and severe drought stress at an elevated (700 ppm) CO_2 level. We also found increased WUE values when the CO_2 level was elevated, but only at 1000 ppm (compared to ambient CO_2 levels and different treatments).

5. Conclusions

Genotypic differences related to elevated atmospheric CO_2 concentration were confirmed in this study, however, the role of environmental factors was significant in this regard. Generally, positive CO_2 fertilization effects were found for barley (Mv Initium); CO_2 enrichment induced higher biomass and grain yield, decreased water uptake and better water-use efficiency. Low CO_2 sink ability was determined for the oat (Mv Hópehely) because an elevated CO_2 concentration resulted in a reduction in grain yield and harvest index, as well as in water-use efficiency, regardless of the gas concentration. High variability was observed in the CO_2 responses of the four winter wheat varieties. The biomass and grain yield of Mv Kolompos were increased by CO_2 fertilization. The water uptake of all wheat varieties decreased significantly, and at the same time, their water-use efficiency improved, but only under 1000 ppm CO_2. Mv Ikva was not susceptible to CO_2 fertilization. The reason behind this phenomenon might be that this cultivar was an early-ripening variety among the tested plants. No consistent CO_2 reactions were observed for Mv Nádor and Mv Nemere. Positive CO_2 fertilization effects were found for Mv Kolompos; CO_2 enrichment induced higher biomass, grain yield and harvest index, decreased water uptake and improved water-use efficiency. The present study suggests that the effects of different CO_2 concentrations should be tested by taking into consideration the determined genotypic differences of cereal species and varieties.

The CO_2 responses of genotypes can differ significantly, as confirmed by our study. However, the number of genotypes was not high enough to be able to confirm correlation between the ripening times of the varieties and their CO_2 reactions. Our study may be a good base for a more detailed examination in which the physiological background of the CO_2 responses can be determined.

Author Contributions: Conceptualization, B.V.; methodology, Z.F and B.V.; software, Z.F. and B.V.; validation, Z.F. and B.V.; formal analysis, Z.F. and B.V.; investigation, Z.F. and B.V.; resources, G.V. and O.V.; data curation, Z.F. and B.V.; writing—original draft preparation, Z.F.; writing—review and editing, B.V.; visualization, Z.F.; supervision, A.A., G.V. and O.V.; project administration, Z.F. and B.V.; funding acquisition, G.V. and O.V. All authors have read and agreed to the published version of the manuscript.

Funding: This research was funded by the GINOP-2.3.2.-15-2016-00029 (Multifunkcionálisan hasznos ítható növények, mint alternatívák a fenntartható mezőgazdaság szolgálatában) project.

Institutional Review Board Statement: Not applicable.

Informed Consent Statement: Not applicable.

Data Availability Statement: The data presented in this study are available on request from the corresponding author.

Acknowledgments: Not applicable.

Conflicts of Interest: The authors declare no conflict of interest. The funders had no role in the design of the study, in the collection, analyses, or interpretation of data, in the writing of the manuscript, or in the decision to publish the results.

Appendix A

Table A1. Effects of the examined factors on biomass.

	Df	Sum Sq	Mean Sq	F Value	Pr(>F)	Signif.
CO_2_level	2	976	488.1	351.26	$<2*10^{-16}$	***
Watering	2	1801	900.6	648.13	$<2*10^{-16}$	***
Variety	6	8191	1365.2	982.49	$<2*10^{-16}$	***
CO_2_level:Watering	4	95	23.8	17.14	$7.59*10^{-11}$	***
CO_2_level:Variety	10	561	56.1	40.38	$<2*10^{-16}$	***
Watering:Variety	12	1436	119.7	86.14	$<2*10^{-16}$	***
CO_2_level:Watering:Variety	20	758	37.9	27.29	$<2*10^{-16}$	***
Residuals	105	146	1.4			

Signif. codes: '***' 0.001 '**' 0.01 '*' 0.05 '.' 0.1 ' ' 1

Table A2. Effects of the examined factors on spike number per pot.

	Df	Sum Sq	Mean Sq	F Value	Pr(>F)	Signif.
CO_2_level	2	101.26	50.63	49.782	$6.22*10^{-16}$	***
Watering	2	38.65	19.33	19.001	$9.05*10^{-08}$	***
Variety	6	214.76	35.79	35.193	$<2*10^{-16}$	***
CO_2_level:Watering	4	10.18	2.55	2.503	0.0467	*
CO_2_level:Variety	10	80.67	8.07	7.932	$2.14*10^{-09}$	***
Watering:Variety	12	76.39	6.37	6.259	$3.15*10^{-08}$	***
CO_2_level:Watering:Variety	20	107.53	5.38	5.286	$5.92*10^{-09}$	***
Residuals	105	106.79	1.02			

Signif. codes: '***' 0.001 '**' 0.01 '*' 0.05 '.' 0.1 ' ' 1

Table A3. Effects of the examined factors on grain yield.

	Df	Sum Sq	Mean Sq	F Values	Pr(>F)	Signif.
CO2_level	2	117	58.5	73.69	$<2*10^{-16}$	***
Watering	2	987.3	493.6	21.76	$<2*10^{-16}$	***
Variety	6	402.2	67	84.44	$<2*10^{-16}$	***
CO2_level:Watering	4	108.3	27.1	34.11	$<2*10^{-16}$	***
CO2_level: Variety	10	319.5	32	40.24	$<2*10^{-16}$	***
Watering: Variety	12	633.4	52.8	66.49	$<2*10^{-16}$	***
CO2_level:Watering: Variety	20	198	9.9	12.47	$<2*10^{-16}$	***
Residuals	105	83.4	0.8			

Signif. codes: '***' 0.001 '**' 0.01 '*' 0.05 '.' 0.1 ' ' 1

Table A4. Effects of the examined factors on harvest index.

	Df	Sum Sq	Mean Sq	F Value	Pr(>F)	Signif.
CO2_level	2	37	18.7	10.075	$9.94*10^{-05}$	***
Watering	2	1116	557.9	300.277	$<210*^{-16}$	***
Variety	6	13650	2274.9	1224.323	$<2*10^{-16}$	***
CO2_level:Watering	4	373	93.3	50.224	$<2*10^{-16}$	***
CO2_level: Variety	10	558	55.8	30.031	$<2*10^{-16}$	***
Watering: Variety	12	1147	95.6	51.457	$<2*10^{-16}$	***
CO2_level:Watering:Variety	20	331	16.5	8.897	$8.02*10^{-15}$	***
Residuals	105	195	1.9			

Signif. codes: '***' 0.001 '**' 0.01 '*' 0.05 '.' 0.1 ' ' 1

Table A5. Effects of the examined factors on water use.

	Df	Sum Sq	Mean Sq	F Value	Pr(>F)	Signif.
CO2_level	2	501	250.6	329.001	$<210*^{-16}$	***
Watering	2	61	30.6	40.108	$1.15*10^{-13}$	***
Variety	6	3229	538.2	706.466	$<2*10^{-16}$	***
CO2_level:Watering	4	5	1.3	1.681	0.159995	
CO2_level: Variety	10	27	2.7	3.491	0.000533	***
Watering: Variety	12	47	3.9	5.176	$9.3210*^{-07}$	***
CO2_level:Watering: Variety	20	22	1.1	1.435	0.122577	
Residuals	105	80	0.8			

Signif. codes: '***' 0.001 '**' 0.01 '*' 0.05 '.' 0.1 ' ' 1

Table A6. Effects of the examined factors on water-use efficiency.

	Df	Sum Sq	Mean Sq	F Value	Pr(>F)	Signif.
CO2_level	2	3.459	1.73	195.49	$<2*10^{-16}$	***
Watering	2	1.157	0.579	65.397	$<2*10^{-16}$	***
Variety	6	25.725	4.287	484.559	$<2*10^{-16}$	***
CO2_level:Watering	4	0.189	0.047	5.336	0.000596	***
CO2_level: Variety	10	1.57	0.157	17.749	$<2*10^{-16}$	***
Watering: Variety	12	1.19	0.099	11.205	$4.20*10^{-14}$	***
CO2_level:Watering: Variety	20	0.446	0.022	2.519	0.001285	**
Residuals	105	0.929	0.009			

Signif. codes: '***' 0.001 '**' 0.01 '*' 0.05 '.' 0.1 ' ' 1

References

1. UN (United Nations) Department of Economic and Social Affairs Population Dynamics: World Population Prospects 2019. Available online: https://population.un.org/wpp/ (accessed on 9 June 2021).
2. Long, S.P.; Ainsworth, E.A.; Rogers, A.; Ort, D.R. Rising atmospheric carbon dioxide: Plants face the future. *Annu. Rev. Plant Biol.* **2004**, *55*, 591–628. [CrossRef]
3. NOAA. National Oceanic and Atmospheric Administration. Available online: https://www.esrl.noaa.gov/gmd/ccgg/trends/monthly.html (accessed on 9 June 2021).
4. IPCC. Technical summary. In *Climate Change 2013: The Physical Science Basis. Contribution of Working Group I to the Fifth Assessment Report of the Intergovernmental Panel on Climate Change*; Qin, D., Plattner, G.-K., Tignor, M., Allen, S.K., Boschung, J., Nauels, A., Xia, Y., Bex, V., Midgley, P.M., Eds.; Cambridge University Press: Cambridge, UK; New York, NY, USA, 2014.
5. Jiménez-de-la-Cuesta, D.; Mauristen, T. Emergent constraints on Earth's transient and equilibrium response to doubled CO_2 from post-1970 global warming. *Nat. Geosci.* **2019**, *12*, 902–905. [CrossRef]
6. Steffen, W.; Canadell, P. *Carbon Dioxide Fertilisation and Climate Change Policy*; Australian Greenhouse Office, Department of Environment and Heritage: Canberra, Australia, 2005; p. 33.
7. Allen, L.H. Plant responses to rising carbon dioxide and potential interactions with air pollutants. *J. Environ. Qual.* **1990**, *19*, 15–34. [CrossRef]
8. Gleadow, R.M.; Foley, W.J.; Woodrow, I.E. Enhanced CO_2 alters the relationship between photosynthesis and defence in cyanogenic *Eucalyptus cladocalyx* F. Muell. *Plant Cell Environ.* **1998**, *21*, 12–22. [CrossRef]

9. Bender, J.; Weigel, H.-J. Changes in atmospheric chemistry and crop health: A review. *Agron. Sustain. Dev.* **2011**, *31*, 81–89. [CrossRef]
10. Schütz, M.; Fangmeier, A. Growth and yield responses of spring wheat (*Triticum eastivum* L. cv. Minaret) to elevated CO_2 and water limitation. *Environ. Pollut.* **2001**, *114*, 187–194. [CrossRef]
11. Kimball, B.A. Crop responses to elevated CO_2 and interactions with H_2O, N and temperature. *Curr. Opin. Plant Biol.* **2016**, *31*, 36–43. [CrossRef]
12. Högy, P.; Wieser, H.; Koehler, P.; Schwadorf, K.; Breuer, J.; Franzring, J. Effects of elevated CO_2 on grain yield and quality of wheat: Results from a 3-year free-air CO_2 enrichment experiment. *Plant Biol.* **2009**, *11*, 60–69. [CrossRef]
13. Hasegawa, T.; Sakai, H.; Tokida, T.; Nakamura, H.; Zhu, C.W.; Usui, Y. Rice cultivar responses to elevated CO_2 at two free-air CO_2 enrichment (FACE) sites in Japan. *Funct. Plant Biol.* **2013**, *40*, 148–159. [CrossRef]
14. Fitzgerald, G.J.; Tausz, M.; O'Leary, G.; Mollah, M.R.; Tausz-Posch, S.; Seneweera, S. Elevated atmospheric CO_2 can dramatically increase wheat yields in semi-arid environments and buffer against heatwaves. *Glob. Change Biol.* **2016**, *22*, 2269–2284. [CrossRef] [PubMed]
15. Leakey, A.D.B.; Bishop, K.A.; Ainsworth, E.A. A multi-biome gap in understanding of crop and ecosystem responses to elevated CO_2. *Curr. Opin. Plant Biol.* **2012**, *15*, 228–236. [CrossRef]
16. McGrath, J.M.; Lobell, D.B. Regional disparities in the CO_2 fertilization effect and implications for crop yields. *Environ. Res. Lett.* **2013**, *8*, 014054. [CrossRef]
17. Reich, P.B.; Hobbie, S.E.; Lee, T.D. Plant growth enhancement by elevated CO_2 eliminated by joint water and nitrogen limitation. *Nat. Geosci.* **2014**, *7*, 920–924. [CrossRef]
18. IPCC. *Fifth Assessment Report (AR5), Climate Change 2014, Synthesis Report Summary for Policymakers*; Intergovernmental Panel on Climate Change: Washington, DC, USA, 2015.
19. Shahryari, R.; Mollasadeghi, V. Introduction of two principal components for screening of wheat genotypes under end seasonal drought. *Adv. Environ. Biol.* **2011**, *5*, 519–522.
20. Szegletes, Z.; Erdei, L.; Tari, I.; Cseuz, L. Accumulation of osmoprotectants in wheat cultivars of different drought tolerance. *Cereal Res. Commun.* **2000**, *28*, 403–410. [CrossRef]
21. Lawlor, D.W.; Cornic, G. Photosynthetic carbon assimilation and associated metabolism in relation to water deficits in higher plants. *Plant Cell Environ.* **2002**, *25*, 275–294. [CrossRef]
22. Zhu, J.K. Salt and drought stress signal transduction in plants. *Annu. Rev. Plant Biol.* **2002**, *53*, 247–273. [CrossRef]
23. Ma, J.; Li, R.; Wang, H.; Li, D.; Wang, X.; Zhang, Y.; Zhen, W.; Duan, H.; Yan, G.; Li, Y. Transcriptomics analyses reveal wheat responses to drought stress during reproductive stages under field conditions. *Front. Plant Sci.* **2017**, *8*, 592. [CrossRef]
24. Chun, J.A.; Li, S.; Wang, Q. Effects of Elevated Carbon Dioxide and Drought Stress on Agricultural Crops. In *Drought Stress Tolerance in Plants*; Hossain, M., Wani, S., Bhattacharjee, S., Burritt, D., Tran, L.S., Eds.; Springer: Berlin/Heidelberg, Germany, 2016; Volume 1.
25. Deuthmann, D.; Blöschl, G. Why has catchment evaporation increased in the past 40 years? A data-based study in Austria. *Hydrol. Earth Syst. Sci.* **2018**, *22*, 5143–5158. [CrossRef]
26. Konapala, G.; Mishra, A.K.; Wada, Y.; Mann, M.E. Climate change will affect global water availability through compounding changes in seasonal precipitation and evaporation. *Nat. Commun.* **2020**, *11*, 3040. [CrossRef] [PubMed]
27. Farooq, M.; Hussain, M.; Siddique, K.H. Drought stress in wheat during flowering and grain-filling periods. *Crit. Rev. Plant Sci.* **2014**, *33*, 331–349. [CrossRef]
28. Varga, B.; Vida, G.; Varga-László, E.; Hoffman, B.; Veisz, O. Combined effect of drought stress and elevated atmospheric CO_2 concentration on the yield parameters and water use properties of winter wheat (*Triticum aestivum* L.) genotypes. *J. Agron. Crop Sci.* **2017**, *203*, 192–205. [CrossRef]
29. Farkas, Z.; Varga-László, E.; Anda, A.; Veisz, O.; Varga, B. Effects of waterlogging, drought and their combination on yield and water-use efficiency of five Hungarian winter wheat varieties. *Water* **2020**, *12*, 1318. [CrossRef]
30. Setter, T.L.; Waters, I. Review of prospects for germplasm improvement for waterlogging tolerance in wheat, barley and oats. *Plant Soil* **2003**, *253*, 1–34. [CrossRef]
31. Semenov, M.A.; Stratonovitch, P.; Alghabari, F.; Gooding, M.J. Adapting wheat in Europe for climate change. *J. Cereal Sci.* **2014**, *59*, 245–256. [CrossRef] [PubMed]
32. Bányai, J.; Maccaferri, M.; Láng, L.; Mayer, M.; Tóth, V.; Cséplő, M.; Pál, M.; Mészáros, K.; Vida, G. Abiotic Stress Response of Near-Isogenic Spring Durum Wheat Lines under Different Sowing Densities. *Int. J. Mol. Sci.* **2021**, *22*, 2053.
33. Bányai, J.; Kiss, T.; Gizaw, S.A.; Mayer, M.; Spitkó, T.; Tóth, V.; Kuti, C.; Mészáros, K.; Láng, L.; Karsai, I.; et al. Identification of superior spring durum wheat genotypes under irrigated and rain-fed conditions. *Cereal Res. Commun.* **2020**, *48*, 355–364. [CrossRef]
34. Yang, J.; Zahng, J.; Huang, Z.; Zhu, Q.; Wang, L. Remobilization of carbon reserves is improved by controlled soil-drying during grain filling of wheat. *Crop Sci.* **2000**, *40*, 1645–1655. [CrossRef]
35. Wu, J.; Li, J.-C.; Wei, F.-Z.; Wang, C.-Y.; Zhang, Y.; Sun, G. Effects of nitrogen spraying on the post-anthesis stage of winter wheat under waterlogging stress. *Acta Physiol. Plant.* **2014**, *36*, 207–216. [CrossRef]
36. Igrejas, G.; Branlard, G. The importance of wheat. In *Wheat Quality for Improving Processing and Human Health*; Igrejas, G., Ikeda, T., Guzmán, C., Eds.; Springer: Berlin/Heidelberg, Germany, 2020; pp. 1–7.

37. FAO. *World Food and Agriculture—Statistical Pocketbook*; FAO: Rome, Italy, 2018; p. 254.
38. World Agricultural Production. Available online: http://www.worldagriculturalproduction.com/crops/barley.aspx (accessed on 19 August 2021).
39. Vacizi, B.; Bavci, V.; Shiran, B. Screening of barley genotypes for drought tolerance by agro-physiological traits in field condition. *Afr. J. Agric. Res.* **2010**, *5*, 881–892.
40. Hoffmann, L.A. World production and use of oats. In *The Oat Crop—Production and Utilization*; Welch, R.W., Ed.; Chapman ad Hall: London, UK, 1995; pp. 34–61.
41. Buerstmayr, H.; Krenn, N.; Stephan, U.; Grasgruber, H.; Zechner, E. Agronomic performance and quality of oat (*Avena sativa* L.) genotypes of worldwide origin produced under Central European growing conditions. *Field. Crop. Res.* **2007**, *101*, 343–351. [CrossRef]
42. Varga, B.; Varga-László, E.; Bencze, S.; Balla, K.; Veisz, O. Water use of winter cereals under well-watered and drought-stressed conditions. *Plant Soil Environ.* **2013**, *59*, 150–155. [CrossRef]
43. Ullah, H.; Santiago-Arenas, R.; Ferdous, Z.; Attia, A.; Datta, A. Chapter Two—Improving water use efficiency, nitrogen use efficiency, and radiation use efficiency in field crops under drought stress: A review. *Adv. Agron.* **2019**, *156*, 109–157.
44. Lancashire, P.D.; Bleiholder, H.; Boom, T.V.D.; Langelüddeke, P.; Stauss, R.; Weber, E.; Witzenberger, A. A uniform decimal code for growth stages of crops and weeds. *Ann. Appl. Biol.* **1991**, *119*, 561–601. [CrossRef]
45. Tischner, T.; Kőszegi, B.; Veisz, O. Climatic programs used in the Martonvasar phytotron most frequently in recent years. *Acta. Agric. Hung.* **1997**, *45*, 85–104.
46. Dong, B.; Zheng, X.; Liu, H.; Able, J.A.; Yang, H.; Zhao, H.; Zhang, M.; Qiao, Y.; Wnag, Y.; Liu, M. Effects of drought stress on pollen sterility, grain yield, abscisic acid and protective enzymes in two winter wheat cultivars. *Front. Plant Sci.* **2017**, *8*, 1008. [CrossRef] [PubMed]
47. Ding, J.; Huang, Z.; Zhu, M.; Li, C.; Zhu, X.; Guo, W. Does cyclic water stress damage wheat yield more than single stress? *PLoS ONE* **2018**, *13*, e0195535. [CrossRef]
48. Zhao, B.; Ma, B.-L.; Hu, Y.; Liu, J. Source-sink adjustment: A mechanistic understanding of timing and severity of drought stress on photosynthesis and grain yields of two contrasting oat (*Avena sativa* L.) genotypes. *J. Plant Growth Regul.* **2021**, *40*, 263–276. [CrossRef]
49. Manderscheid, R.; Weigel, J.-H. Drought stress effects on wheat are mitigated by atmospheric CO_2 enrichment. *Agron. Sustain. Dev.* **2007**, *27*, 79–87. [CrossRef]
50. Li, Y.; Li, X.; Yu, J.; Liu, F. Effect of the transgenerational exposure to elevated CO_2 on the drought response of winter wheat: Stomatal control and water use efficiency. *Environ. Exp. Bot.* **2017**, *136*, 78–84. [CrossRef]
51. Garmendia, I.; Gogorcena, Y.; Aranjuelo, I.; Goicoechea, N. Responsiveness of durum wheat to mycorrhizal inoculation under different environmental scenarios. *J. Plant Growth Regul.* **2017**, *36*, 855–867. [CrossRef]
52. Bista, D.R.; Heckathorn, S.A.; Jayawardena, D.M.; Boldt, J.K. Effect of drought and carbon dioxide on nutrient uptake and levels of nutrient-uptake proteins in roots of barley. *Am. J. Bot.* **2020**, *107*, 1401–1409. [CrossRef] [PubMed]
53. Ulfat, A.; Shokat, S.; Li, X.; Fang, L.; Großkinsky, D.K.; Majid, S.A.; Roitsch, T.; Liu, F. Elevated carbon dioxide alleviates the negative impact of drought on wheat by modulating plant metabolism and physiology. *Agric. Water Manag.* **2021**, *250*, 106804. [CrossRef]
54. Samarah, N.H.; Alqudah, A.M.; Amayreh, J.A.; McAndrews, G.M. The effects of late-terminal drought stress on yield components of four barley cultivars. *J. Agron. Crop Sci.* **2009**, *195*, 427–441. [CrossRef]
55. Khakwani, A.A.; Dennett, M.D.; Minur, M.; Abid, M. Growth and yield response of varieties to water stress at booting and anthesis stages of development. *Pak. J. Bot.* **2012**, *44*, 879–886.
56. Rollins, J.A.; Habte, E.; Templer, S.E.; Colby, T.; Schmidt, J.; von Korff, M. Leaf proteome alterations in the context of physiological and morphological responses to drought and heat stress in barley (*Hordeum vulgare* L.). *J. Exp. Bot.* **2013**, *64*, 3201–3212. [CrossRef] [PubMed]
57. Sionit, N.; Hellmers, H.; Strain, B.R. Growth and yield of wheat under CO_2 enrichment and water stress. *Crop Sci.* **1980**, *20*, 677–690. [CrossRef]
58. Thilakarathne, C.L.; Tausz-Posch, S.; Cane, K.; Norton, R.M.; Tausz, M.; Seneweera, S. Intraspecific variation in growth and yield response to elevated CO_2 in wheat depends on the differences of leaf mass per unit area. *Funct. Plant Biol.* **2013**, *40*, 189–194. [CrossRef]
59. Quaseem, M.F.; Qureshi, R.; Shaheen, H. Effects of pre-anthesis droughts, heat and their combination on the growth, yield and physiology of diverse wheat (*Triticum aestivum* L.) genotypes varying in sensitivity to heat and drought stress. *Sci. Rep.* **2019**, *9*, 6955. [CrossRef]
60. Wu, D.-X.; Wang, G.-X.; Bai, Y.-F.; Liao, J.-X. Effects of elevated CO_2 concentration on growth, water use, yield and grain quality of wheat under two soil water levels. *Agric. Ecosyst. Environ.* **2004**, *104*, 493–507. [CrossRef]
61. Liu, J.-C.; Temme, A.A.; Cornwell, W.K.; van Logtestijn, R.S.P.; Aerts, R.; Cornelissen, J.H.C. Does plant size effect growth responses to water availability at glacial, modern and future CO_2 concentration? *Ecol. Res.* **2016**, *31*, 213–227. [CrossRef]

62. Robredo, A.; Pérez-López, U.; Sainz de la Manza, H.; González-Moro, B.; Lacuesta, M.; Mena-Petite, A.; Muñoz-Rueda, A. Elevated CO_2 alleviates the impact of drought on barley improving water status by lowering stomatal conductance and delaying its effects on photosynthesis. *Environ. Exp. Bot.* **2007**, *59*, 252–263. [CrossRef]

63. Medeiros, J.S.; Ward, J.K. Increasing atmospheric [CO_2] from glacial through future levels affects drought tolerance via impacts on leaves, xylem and their integrated function. *New Phytol.* **2013**, *199*, 738–748. [CrossRef] [PubMed]

 *sustainability*

Article

Pro197Thr Substitution in *Ahas* Gene Causing Resistance to Pyroxsulam Herbicide in Rigid Ryegrass (*Lolium Rigidum* Gaud.)

Barbara Kutasy [1], Zsolt Takács [2], Judit Kovács [3], Verëlindë Bogaj [4], Syafiq A. Razak [4], Géza Hegedűs [5], Kincső Decsi [1], Kinga Székvári [6] and Eszter Virág [5,7,*]

1 Campus Keszthely, Department of Plant Physiology and Plant Ecology, Hungarian University of Agriculture and Life Sciences Georgikon, 7 Festetics Str., 8360 Keszthely, Hungary; kutasybarbara@gmail.com (B.K.); szaszkone.decsi.kincso@uni-mate.hu (K.D.)
2 Department of Agricultural Mechanization, University of Pannonia, 16 Deák F. Str., 8360 Keszthely, Hungary; permetezo.takacs@gmail.com
3 Gregor Mendel Institute of Molecular Plant Biology, Austrian Academy of Sciences, Vienna Biocenter, Dr. Bohr-Gasse 3, 1030 Vienna, Austria; juditkovacsbio@gmail.com
4 Institute of Plant Protection, University of Pannonia, 16 Deák F. Str., 8360 Keszthely, Hungary; verelindebogaj@gmail.com (V.B.); nanosyafiq92@gmail.com (S.A.R.)
5 Educomat Ltd., 12/A Iskola Str., 8360 Keszthely, Hungary; hegedus@educomat.hu
6 Festetics Doctoral School, Hungarian University of Agriculture and Life Sciences, 7 Festetics Str., 8360 Keszthely, Hungary; kinga.szkvr@gmail.com
7 Department of Molecular Biotechnology and Microbiology, Institute of Biotechnology, Faculty of Science and Technology, University of Debrecen, 1 Egyetem Square, 4032 Debrecen, Hungary
* Correspondence: eszterandreavirag@gmail.com

Citation: Kutasy, B.; Takács, Z.; Kovács, J.; Bogaj, V.; Razak, S.A.; Hegedűs, G.; Decsi, K.; Székvári, K.; Virág, E. Pro197Thr Substitution in *Ahas* Gene Causing Resistance to Pyroxsulam Herbicide in Rigid Ryegrass (*Lolium Rigidum* Gaud.). *Sustainability* **2021**, *13*, 6648. https://doi.org/10.3390/su13126648

Academic Editor: Balázs Varga

Received: 3 May 2021
Accepted: 9 June 2021
Published: 10 June 2021

Publisher's Note: MDPI stays neutral with regard to jurisdictional claims in published maps and institutional affiliations.

Abstract: *Lolium rigidum* Gaud. is a cross-pollinated species characterized by high genetic diversity and it was detected as one of the most herbicide resistance-prone weeds, globally. Acetohydroxyacid synthase (AHAS) resistant populations cause significant problems in cereal production; therefore, monitoring the development of AHAS resistance is widely recommended. Using next-generation sequencing (NGS), a de novo transcriptome sequencing dataset was presented to identify the complete open reading frame (ORF) of AHAS enzyme in *L. rigidum* and design markers to amplify fragments consisting of all of the eight resistance-conferring amino acid mutation sites. Pro197Thr, Pro197Ala, Pro197Ser, Pro197Gln, and Trp574Leu amino acid substitutions have been observed in samples. Although the Pro197Thr amino acid substitution was already described in SU and IMI resistant populations, this is the first report to reveal that the Pro197Thr in AHAS enzyme confers a high level of resistance (ED_{50} 3.569) to pyroxsulam herbicide (Triazolopyrimidine).

Keywords: target-site resistance; *ahas*; *als*; Triazolopyrimidine herbicide; *Lolium rigidum*

1. Introduction

Rigid ryegrass (*Lolium rigidum* Gaud.) is a diploid annual grass species distributed worldwide [1]. *L. rigidum* is one of the most widespread and harmful weeds in the Mediterranean area and it was introduced to North and South America, South Africa, and Australia [2]. Globally, *L. rigidum* occurs as weeds of cereal crops, vineyards, and orchards [3]. Though it was intentionally introduced to south-western Australia as a pasture plant species, it is considered as one of the most problematic weeds [4,5].

L. rigidum is one of the most important herbicide-resistant weeds causing economic problems worldwide [6]. Since the 1980s, the development of broad and complex herbicide resistance patterns, such as multiple and cross resistance, has been detected across *L. rigidum* populations [7–9].

However, in weed control the dominant strategy is to use different herbicides that are highly effective on weeds, within these populations, several plants were documented carrying multiple resistance to herbicides including ten different chemical classes [10,11]. Multiple resistance to PSII, acetyl coenzyme A carboxylase (ACCase) and acetolactate synthase (ALS) inhibitors has been demonstrated in Australian and Spanish populations [5,12,13]. This observation has significant consequences on the treatment management for arable lands, as it indicates that the selection with a single herbicide can cause a complex and highly unpredictable pattern of cross-resistance [14].

An effective herbicide is able to penetrate the plants to their site of action at a lethal dose. Herbicides can inhibit plant specific enzymes on a specific target site, thereby altering plant metabolic mechanisms. If the structure of the target site is changing, then the effect of the herbicide is limited, therefore, plants can develop target-site resistance (TSR) against several herbicides. The mechanism of TSR includes point mutations in a target gene causing amino acid changes and lower herbicide binding capacity to the target enzyme. TSR often involves mutations in genes encoding protein targets of herbicides that affect herbicide binding either at or near the catalytic domains or in regions that affect access to them [15]. The mutant target protein shows increased expression in response to herbicide exposure [16]. Because of the rapid evolution of TSR mutations triggered by genetic and other factors, an increased number of herbicide resistant weed populations can be observed on arable fields [17]. This problem will endanger the sustainability of weed control [18], therefore, it is necessary to detect herbicide resistance in agricultural fields. Determining herbicide resistance in surviving weeds after chemical treatment facilitates the planning and effectiveness of subsequent herbicide treatment. In the next-generation sequencing-targeted amplicon sequencing (NGS-TAS) method, TSR is determined by single nucleotide polymorphism (SNP) genotyping and allows identifying the formed multiple TSR (MTSR) in weed population [19].

A remarkable phenomenon of herbicide resistance in *L. rigidum* cross-pollinated weed species, is the high rate of occurrence of cross-resistance that has been confirmed globally. For instance, resistance to AHAS (acetohydroxyacid synthase, EC 4.1.3.18, also known as ALS) inhibitors has been found in Australia, Chile, France, Greece, Israel, and South Africa [20]. Moreover, triasulfuron-resistant populations were found to be resistant also to imidazolinone herbicides [21]. As a consequence of amino acid substitutions, resistant plants were detected in the genetically diverse and resistance-prone *L. rigidum* genus. Examining a single population, 1–6 different mutations were found in the *ahas* gene sequence causing herbicide resistance [22]. So far, 13 Sites of Action were detected in resistant populations in the case of 35 herbicides, therefore, *L. rigidum* was ranked as one of the most herbicide resistance-prone weeds [20].

Resistance to AHAS targeting herbicides is the most common problem among weed species. Most weed species are more resistant to AHAS-inhibiting herbicides than to any other herbicides [8,23]. AHAS catalyzes the first common step in the pathway for synthesis of the branched-chain amino acids such as leucine, isoleucine, and valine in plants. This herbicide group consists of five different chemical families: sulfonylureas (SU), imidazolinones (IMI), sulfonylaminocarbonyl-triazolinones (SCT), pyrimidinylthiobenzoates (PTB), and triazolopyrimidines (TP). AHAS inhibitor herbicides have been widely used as a result of their effectiveness against several weeds at low doses with low mammalian toxicity and their selectivity for major crops [24,25]. Pyroxsulam (molecular formula: C14H13F3N6O5S) (IUPAC name: N-(5,7-dimethoxy[1,2,4] triazolo[1,5-a]pyrimidin-2-yl)-2-methoxy-4- (trifluoromethyl) pyridine-3-sulfonamide) is a member of the TP chemical family. Pyroxsulam herbicide inhibits the production of the AHAS enzyme in plants. Therefore, it can control annual grasses and broadleaf weeds in cereal crops, corn, soybean, sunflower, cotton, tree, and vine crops.

A few point mutations may occur in the sequence of a target enzyme gene, evolving resistance to herbicides. In plants, point mutations causing amino acid substitutions at eight different codons of the *ahas* gene, have been confirmed to cause resistance in field-evolved resistant weed species [17,26]. So far in AHAS enzyme, two amino acid substitutions at the position of Pro197 and Trp574 (according to *Arabidopsis thaliana* numbering) have been reported in *L. rigidum* samples as a consequence of SU and IMI. Considering

AHAS enzyme, six resistance-endowing amino acid substitutions were identified in one *L. rigidum* population: Pro197Ala, Pro197Arg, Pro197Gln, Pro197Leu, Pro197Ser, and Trp574Leu [20,22,27]. Furthermore, Ala122Gly, Pro197Thr, Ala205, Asp376Asn, Asp376Glu amino acid changes were observed in Greek and Italian populations after SU and IMI treatment [28,29]. Although AHAS has a crucial role in the development of herbicide resistance, interestingly, its whole gene sequence is still unknown. It would be essential to identify it to be able to discover new amino acid substitutions in order to fully understand its role in herbicide resistance.

L. rigidum, as a highly susceptible weed, poses a serious economic threat to intensive cropping systems. The use of herbicides is a dominant part of integrated weed management practices, so the knowledge of developed herbicide resistance is important (Beckie 2006). Our aim was to determine the degree of AHAS resistance and the type of resistance (TSR or NTSR) in the suspected Tunisian population of *L. rigidum*. The site resistance can be confirmed by examining mutations that cause known resistance.

In this study, the complete AHAS enzyme mRNA sequence in sensitive *L. rigidum* was identified using Illumina de novo sequencing technology, which sequence information was gap-filling information in this area. After pyroxsulam treatment resistant populations were produced in order to analyze the mutations in the *ahas* gene coding sequences. Based on Illumina data, point mutations were identified using Sanger sequencing (Figure 1).

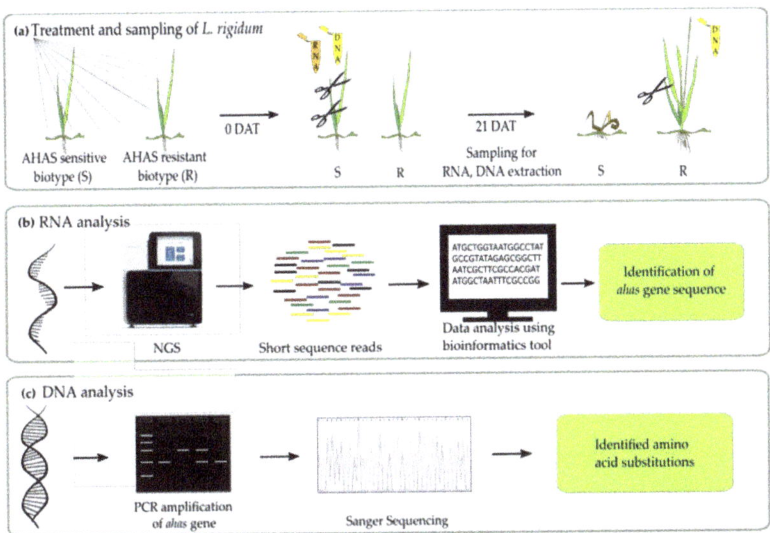

Figure 1. The investigation of the *ahas* gene in *L. rigidum* population. (**a**) Pyroxsulam (TP herbicide) treatment and collection of leaf tissue for RNA and DNA isolation at 0 DAT and for DNA isolation at 21 DAT. (**b**) Process of RNA analysis. After RNA extraction, library preparation was prepared for Illumina NextSeq500 paired-end sequencing. The quality of the raw reads was checked, and de novo assembly of transcripts was performed using Trinity software and the ORF of the *ahas* gene was identified. (**c**) DNA analysis. After DNA extraction, gene fragments of the *ahas* gene were amplified by PCR and their nucleotide sequence was identified by Sanger sequencing. As a result, amino acid substitutions were identified.

2. Materials and Methods

2.1. Plant Materials and Growth Conditions

Seeds of 50 putative pyroxsulam-resistant *L. rigidum* mature plants were collected from a wheat field beside Ben Arous city, in north-eastern Tunisia where pyroxsulam (Pallas[TM] 45 OD, Dow AgroSciences) was used at a normal field dose of 18.4 g ai/ha. Immediately, after collection of all seeds, they were air-dried and stored at 4 °C until used in consecutive experiments. Sensitive seeds of *L. rigidum* that have never been treated with herbicides were obtained from Seed Bank of Hungarian University of Agriculture and Life Sciences Georgikon Campus, Institute of Plant Protection. Seeds were germinated in Petri dishes at 30 °C/20 °C 16 h photoperiod and were planted in pots and were stored in a greenhouse.

2.2. De Novo Transcriptome Sequencing and Assembly

Leaves of five herbicide-susceptible samples were collected and frozen in liquid nitrogen immediately after collection and were stored at −80 °C. The total RNA from plant tissues were isolated using TaKaRa Plant RNA Extraction Kit following the manufacturer's instruction (Takara Bio Inc.; Shiga, Japan). The purity and concentration of the RNA sample were assessed on the Agilent TapeStation system (Agilent Technologies Inc.; Santa Clara, CA, USA) and Qubit 4 Fluorometer (Thermo Fisher Scientific; Waltham, MA, USA) (Table S1, Figure S1). As starting material, 2.0 μg of total RNA was used from which cDNA synthesis and library construction were performed. TruSeq™ RNA sample preparation kit (Low-Throughput protocol) was used to enrich the mRNA, synthesize cDNA, and prepare the library for Illumina NextSeq paired-end sequencing (Figure S2). The quality of the raw (2 × 80 base long) reads was checked by using FastQC quality control software [30]. Quality trimming at a Phred score $\geq$30 was performed by using self-developed software [31]. De novo assembly of transcripts was performed using Trinity [32] with 25 k-mer size. Assembled contigs were used as *L. rigidum* reference transcriptome. The transcript abundances were estimated using the RSEMprogram (http://deweylab.github.io/RSEM/ accessed on 10 June 2021) with the scripts provided in the Trinity package. Raw reads and assembled transcript dataset were deposited in the NCBI Sequence Read Archive (SRA: SRR8469510) and Transcriptome Shotgun Assembly (TSA: GHHB00000000.1; BioProject PRJNA512627).

2.3. Identification of Ahas mRNA in L. Rigidum

Ahas mRNA of *Lolium multiflorum* (GenBank: AF310684.2) was used as a reference. To dynamically translate all reading frames, Blastn algorithm was used [33]. The mRNA sequences were identified from BLAST and aligned into a single gene by using ClustalW [34]. Open reading frames (ORF) were identified using NCBI ORF finder (https://www.ncbi.nlm.nih.gov/orffinder/ accessed on 10 June 2021). The investigated sequence of *L. rigidum* was deposited in the NCBI GenBank MK492446.

2.4. Dose-Response Analysis

Twenty putative pyroxsulam-resistant plants at the three-leaf stage in each pot were treated with pyroxsulam herbicide as commercial herbicide formulation PallasTM 45 OD, using a plot sprayer delivering 250 L ha^{-1}, at a pressure of 300 kPa and a speed of 1.2 ms^{-1}, with a boom equipped multirange flat fan hydraulic spray nozzle (TeeJetR© XR11002). Relative biomass reduction bioassay was performed using six application rates: 0, 4.6, 9.2, 18.4, 36.8, 73.6, and 147.2 (18.4 g ai/ha as the recommended field dose). The treatment was applied on sensitive samples, as control and on resistant samples. Leaves of sensitive samples were collected at 0 day after treatment (DAT) for DNA extraction. The pots were placed in a greenhouse and survivors were assessed at 21 DAT. The herbicide treatment was applied in three technical replicates with a complete randomized block. The dose-response experiment was repeated once more. The nonlinear regression analysis was done using the "drc" package in R v4.0.4 (https://cran.r-project.org/web/packages/drc/drc.pdf accessed on 10 June 2021) (Figure S3). According to the biomass weight of the resistant population, effective dose (ED$_{50}$) was estimated using the four-parameter logistic Equation (1) [35,36]:

$$y = C + (D - C)/[1 + (x/ED_{50})^{b}] \tag{1}$$

ED$_{50}$, the effective dose causing 50% reduction in biomass, indicates the level of resistance (R/S ratios).

2.5. Ahas Gene of L. Rigidum Amplification and Sequencing

Total DNA was extracted from leaf tissues of 50–50 putative pyroxsulam-resistant and sensitive samples [37]. Three primer pairs were designed using the software Primer3Plus to amplify the earlier described resistance-conferring mutation sites based on the *ahas* mRNA in *L. rigidum* (Table S1). The PCR was performed in a final volume of 15 µL (Thermo Fisher Scientific; Waltham, MA, USA). Amplification was carried out in an Eppendorf Mastercycler ep384 (Eppendorf AG; Hamburg, Germany). PCR products were controlled with Agilent 2100 Bioanalyzer (Agilent Technologies Inc.; Santa Clara, CA, USA) using Agilent DNA 1000 Kit and were purified using NucleoSpin Gel and PCR Clean-up System (Macharey-Nagel GmbH &Co; Düren, Germany). The pGEM-T Vector System II (Promega; Madison, WI, USA) was used for cloning according to the manufacturer's instructions and fragments were sequenced ABI 3130x1 Genetic Analyzer (Applied Biosystems; Carlsbad, CA, USA). This 16-capillary array genetic analyzer was used with 50 cm capillary and 3130 POP-7 separation polymers.

3. Results

In this study, an RNA-seq approach is presented to identify complete coding sequences (cds) of the *ahas* gene in *L. rigidum* and mutation points involved resistance to TP herbicide, pyroxsulam. A novel amino acid substitution was identified among the earlier described resistance-conferring mutation sites using Sanger sequencing.

3.1. Transcriptome Reconstruction and Identification of Ahas Cds

The *L. rigidum* mRNA library was sequenced using NextSeq500 platform. After quality filtering, 18,636,868 cleaned reads were obtained from the total raw reads of 18,642,689 (2 × 80 bp paired-end) showing Phred score 34 and GC% 55 on average. Based on FastQC report, sequences represented more than 0.1% of the total and low-quality bases (corresponding to a 0.1% sequencing error rate) were removed. After transcriptome assembly by Trinity, de novo transcript dataset resulted in 74,279 contigs with lengths from 201 to 15,531bp (GC% 52.5). This transcript dataset was used to identify mRNA of *ahas* gene in *L. rigidum*. The DN19950_c0_g2_i3 contig was used to determine the ORF of *ahas* gene. The similarity between *L. multiflorum* as the reference and the identified cds in *L. rigidum* was 98%. The determined *ahas* mRNA sequence was 1929bp long.

3.2. Dose–Response Experiment

The population of *L. rigidum* contains different amino acid substitutions. Each sample contains one of the amino acid substitutions of Pro197Thr, Pro197Ala, Pro197Ser, Pro197Gln, or Trp574Leu. The resistance patterns of the combined substitution population were characterized. The estimated ED_{50} is 3.569 that indicates the 50% survival of the resistance population to pyroxsulam that showed Pro197 or Trp574 substitution (Figure 2). Following a pyroxsulam field application rate of 18.4 g ha^{-1}, *L. rigidum* putative pyroxsulam-resistant population survived with 98.9% growth reduction.

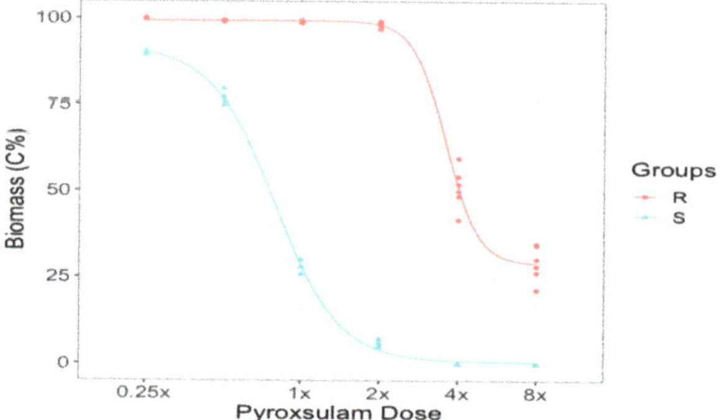

Figure 2. Effect of pyroxsulam herbicides on biomass reduction of *L. rigidum* population. Dose-response curve of biomass weight after pyroxsulam treatment. Data points indicate mean values of 20 independent replicate experiments (average n = 120 plants per dose). Red line represents a 4-parameter logistic regression of pyroxsulam-resistant population. Green line represents sensitive samples.

3.3. Determination of Mutation Points of Ahas Gene in L. Rigidum

Three parts of the *ahas* gene were amplified and sequenced by Sanger sequencing. The primer pairs covered all of eight resistance-conferring amino acid mutation sites at three different regions: 418, 259, and 399 bp (Figure S4, Table S2). The *Ahas* gene fragments were amplified by PCR from 50–50 resistant and sensitive samples and sequenced by the Sanger method (Figure S5). Results of sequence alignment analysis showed single nucleotide substitutions in Pro197 and Trp574 positions in resistant samples (Figure 3). Wild type of *L. rigidum* samples showed no mutations at the eight known nucleotide sequences of the *ahas* gene that correspond to the resistance.

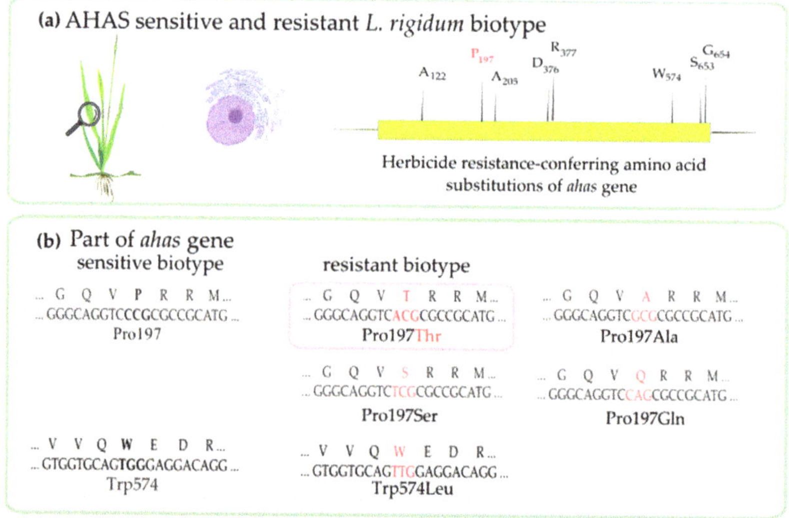

Figure 3. Analysis of the *ahas* gene nucleotide sequence localized in the nucleus in AHAS sensitive and resistant *L. rigidum* biotype. (**a**) The eight known amino acid position in the *ahas* gene responsible for resistance in *L. rigidum*. (**b**) Collection of the amino acid substitutions (red) at the 197 and 574 positions in AHAS sensitive and resistant biotypes. Box highlighted in red indicates the newly discovered amino acid substitution (Pro197Thr).

Sequence alignment analysis showed a single nucleotide change of CCG to ACG at the 197 positions that resulted in a Pro/Thr substitution in the resistant samples (Figures S6 and S7). This is the first Pro197Thr substitution in an *ahas* resistant *L. rigidum* population that was reported after triazolopyrimidines herbicide treatment. Moreover, amino acid mutations were detected at Pro197Ala, Pro197Ser, Pro197Gln, and Trp574Leu (Figure S8) in resistant samples as previously reported [22]. A nucleotide substitution (GAT/GAC) was observed at the Asp376 amino acid position that did not evolve amino acid substitution (Figure S9).

4. Discussion

In arable lands, chemical weed control is one of the most common and unavoidable methods of controlling weeds, which means high selection pressure for resistant weeds. The presence of high levels of herbicide resistance risks the sustainability of current weed control strategies, the integrated weed control [38,39]. In this study, Illumina RNA-sequencing and the "gold standard" Sanger sequencing were used to investigate the sequence of the ORF *ahas* gene in *L. rigidum*. Rigid ryegrass was noticed as a highly resistance-prone weed species and numerous populations were described as resistant to AHAS inhibitors, including SU and IMI. Populations containing Pro197 Ala122, Ala205, Asp376, and Leu 574 amino acid mutations showed high resistance (RI from 8 to 70) after SU treatment. Types of mutant ALS alleles, such as Pro197Ala, Pro197Gln, Pro197Leu, Pro197Ser, and Pro197Thr were observed in these samples [28]. The same biotypes were detected in our samples after pyroxsulam treatment. This is the first report that reveals that the Pro197Thr amino acid change in the AHAS enzyme can confer a high level of resistance (ED_{50} is 3.569) to pyroxsulam (TP) herbicide in *L. rigidum*. The different AHAS-mutations confer different resistance against the AHAS inhibitor herbicide chemical families [20]. It has been already shown that several weed species carrying the Pro197 and Thr574 amino acid substitutions developed high levels of resistance after TP herbicide treatment. In this study, the investigated *L. rigidum* population that contained amino acid substitutions at the 197 and 574 positions, also showed high resistance against the pyroxsulam herbicide. So far, the Pro197Thr resistance-endowing mutation position has been reported in four weed species after TP treatment [20]. Substitutions at Pro197Thr are known to cause a high level of resistance to TP in *Raphanus raphanistrum* [40], in *Kochia scoparia* [41] and moderated resistance to TP in *Apera spica-venti* [26] and *Alopecurus aequalis* [42]. In this study, fifty *L. rigidum* resistant plants were examined and every individual carried only one mutation point out of the eight in the *ahas* gene at the known positions of resistance-endowing amino acid substitutions. Single Pro197Thr substitution was observed in 16 samples (Table S3). As the pyroxsulam resistant biotypes contained different single amino acid substitution in the *ahas* gene, the statistical analysis describes the mixed population. To obtain more information about the Pro197Thr amino acid substitution in the *ahas* gene, larger populations need to be examined. Moreover, the Thr197Pro mutation was described, after SU and IMI treatment, as the same mutation that occurs after pyroxsulam treatment; therefore, presence of cross-resistance is presumed. The existence of cross-resistance in this allelic variant should be demonstrated by treatments with other herbicides belonging to other AHAS inhibitor families (e.g., SU, IMI). Additional herbicides that demonstrate the presence or absence of cross-resistance will be included in our further experiments. Indeed, the detected mutation would provide important information about possible cross-resistance and can help to plan treatment with AHAS inhibitors.

5. Conclusions

Due to the rapid evolution of herbicide resistance, it is necessary to constantly monitor resistant populations on the arable fields in order to maintain weed control considering environmental protection and sustainable agriculture. The pyroxsulam-resistant *L. rigidum* population showed a TSR mutation detected at positions Pro197 and Leu574. The knowledge of resistance-inducing effects of compounds containing AHAS, including pyroxsulam-triggered TSR, would help to develop diagnostic methods, such as NGS-TAS. By diagnosing the presence of TSR and MTSR, chemical weed control would become more predictable, and it will guide farmers on how to plan integrated and resistant weed management strategies that are less harmful to the environment.

Supplementary Materials: The following are available online at https://www.mdpi.com/article/10 .3390/su13126648/s1, Figures S1–S9: Data of RNA sample, Statistic data of *drc* analysis, Primers, Gel pictures and Electropherogram of Sanger sequencing. Tables S1–S3: Data of RNA sample, Primers, Data of amino acid substitutions.

Author Contributions: Conceptualization, B.K. and Z.T.; methodology, B.K., Z.T., V.B., S.A.R., K.D. and K.S.; software, G.H.; validation, B.K., E.V. and G.H.; formal analysis, B.K.; investigation, B.K. and E.V.; resources B.K.; data curation, E.V.; writing—original draft preparation, B.K., J.K. and E.V.; writing—review and editing, J.K. and E.V.; visualization B.K. and J.K.; supervision, E.V. and J.K.; project administration, B.K., E.V.; funding acquisition, B.K. and K.D. All authors have read and agreed to the published version of the manuscript.

Funding: This research received no external funding. The work was supported by Hungarian University of Agriculture and Life Sciences, Institute of Agronomy, Department of Plant Physiology and Plant Ecology, Gödöllő, 1 Páter Károly str., 2100 Gödöllő, Hungary.

Institutional Review Board Statement: Not applicable.

Informed Consent Statement: Not applicable.

Data Availability Statement: Campus Keszthely, Department of Plant Physiology and Plant Ecology, Institute of Agronomy, Hungarian University of Agriculture and Life Sciences Georgikon, 7 Festetics str., 8360 Keszthely, Hungary.

Acknowledgments: We express our thanks to Peter Nagy, DowDuPont Inc. Mougins, France and Janos Taller, Hungarian University of Agriculture and Life Sciences (MATE) to release available the resistant *L. rigidum* seeds (biological samples) to our experiments. Moreover, we express our thanks to László Kocsis, Government Office of Fejér County, Velence to perform the chemical spraying. We are grateful to the editor and to two anonymous reviewers for valuable comments that have helped to improve the manuscript.

Conflicts of Interest: The authors declare no conflict of interest. The funders had no role in the design of the study; in the collection, analyses, or interpretation of data; in the writing of the manuscript, or in the decision to publish the results.

References

1. Bowden, W.M. Chromosome numbers and taxonomic notes on northern grasses: iii. twenty-five genera. *Can. J. Bot.* **1960**, *38*, 541–557. [CrossRef]
2. Castellanos-Frías, E.; De León, D.G.; Bastida, F.; Gonzalez-Andujar, J.L. Predicting global geographical distribution of *Lolium rigidum* (rigid ryegrass) under climate change. *J. Agric. Sci.* **2016**, *154*, 755–764. [CrossRef]
3. Preston, C.; Wakelin, A.M.; Dolman, F.C.; Bostamam, Y.; Boutsalis, P. A Decade of Glyphosate-Resistant Lolium around the World: Mechanisms, Genes, Fitness, and Agronomic Management. *Weed Sci.* **2009**, *57*, 435–441. [CrossRef]
4. Niknam, S.; Moerkerk, M.; Cousens, R. Weed seed contamination in cereal and pulse crops. In Proceedings of the 13rd Australian Weeds Conference, Perth, Australia, 8–13 September 2002; Jacob, H.S., Dodd, J., Moore, J.H., Eds.; Plant Protection Society of Western Australia Inc.: Victoria, Australia, 2002; pp. 59–62.
5. Preston, C.; Tardif, F.J.; Christopher, J.T.; Powles, S.B. Multiple Resistance to Dissimilar Herbicide Chemistries in a Biotype of Lolium rigidum Due to Enhanced Activity of Several Herbicide Degrading Enzymes. *Pestic. Biochem. Physiol.* **1996**, *54*, 123–134. [CrossRef]
6. Beckie, H. Herbicide-Resistant Weeds: Management Tactics and Practices. *Weed Technol.* **2006**, *20*, 793–814. [CrossRef]
7. Christopher, J.T.; Powles, S.B.; Liljegren, D.R.; Holtum, J.A. Cross-resistance to herbicides in annual ryegrass (*Lolium rigidum*): II. Chlorsulfuron resistance involves a wheat-like detoxification system. *Plant Physiol.* **1991**, *95*, 1036–1043. [CrossRef] [PubMed]
8. Owen, M.J.; Walsh, M.J.; Llewellyn, R.S.; Powles, S.B. Widespread occurrence of multiple herbicide resistance in Western Australian annual ryegrass (Lolium rigidum) populations. *Aust. J. Agric. Res.* **2007**, *58*, 711–718. [CrossRef]
9. Han, H.; Yu, Q.; Owen, M.J.; Cawthray, G.R.; Powles, S.D. Widespread occurrence of both metabolic and target-site herbicide resistance mechanisms in Lolium rigidum populations. *Pest Manag. Sci.* **2016**, *72*, 255–263. [CrossRef]

10. Powles, S.B.; Matthews, J.M. Multiple herbicide resistance in annual ryegrass (*Lolium rigidum*): A driving force for the adoption of integrated weed management. In *Resistance'91: Achievements and Developments in Combating Pesticide Resistance*; Springer: Berlin/Heidelberg, Germany, 1992; pp. 75–87.

11. Burnet, M.W.M.; Hart, Q.; Holtum, J.A.M.; Powles, S.B. Resistance to Nine Herbicide Classes in a Population of Rigid Ryegrass (*Lolium rigidum*). *Weed Sci.* **1994**, *42*, 369–377. [CrossRef]

12. Torra, J.; Montull, J.M.; Taberner, A.; Onkokesung, N.; Boonham, N.; Edwards, R. Target-Site and Non-target-Site Resistance Mechanisms Confer Multiple and Cross- Resistance to ALS and ACCase Inhibiting Herbicides in Lolium rigidum From Spain. *Front. Plant Sci.* **2021**, *12*, 88. [CrossRef] [PubMed]

13. Owen, M.J.; Martinez, N.J.; Powles, S.B. Multiple herbicide-resistant *Lolium rigidum* (annual ryegrass) now dominates across the Western Australian grain belt. *Weed Res.* **2014**, *54*, 314–324. [CrossRef]

14. Neve, P.; Powles, S.B. High survival frequencies at low herbicide use rates in populations of *Lolium rigidum* result in rapid evolution of herbicide resistance. *Heredity* **2005**, *95*, 485–492. [CrossRef] [PubMed]

15. Gaines, T.A.; Duke, S.O.; Morran, S.; Rigon, C.A.G.; Tranel, P.J.; Küpper, A.; Dayan, F.E. Mechanisms of evolved herbicide resistance. *J. Biol. Chem.* **2020**, *295*, 10307–10330. [CrossRef]

16. Délye, C.; Jasieniuk, M.; Le Corre, V. Deciphering the evolution of herbicide resistance in weeds. *Trends Genet.* **2013**, *29*, 649–658. [CrossRef]

17. Powles, S.B.; Yu, Q. Evolution in Action: Plants Resistant to Herbicides. *Annu. Rev. Plant Biol.* **2010**, *61*, 317–347. [CrossRef]

18. Harker, K.N.; O'Donovan, J.T. Recent Weed Control, Weed Management, and Integrated Weed Management. *Weed Technol.* **2013**, *27*, 1–11. [CrossRef]

19. Kutasy, B.; Farkas, Z.; Kolics, B.; Decsi, K.; Hegedűs, G.; Kovács, J.; Taller, J.; Tóth, Z.; Kálmán, N.; Kazinczi, G.; et al. Detection of Target-Site Herbicide Resistance in the Common Ragweed: Nucleotide Polymorphism Genotyping by Targeted Amplicon Sequencing. *Diversity* **2021**, *13*, 118. [CrossRef]

20. Heap, I. The International Survey of Herbicide Resistant Weeds. 2021. Available online: www.weedscience.org (accessed on 10 June 2021).

21. Gill, G. Development of herbicide resistance in annual ryegrass populations (*Lolium rigidum* Gaud.) in the cropping belt of Western Australia. *Aust. J. Exp. Agric.* **1995**, *35*, 67–72. [CrossRef]

22. Yu, Q.; Han, H.; Powles, S.B. Mutations of the ALS gene endowing resistance to ALS-inhibiting herbicides in *Lolium rigidum* populations. *Pest Manag. Sci.* **2008**, *64*, 1229–1236. [CrossRef]

23. Tranel, P.J.; Wright, T.R. Resistance of weeds to ALS-inhibiting herbicides: What have we learned? *Weed Sci.* **2002**, *50*, 700–712. [CrossRef]

24. Délye, C.; Causse, R.; Michel, S. Genetic basis, evolutionary origin and spread of resistance to herbicides inhibiting acetolactate synthase in common groundsel (*Senecio vulgaris*). *Pest Manag. Sci.* **2016**, *57*, 1119–1126. [CrossRef]

25. Yu, Q.; Powles, S.B. Resistance to AHAS inhibitor herbicides: Current understanding. *Pest Manag. Sci.* **2014**, *70*, 1340–1350. [CrossRef] [PubMed]

26. Massa, D.; Krenz, B.; Gerhards, R. Target-site resistance to ALS-inhibiting herbicides in Apera spica-venti populations is conferred by documented and previously unknown mutations. *Weed Res.* **2011**, *51*, 294–303. [CrossRef]

27. Menegat, A.; Bailly, G.C.; Aponte, R.; Heinrich, G.M.T.; Sievernich, B.; Gerhards, R. Acetohydroxyacid synthase (AHAS) amino acid substitution Asp376Glu in Lolium perenne: Effect on herbicide efficacy and plant growth. *J. Plant Dis. Prot.* **2016**, *123*, 145–153. [CrossRef]

28. Scarabel, L.; Panozzo, S.; Loddo, D.; Mathiassen, S.K.; Kristensen, M.; Kudsk, P.; Gitsopoulos, T.; Travlos, I.; Tani, E.; Chachalis, D.; et al. Diversified Resistance Mechanisms in Multi-Resistant Lolium spp. in Three European Countries. *Front. Plant Sci.* **2020**, *11*. [CrossRef] [PubMed]

29. Anthimidou, E.; Ntoanidou, S.; Madesis, P.; Eleftherohorinos, I. Mechanisms of *Lolium rigidum* multiple resistance to ALS- and ACCase-inhibiting herbicides and their impact on plant fitness. *Pestic. Biochem. Physiol.* **2020**, *164*, 65–72. [CrossRef] [PubMed]

30. Andrews, S. FastQC: A Quality Control Tool for High Throughput Sequence Data. 2010. Available online: http://www.bioinformatics.babraham.ac.uk/projects/fastqc (accessed on 10 June 2021).

31. Nagy, E.; Hegedűs, G.; Taller, J.; Kutasy, B.; Virág, E. Illumina sequencing of the chloroplast genome of common ragweed (*Ambrosia artemisiifolia* L.). *Data Brief* **2017**, *15*, 606–611. [CrossRef]

32. Haas, B.J.; Papanicolaou, A.; Yassour, M.; Grabherr, M.; Blood, P.D.; Bowden, J.; Couger, M.B.; Eccles, D.; Li, B.; Lieber, M.; et al. De novo transcript sequence reconstruction from RNA-seq using the Trinity platform for reference generation and analysis. *Nat. Protoc.* **2013**, *8*, 1494–1512. [CrossRef] [PubMed]

33. Altschul, S.F.; Gish, W.; Miller, W.; Myers, E.W.; Lipman, D.J. Basic local alignment search tool. *J. Mol. Biol.* **1990**, *215*, 403–410. [CrossRef]

34. Thompson, J.D.; Higgins, D.; Gibson, T.J. CLUSTAL W: Improving the sensitivity of progressive multiple sequence alignment through sequence weighting, position-specific gap penalties and weight matrix choice. *Nucleic Acids Res.* **1994**, *22*, 4673–4680. [CrossRef]

35. Seefeldt, S.S.; Jensen, J.E.; Fuerst, E.P. Log-Logistic Analysis of Herbicide Dose-Response Relationships. *Weed Technol.* **1995**, *9*, 218–227. [CrossRef]

36. Knezevic, S.Z.; Streibig, J.C.; Ritz, C. Utilizing R Software Package for Dose-Response Studies: The Concept and Data Analysis. *Weed Technol.* **2007**, *21*, 840–848. [CrossRef]
37. Doyle, J.J.; Doyle, J. LA rapid DNA isolation procedure for small quantities of fresh leaf tissue. *Phytochem. Bull.* **1987**, *19*, 11–15.
38. Westwood, J.H.; Charudattan, R.; Duke, S.O.; Fennimore, S.A.; Marrone, P.; Slaughter, D.C.; Swanton, C.; Zollinger, R. Weed Management in 2050: Perspectives on the Future of Weed Science. *Weed Sci.* **2018**, *66*, 275–285. [CrossRef]
39. Bajwa, A.; Mahajan, G.; Chauhan, B.S. Nonconventional Weed Management Strategies for Modern Agriculture. *Weed Sci.* **2015**, *63*, 723–747. [CrossRef]
40. Yu, Q.; Han, H.; Li, M.; Purba, E.; Walsh, M.J.; Powles, S.B. Resistance evaluation for herbicide resistance-endowing acetolactate synthase (ALS) gene mutations using *Raphanus raphanistrum* populations homozygous for specific ALS mutations. *Weed Res.* **2012**, *52*, 178–186. [CrossRef]
41. Saari, L.L.; Cotterman, J.C.; Primiani, M.M. Mechanism of Sulfonylurea Herbicide Resistance in the Broadleaf Weed, *Kochia scoparia*. *Plant Physiol.* **1990**, *93*, 55–61. [CrossRef] [PubMed]
42. Xia, W.; Pan, L.; Li, J.; Wang, Q.; Feng, Y.; Dong, L. Molecular basis of ALS- and/or ACCase-inhibitor resistance in shortawn foxtail (*Alopecurus aequalis* Sobol.). *Pestic. Biochem. Physiol.* **2015**, *122*, 76–80. [CrossRef]

MDPI

St. Alban-Anlage 66

4052 Basel

Switzerland

Tel. +41 61 683 77 34

Fax +41 61 302 89 18

www.mdpi.com

Sustainability Editorial Office

E-mail: sustainability@mdpi.com

www.mdpi.com/journal/sustainability